ADVANCED COMPOSITE MATERIALS FOR AUTOMOTIVE APPLICATIONS

ADVANCED COMPOSITE MATERIALS FOR AUTOMOTIVE APPLICATIONS

STRUCTURAL INTEGRITY AND CRASHWORTHINESS

Editor

Ahmed Elmarakbi
University of Sunderland, UK

WILEY

This edition first published 2014
© 2014 John Wiley & Sons, Ltd

Registered office
John Wiley & Sons Ltd, The Atrium, Southern Gate, Chichester, West Sussex, PO19 8SQ, United Kingdom

For details of our global editorial offices, for customer services and for information about how to apply for permission to reuse the copyright material in this book please see our website at www.wiley.com.

Library of Congress Cataloging-in-Publication Data

Advanced composite materials for automotive applications : structural integrity and crashworthiness / [compiled by] Ahmed Elmarakbi.
 pages cm
Includes bibliographical references and index.
ISBN 978-1-118-42386-8 (cloth)
1. Composite materials in automobiles. 2. Automobiles–Crashworthiness. I. Elmarakbi, Ahmed.
TL240.5.C65A38 2014
629.2′32–dc23

                         2013023086

A catalogue record for this book is available from the British Library.

ISBN: 978-1-118-42386-8

Typeset in 10/12pt Times by Aptara Inc., New Delhi, India

1 2014

Contents

About the Editor

Ahmed Elmarakbi is Professor of Automotive Engineering in the Department of Computing, Engineering and Technology at the University of Sunderland, UK. He obtained his PhD in Mechanical Engineering from the University of Toronto, Canada, in September 2004. Then, he began a prestigious postdoctoral research fellowship supported by NSERC/JSPS in the Department of Aeronautics and Space Engineering at the Tohoku University, Japan. His research interests lie in the area of energy efficient vehicles, including lightweight materials for low carbon vehicles, advanced composite materials, automotive composites, vehicle safety and crashworthiness. His research outcomes are recognized both nationally and internationally, as evident from his over 120 publications, many of which are published in high-impact journals and well cited. He has presented papers and delivered scientific talks and seminars in many countries worldwide. He has expertise in gaining national and international funding, has established a number of fruitful national and international collaborations and has worked with a number of highly respected researchers in world-leading laboratories in the United States, Japan, Canada and Europe.

List of Contributors

Alamusi, Department of Mechanical Engineering, Chiba University, Yayoi-cho 1-33, Inage-ku, Chiba 263-8522, Japan

Othman Al-Khudairi, Material Research Centre, SEC Faculty, Kingston University, London, UK

Antonio Argüelles, Polytechnic School of Engineering, University of Oviedo, Gijón, Spain

Satoshi Atobe, Department of Aerospace Engineering, Tohoku University, Aramaki-Aza-Aoba 6-6-01, Aoba-ku, Sendai 980-8579, Japan

David C. Barton, School of Mechanical Engineering, University of Leeds, Leeds, LS2 9JT, UK

Giovanni Belingardi, Dipartimento di Meccanica, Politecnico di Torino, Corso Duca degli Abruzzi, 24, 10129, Torino, Italy

Jorge Bonhomme, Polytechnic School of Engineering, University of Oviedo, Gijón, Spain

Simonetta Boria, School of Science and Technology, Mathematics Division, University of Camerino, Camerino, Italy

Lee S. Bryars, Material Research Centre, SEC Faculty, Kingston University, London, UK

Andreas Büter, Fraunhofer Institute for Structural Durability and System Reliability (LBF), Bartningstrasse 47, 64295 Darmstadt, Germany

Sujit Das, Oak Ridge National Laboratory, Oak Ridge, USA

Hicham El-Hage, Department of Mechanical Engineering, LIU, Beirut, Lebanon

Ahmed Elmarakbi, Department of Computing, Engineering and Technology, Faculty of Applied Sciences, University of Sunderland, Sunderland, SR6 0DD, UK

Hisao Fukunaga, Department of Aerospace Engineering, Tohoku University, Aramaki-Aza-Aoba 6-6-01, Aoba-ku, Sendai 980-8579, Japan

Nikhil Gupta, Composite Materials and Mechanics Laboratory, Mechanical and Aerospace Engineering Department, Polytechnic Institute of New York University, Brooklyn, NY 11201 USA

Homayoun Hadavinia, Material Research Centre, SEC Faculty, Kingston University, London, UK

Ali Hallal, Department of Mechanical Engineering, LIU, Beirut, Lebanon

Peyman Honarmandi, Mechanical Engineering Department, The City College of the City University of New York, New York, USA

Ning Hu, Department of Mechanical Engineering, Chiba University, Yayoi-cho 1-33, Inage-ku, Chiba 263-8522, Japan

Ermias Koricho, Dipartimento di Meccanica, Politecnico di Torino, Corso Duca degli Abruzzi, 24, 10129, Torino, Italy

Lovre Krstulović-Opara, Faculty of Electrical Engineering, Mechanical Engineering and Naval Architecture, University of Split, Split, Croatia

Christian Lauter, Automotive Lightweight Construction, University of Paderborn, Paderborn, Germany

Eoin Lewis, Material Research Centre, SEC Faculty, Kingston University, London, UK

Yaolu Liu, Department of Mechanical Engineering, Chiba University, Yayoi-cho 1-33, Inage-ku, Chiba 263-8522, Japan

Dirk H.-J.A. Lukaszewicz, Research and Innovation Centre, BMW AG, Knorrstrasse 147, 80788, Munich, Germany

Anthony Macke, Center for Composite Materials, Materials Engineering Department, University of Wisconsin-Milwaukee, Milwaukee, WI 53201, USA

Brunetto Martorana, Centro Ricerche FIAT, Strada Torino, 50, 10043, Orbassano, Italy

Victoria Mollón, Polytechnic School of Engineering, University of Oviedo, Gijón, Spain

Barnaby Osborne, Material Research Centre, SEC Faculty, Kingston University, London, UK

Arun Balan Ramamohan, WMG, University of Warwick, Warwick, UK

Zoran Ren, Faculty of Mechanical Engineering, University of Maribor, Maribor, Slovenia

Corin Reuter, Automotive Lightweight Construction, University of Paderborn, Paderborn, Germany

Neil Reynolds, WMG, University of Warwick, Warwick, UK

Pradeep K. Rohatgi, Center for Composite Materials, Materials Engineering Department, University of Wisconsin-Milwaukee, Milwaukee, WI 53201, USA

Davide Roncato, Centro Ricerche FIAT, Strada Torino, 50, 10043, Orbassano, Italy

Nicole Schweizer, Fraunhofer Institute for Structural Durability and System Reliability (LBF), Bartningstrasse 47, 64295 Darmstadt, Germany

Ali Shaito, Department of Mechanical Engineering, LIU, Beirut, Lebanon

Benjamin F. Schultz, Center for Composite Materials, Materials Engineering Department, University of Wisconsin-Milwaukee, Milwaukee, WI 53201, USA

Ala Tabiei, School of Advance Structures, College of Engineering and Applied Science, University of Cincinnati, Cincinnati, OH 45221-0071, USA

Alem Tekalign, Dipartimento di Meccanica, Politecnico di Torino, Corso Duca degli Abruzzi, 24, 10129, Torino, Italy

Thomas Tröster, Automotive Lightweight Construction, University of Paderborn, Paderborn, Germany

Matej Vesenjak, Faculty of Mechanical Engineering, University of Maribor, Maribor, Slovenia

Jaime Viña, Polytechnic School of Engineering, University of Oviedo, Gijón, Spain

David Warren, Oak Ridge National Laboratory, Oak Ridge, USA

Tomonori Watanabe, Department of Mechanical Engineering, Chiba University Yayoi-cho 1-33, Inage-ku, Chiba 263-8522, Japan

Alan Wheatley, Department of Computing, Engineering and Technology, University of Sunderland, Sunderland, SR6 0DD, UK

Series Preface

One of the major challenges that the automotive sector will face in both the near and long-term future is the need for higher fuel efficiency. This is being driven by international requirements targeting reduced fuel consumption and carbon emissions, in a quest for sustainability. One of the most significant methods by which fuel economy can be achieved is by reducing the weight of the vehicle, or lightweighting the car. Composite materials, with their high strength to weight ratio provide an excellent platform upon which to develop the next generation of lightweight vehicles. Significant successes in the aerospace sector have led to the initial integration of carbon fiber composites into specialized vehicles such as Formula 1 racing systems, demonstrating the viability of composites in the ground vehicle. This viability is not only related to a successful lightweight vehicle that is more fuel efficient, but one that possesses both significant crashworthiness and is highly durable.

Based on initial successes, the promise of successfully integrating composites into commercial vehicles that are mass produced is within reach. However, the integration of composites into the vehicle and many of its components requires significant modifications to many of the vehicle design and analysis practices, where new material models and design characteristics must be considered. *Advanced Composite Materials for Automotive Applications* captures the basic and pragmatic concepts necessary to rethink the automobile's design to incorporate composite materials. It is part of the *Automotive Series* whose primary goal is to publish practical and topical books for researchers and practitioners in the industry and postgraduate/advanced undergraduates in automotive engineering. The series addresses new and emerging technologies in automotive engineering supporting the development of more fuel efficient, safer and more environmentally friendly vehicles. It covers a wide range of topics, including design, manufacture and operation, and the intention is to provide a source of relevant information that will be of interest and benefit to people working in the field of automotive engineering.

Advanced Composite Materials for Automotive Applications presents a number of different design and analysis considerations related to the integration and use of composites in the vehicle and its various components, including manufacturing methods, crash, impact and load analysis, multi-material integration, damage, curability and failure analysis. Also, the text provides a number of excellent real-world examples that punctuate the fundamental concepts developed in the book. It is a state of the art text, written by recognized experts in the field providing both fundamental and pragmatic information to the reader, and it is a welcome addition to the *Automotive Series*.

Thomas Kurfess
August 2013

Preface

The automotive industry faces many challenges, including increased global competition, the need for higher performance vehicles, a reduction in costs and tighter environmental and safety requirements. The materials used in automotive engineering play key roles in overcoming these challenges. However, the development of materials and processes to facilitate the use of composites in high-volume automotive applications is also still a big challenge. Thermoplastics and thermoset composites are being heavily considered by many automotive companies. Nowadays, there is a clear direction within car industries to replace metal parts by polymer composites in order to improve fuel consumption and produce lighter vehicles. The main advantages that composites offer to automotive applications are in cost reduction, weight reduction, recyclability and excellent crash performance compared with traditional steels.

This book provides a comprehensive explanation of how advanced composite materials, including FRPs, reinforced thermoplastics, carbon-based composites and many others are designed, processed and utilized in vehicles. The book includes a technical explanation of composite materials in vehicle design and analysis and covers all phases of composites design, modelling, testing and failure analysis. It also sheds light on the performance of existing materials, including carbon composites and future developments in automotive material technology which work towards reducing the weight of the vehicle structure.

A lot of case studies and examples covering all aspects of composite materials and their application in automotive industries are provided and explained in detail by the authors.

The initial chapters of the book focus on the fundamental background, providing a detailed overview of composite materials, their technology and their automotive applications. Impact, crash analysis, composite responses, damage and failure behaviour are presented and discussed in detail in Chapters 4–12. In addition, detailed work on metal matrix composites and their automotive applications are presented in Chapter 13. Finally, several case studies and designs are then covered in Chapters 14–17, including a wheel with integrated hub motor, safety components in composite body panels, noise and vibration analysis, braking systems and using low cost carbon fibre, together with performance and cost models.

A book covering such vital topics definitely would be attractive to the entire scientific community. The book will be valuable for those already working with composites and for those who are considering their use in the future for automotive applications. This book is proposed to give readers an appreciation of composite materials and their characteristics. The book will also provide the reader with the state of the art in the failure analysis of composite materials and their implications in the automotive industry. It will provide many technical advantages on the current and future uses of composites and the development and specific characteristics of composites and their energy absorption capabilities for crash safety.

This book is aimed at engineers, researchers and professionals who have been working in composites or are considering their use in the future in automotive applications. This book would be described as advanced/specialist.

The book is unique, with valuable contributions from renowned world-class experts from all over the world. The Editor would like to express his gratitude and appreciation to all contributors of this book for their efforts and decent work and to all my colleagues who served as reviewers for their comments, opinions and suggestions. The Editor would also like to thank John Wiley & Sons for this opportunity and for their enthusiastic and professional support.

Part One

Fundamental Background

1

Overview of Composite Materials and their Automotive Applications

Ali Hallal[1], Ahmed Elmarakbi[2], Ali Shaito[1] and Hicham El-Hage[1]

[1]*Department of Mechanical Engineering, LIU, Beirut, Lebanon*
[2]*Department of Computing, Engineering and Technology, University of Sunderland, Sunderland, SR6 0DD, UK*

1.1 Introduction

This chapter presents an overview of recent automotive applications of advanced composites. A summary of available composites that could be used in automotive industries is presented. This work mainly deals with new research and studies done in order to investigate the present and potential use of composites for automotive structural components (e.g. tubes, plates, driveshafts, springs, brake discs, etc.). The important conclusions of these experimental and numerical simulation studies are shown in detail. It is important to note that most studies have an interest in enhancing the mechanical properties of automotive parts as well as providing better ecological and economical solutions. The influence of reinforcement types and architecture on the mechanical behaviour of automotive parts is investigated.

It is remarked that unidirectional composites and composite laminates are the most used composites, with a domination of glass fibres. However, carbon reinforced polymers and carbon ceramic composites along with nanocomposites could be considered as the most advanced composites currently in use for the automotive industry. Moreover, the emergence of natural fibre reinforced polymers, green composites, as a replacement of glass fibre reinforced polymers is discussed.

Recently, the use of composite materials has increased rapidly in automotive domains. As reported, according to [1], it is remarked that the total global consumption of lightweight materials used in transportation equipment will increase at a compound annual growth rate (CAGR) of 9.9% in tonnage terms and 5.7% in value terms between 2006 and 2011 (from

Advanced Composite Materials for Automotive Applications: Structural Integrity and Crashworthiness,
First Edition. Edited by Ahmed Elmarakbi.
© 2014 John Wiley & Sons, Ltd. Published 2014 by John Wiley & Sons, Ltd.

42.8 million tons/US\$80.5 billion in 2006 to 68.5 million tons/US\$106.4 billion in 2011) [2]. The use of composites consists of chassis parts, bumpers, driveshafts, brake discs, springs, fuel tanks, and so on.

From a historical point of view, it should be noticed that the first car body made from (glass fibre reinforced polymer, GFRP) composites was for the Chevrolet Corvette, which was introduced to the public at Motorama show at New York in 1953 [3]. For these days, the Corvette series still use composite materials in its design. In motor sports, the use of carbon fibre reinforced polymers has been shown in Formula 1, with the McLaren MP4 in 1981. The open wheel car benefits from lighter body, which leads to a well distributed weight in order to achieve more mechanical grip on the track which significantly increases the overall performance of the car. Nowadays, all Formula series cars and other racing touring cars use composites in huge amounts in almost all of their body parts.

Composites have many advantages over traditional materials, such as their relatively high strength and low weight, excellent corrosion resistance, thermal properties and dimensional stability and more resistance to impact, fatigue and other static and dynamic loads that car structures could be subjected. These advantages increase the performance of cars and lead to safer and lower energy consumption. It should be noticed that car performance is affected not only by the engine horsepower, but also by other important parameters such as the weight/horsepower ratio and the good distribution of the weight. Moreover, lighter vehicles lead to a reduction of fuel consumption. It has been estimated that the fuel economy improves by 7% for every 10% of weight reduction from a vehicle's total weight [1,2]. It is reported that using carbon fibre composites instead of traditional materials in body and chassis car parts could save 50% of weight [1,2]. In addition, it means for every kilogram of weight reduced in a vehicle, there is about 20 kg of carbon dioxide reduction [2].

The major problems still facing the large use of composites in automotive domains are: the high cost in comparison with traditional materials (steel, alloy, aluminium), the complex and expensive manufacturing process for a large number of parts, the unknown physical (mechanical, thermal) behaviour of some kind of composites. Thus, many studies and research are conducted to solve these problems in order to extend the use of composites in large mass. Ford, with a collaboration with materials experts through the Hightech NRW research project, leads the search for a solution of a cost efficient manufacturing of carbon fibre composite components [4]. As estimated by Ford, the use of carbon fibre composites in addition to other advanced materials in the manufacturing of many automotive parts will reduce the weight of their cars by 340 kg at the end of the decade [4]. Another example is the consortium, led by Umeco and partnered by Aston Martin Lagonda, Delta Motorsport Ltd, ABB Robotics and Pentangle Engineering Services Ltd, that has been created to look into the potential for using high-performance composites. The project aims to reduce the cost of composite body in white vehicle structures for the mainstream automotive sector [5].

Many types of composites exist, which give the opportunity to select the optimum material design for any structure. However, this leads to many studies that deal with the mechanical behaviour of composites. The most used composites are composite laminates which consist of several plies with unidirectional long fibres. More developed kinds of composites known as textile composites (woven, braided and knitted fabrics) has emerged recently to be adopted in automotive applications. Moreover, nanocomposites have been used in order to enhance the performance of car structures. Hybrid composites also have been adopted especially in designing tubes and beams. Hybrids consist of several layers of composites and other types of materials, such as aluminium. The aluminium layer is reinforced by laminated or unidirectional (UD) composites composites. These kinds of lightweight materials are used to resist impact loading, as will be shown in the section below.

In this chapter, a brief general introduction of composites type is presented in the second section, while the automotive applications of advanced composites are discussed in the third section. In the fourth section, the potential of analytical and numerical analysis is presented.

1.2   Polymer Composite Materials

In general, composite materials are composed from at least two materials, where one is the reinforcing phase and the other is the matrix. Many combinations can be shown, with different kinds of materials and architectures.

There are two classification systems of composite materials. One is based on the matrix material (metal matrix composites, MMC; ceramic matrix composites, CMC; polymer matrix composites, PMC) and the second is based on the material structure: particulate (random orientation of particles; preferred orientation of particles), fibrous (short-fibre reinforced composites; long-fibre reinforced composites) and laminate composites.

The various PMC are classified as thermoplastic and thermosets and can be reinforced with various types of fibres depending upon the applications. The PMC are used in various automotive applications like crashworthiness, body panels, bumpers and so on. The CMC are used in elevated temperatures of various engine components and braking systems. The MMC use magnesium, copper and aluminium as their matrix with fibres to be used in various engine and crash absorbing components. Moreover, MMC with aluminium matrix and ceramic based composites find some automotive applications with supercar brake discs.

The PMC have heavily been used in the automotive industry. Polymers used in automotive applications are divided into thermoplastics and thermosets. Thermoplastics are high molecular weight materials that soften or melt on the application of heat. Thermoset processing requires the non-reversible conversion of a low molecular weight base resin to a polymerised structure. The resultant material cannot be re-melted or re-formed.

In automotive applications, reinforced plastics are the major composite material. For polymer composites, common fillers used include calcium carbonate ($CaCO_3$), talc, wollastonite, glass and carbon fibre. Some of the common processing techniques for polymer composites are: injection moulding, sheet moulding compound (SMC), glass-mat thermoplastic (GMT) compression moulding, resin transfer moulding (RTM) and reaction injection moulding (RIM).

Some of the factors affecting the processing and manufacture of polymer composites are fibre distribution in the matrix, compatibility between the matrix and fibres, fibre orientation and thermal stability of the fibre.

A common processing technique for the production of polymer composites is thermoforming. Thermoforming is commonly used to produce fibre-mat thermoplastic composites. The fibre and the polymer are inserted into a heated mould. The thermoplastic then flows into the fibre component. The hybrid material is then combined into a composite in a cold press. Compression moulding using thermoset polymer matrices is another major processing technique used to manufacture large parts for the automotive industry.

The main categories of polymer composites used in automotive applications are as follows [6]:

- **Non-structural composites**, composites composed of short glass fibre-reinforced plastics with the reinforcement in the range of 10–50% by weight used in pedal systems, mirror housing and so on.
- **Semi-structural composites**, composites composed of several layers of the reinforcement, which is in the form of a mat in a matrix. The mat could be a chopped strand mat, a random continuous strand mat, or a unidirectional mat. The matrix could be a thermoplastic or a thermoset. These composites are used in body panels, front end structures, seat backs and so on.
- **Structural composites**, structural thermoplastic composites (TPC), structural reaction injection moulded core parts, bumper systems and so on.

1.2.1   Non-Structural Composites

The use of composite materials has been limited to automotive structural components, however recently there has been a wide use of composites in non-structural functional components. Schouwenaars *et al.* [7] studied the fracture during assembly of a radiator head produced from a nylon/33% short glass fibre composite. The study focused on finding the elastic constants and fracture stresses and on resolving some of the manufacturing problems such as distortion after moulding and deformations induced during assembly. They used a combination of in situ measurements, microscopy and reverse modelling of non-linear material properties to determine the stiffness constants and strength as a function of fibre length distribution and fibre orientation distribution. The integrated approach was used effectively to resolve the manufacturing problems presented.

Lee *et al.* [8] addressed the advantages of using composites in reducing the mass of automotive components and improving the fuel efficiency by developing a hybrid valve lifter to be used in an automotive internal combustion engine. The lifter was made from carbon fibre/phenolic composite and steel. The design and manufacture of the hybrid valve was investigated based on the functional requirements such as durability. The mass of the composite was 35% lighter than the conventional steel valve lifter and showed to be durable for the test loads.

Imihezri *et al.* [9] studied the mould flow and component design of a 30% glass fibre polyamide composite to be used as a clutch pedal. Different profile cross-sections were analysed using finite element analysis for stress and using mould flow software for flow properties.

1.2.2   Semi-Structural Composites

Sheet moulded compound (SMC) panels as exterior body panels have been increasingly used in automotive industry [10]. Among the main factors to consider in the design of the body panels are material cost and mass reduction. Mass reduction is achieved by using materials with the high strength-to-weight ratios which composite materials offer [2]. However, there are some barriers for the use of composites in the production of automotive parts, such their cost and their manufacturing.

General Motors Research Laboratories evaluated different body panel designs for a front wheel drive compact car for equal stiffness requirements [11]. The materials considered as a substitute for steel were aluminium, glass SMC and carbon SMC. The SMC fibre content was in the range of 10–70% by weight. The glass SMC showed 27% reduction in mass, while aluminium and carbon SMC showed a mass reduction of 35 and 45%, respectively. Although the carbon SMC showed the highest mass reduction, it had a higher cost due to the higher selling price of carbon fibres. General Motors also uses continuous fibre glass epoxy composites for front and rear leaf springs in selected passenger cars and for longitudinal leaf springs in the GM mini-van. The glass/epoxy leaf spring had a mass reduction ratio of 4: 1 over the conventional steel leaf.

Feraboli *et al.* [12] studied the effect of fibre architecture on the delamination and flexural behaviour of carbon/epoxy body panels of the Lamborghini Murcielogo. They studied the use of four prepreg tapes, two fabrics and woven laminates over directional tape. The strength, durability, environmental resistance, vibration damping and surface finish of the composite body panels were investigated. The body panels were bonded to the tubular steel chassis by a methacrylate adhesive which was found to be effective than other adhesives like epoxy and polyurethane.

Feuillade *et al.* [13] studied the influence of material formulation and SMC process on the surface quality of SMC body panels. The main parameters considered were the amount of sizing on the fibres, the type of antistatic agent and its deposit method and the type of film former. Two commercially sized glass fibres were studied. The results showed that, during the impregnation process, the natures of the antistatic agent and the film former, the fibre wetting properties and the sizing influenced the surface quality of the SMC moulded panels.

Ning *et al.* [14] designed, analysed and manufactured an air conditioning cover roof door on a mass transit bus. Thermoplastic composites and thermoforming processing technology was used in the manufacture of the part. The composite was found to have weight savings of 39%, as compared to aluminium, with an enhanced rigidity of 42% reduced free-standing deflection.

1.2.3 Structural Composites

In addition to the requirements of light weight and lower cost, materials used in automotive applications should also meet the requirements of safety and the ability to absorb impact energy in what is referred to as crashworthiness. Polymer composites have been replacing metal components due to their reduced weight (which improves fuel consumption), their durability and their crashworthiness. Several studies were done on the impact energy absorption, durability and the crushing behaviour of polymer composites [15–24].

Davoodi *et al.* [25] studied the design of a fibre-reinforced epoxy composite bumper absorber. The study focused on studying the use of composites in energy absorption in a car bumper as a pedestrian energy absorber. A carbon reinforced composite was investigated. It was found that the fibre-reinforced epoxy composite absorber is sufficient for pedestrian impact and can substitute for the existing materials, such as expanded polypropylene foam.

Bisagni *et al.* [26] studied the progressive crushing behaviour of fibre-reinforced composite energy absorbers for Formula One side impact and steering column impact. Two series of tubes with different lamina were investigated. A finite element model using LS-DYNA was also developed. The composite absorbers had a high capacity of energy absorption. The numerical model accurately predicted the overall shape, magnitude of impact, deformation and failure of the composite absorbers with about 10% difference to experimental results.

Launay *et al.* [27] studied the cyclic behaviour of a 35% short glass fibre-reinforced polyamide composite to be used in an automotive application. Mechanical tests were done on two different relative humidity specimens. The creep and stress relaxation behaviour of the materials were studied to predict the fatigue life of the polymer matrix composite under different loadings and environmental conditions.

Ruggles *et al.* [28] studied the fatigue properties of carbon fibre-reinforced epoxy matrix composites. The study focused on developing experimentally based, durability-driven design guidelines for the long-term reliability of carbon reinforced composites for structural automotive components. A temperature-dependence study was also done to study the variation of the fatigue behaviour of the composite with temperature.

It is worth mentioning that composites used in automotive applications are joined in different methods. The main methods are mechanical fastening, adhesive bonding and welding. Mechanical joints such as rivets and bolts have the disadvantage of creating stress concentrations. Adhesive bonding uses a combination of polymer-based adhesive blends and provides some advantages, such as controlled mechanical properties of the adhesive, smooth surface finish between the joined materials, increased life time of the joint and good sealing. Welding, which is commonly used in metallic parts, has limited applications with composites. Welding has some advantages, such as durability and short processing time. For thermoplastic composites, the common welding techniques are: ultrasonic, induction and resistance welding.

Concerning advanced composites used or to be implemented in automotive industries, fibre reinforced polymers (FRP) with glass, aramid, carbon and graphite fibres should receive notice (Table 1.1). The polymeric matrix is in a general epoxy resin or polyester (Table 1.2). However, more recently nanocomposites have found some applications in automotive industries. Moreover, metal matrix composites (MMC) with aluminium matrix and ceramic based composites found some automotive applications with supercar brake discs. Composite architectures found in automotive applications are composite laminates, textiles, hybrids and nanocomposites. Tables 1.1, 1.2 and 1.3 show the mechanical properties of some fibre/polymeric matrices and composites. It is well remarked the advantage that composites bring in terms of high stiffness with a low density material.

Advanced composite materials with long fibres can be categorised into three major categories: laminates, hybrid composites and textiles.

Table 1.1 Mechanical properties of some fibres and metals [29].

Material	Young's modulus (Gpa)	Shear modulus (Gpa)	Axial Poisson's ratio	Ultimate strength (Mpa) tension	Strain to failure (%)	Density (Kg/m^3)
Carbon fibre HT-T300	230	23	0.23	3530	1.5	1750
Carbon fibre IM-T800	294	23	0.23	5586	1.9	1800
Carbon fibre HM	385	20	0.23	3630	0.4	2170
E-glass fibre glass	72	27.7	0.3	3450	4.7	2580
S-glass fibre	87	33.5	0.3	4710	5.6	2460
Kevlar 49 fibre	124	5	0.3	3850	2.8	1440
Steel	206	81	0.27	648	4	7800
Aluminium	69	25.6	0.35	234	3.5	2600

Table 1.2 Mechanical properties of some polymeric matrices [29, 30].

Material	Young's modulus (Gpa)	Shear modulus (Gpa)	Axial Poisson's ratio	Ultimate strength (Mpa) tension	Strain to failure (%)	Density (Kg/m^3)
Epoxy	3.1	1.2	0.3	70	4.0	1200
Polyester	3.5	1.4	0.3	70	5.0	1100
Resin RTM 6	2.89	1.08	0.34	75	3.4	1140
Resin RTM 120	2.60	0.96	0.35	77	—	1200

1.2.4 Laminated Composites

Composite laminates, also known as laminated composites, are composed from different plies, where each ply is considered as a UD long fibre lamina. The mechanical behaviour of each lamina is considered to be transverse isotropic, while the behaviour of the laminated composite is orthotropic. This category represents the most used kind of composites. Several kinds of architecture could be found, such as the cross-ply [0, 90] (Figure 1.1), the bi-directional [$-\theta$, $+\theta$] ([-45, $+45$], [-30, $+30$]), in addition to the tri-axial laminates [$-\theta$, 0, $+\theta$]. Laminated composites provide good in-plane mechanical properties: high in plane Young's moduli, shear modulus and in-plane ultimate strength. However, they lack stiffness in the out of plane direction, known as the through thickness direction or Z direction. This implies adding more layers in order to strengthen the through thickness direction, which means more weight and cost and prevents the manufacturing of complex shapes. Moreover, composite laminates are prone to delamination and inetrlaminar shear.

1.2.5 Textile Composites

More advanced long fibre composites have emerged known as textile composites. Textiles are categorised into three major fabric kinds: woven, braided and knitted fabrics. They are introduced to improve the mechanical behaviour of composites and to offer more choices of composite architectures. Textiles are made from interlaced, interlocked, or knitted yarns that

Table 1.3 Mechanical properties of different kinds of composites.

Composite	Longitudinal Young's modulus (Gpa)	Ultimate strength (Mpa) tension	Density (Kg/m^3)
Composite unidirectional graphite/epoxy	181	1500	1600
Composite unidirectional glass/epoxy	38.60	1062	1800
[0,90] graphite/epoxy	95.98	373.0	1600
[0,90] glass/epoxy	23.58	88.25	1800
Textile 2D "Taffetas" carbon/epoxy	59.4	515.05	1500
Textile 3D "interlock"	49.02	672	1400
Textile 3D "orthogonal"	57.5	770	1500

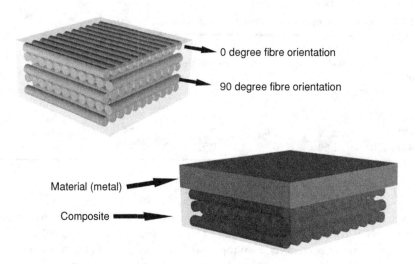

Figure 1.1   Cross-ply laminated composites and hybrid metal-composite material.

undulate above and beneath other yarns to form a complex architecture. Known textiles are composed of yarns with long fibres impregnated in polymer resin, metal, or ceramic matrices. Different 2D and 3D geometries and structures of woven, braided and knitted fabrics could be classified according to the number and shape of yarns. Thus, a large number of different architectures are introduced due to the development and demand needs of industries.

Woven composites represent the biggest category of textiles. Concerning the 2D woven fabrics, the composite is made of two sets of yarns: the warp and weft yarns. They are interlaced in a 90° "in-plane". In this kind of woven fabrics, one type of yarn (warp or weft) goes beneath only one set of the other type of yarn and they have a sinusoidal longitudinal profile. The 2D woven composites can be categorised into three major types: the plain weave, the harness satin weave and the twill weave fabrics (Figure 1.2).

The 3D woven composites introduce a yarn in the through thickness direction. The 3D woven composites are designed in order to reinforce the third direction and to avoid delamination between the layers. They are widely used in highly advanced industries, especially in the aeronautics fields, due to their stability and strength in all three axes. The weaver yarns in these composites go beneath more than one set of layers. They are divided into two types: Orthogonal weave and angle interlock weave fabrics. These two types are also divided into two kinds of fabrics: the layer-to-layer and the through-thickness weaves. 3D orthogonal woven composites are characterised by three set of orthogonal yarns: warp weaver yarns, stuffer yarns and weft yarns. However, the angle interlock fabrics can consist of two or three yarn types. In angle interlock fabrics, the warp weaver yarns have a crimp angle between 0 and 90°, while in orthogonal fabrics these yarns have a 90° angle with the (xy) plane (Figure 1.3).

Concerning the braided fabrics, they have an architecture close to those of 2D woven composites, but in this case the yarns are interlaced by a braider angle which is different from 90°. 2D braided composites are divided in general into: diamond fabrics (Figure 1.4a) and tri-axial fabrics (Figure 1.4b). Diamond fabrics consist of only two sets of braider yarns, while tri-axially braided fabrics have an additional axial yarn. The main advantage of these

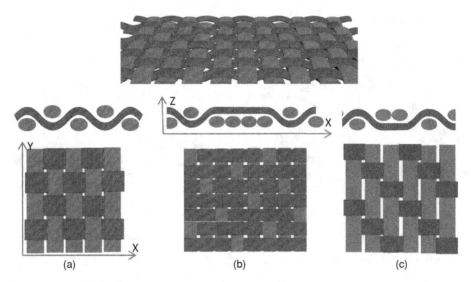

Figure 1.2 2D woven composites: (a) 2D plain weave composite, (b) Five harness satin weave composite, (c) 2D twill weave composite.

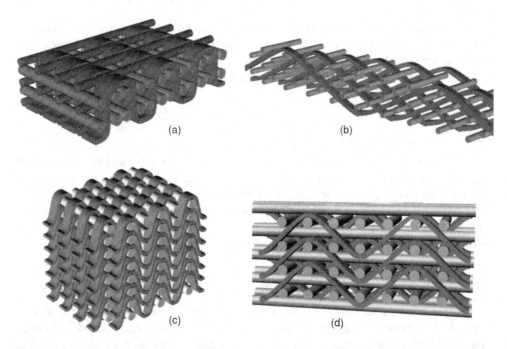

Figure 1.3 Different architectures of 3D woven composites: (a) 3D orthogonal woven composite, (b) 3D through-thickness angle interlock woven composite, (c) 3D layer to layer angle interlock woven composite, (d) 3D layer to layer angle interlock woven composite.

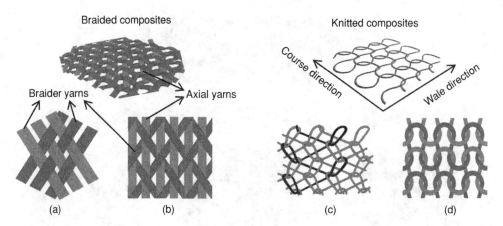

Figure 1.4 (a) 2D diamond braided composite, (b) 2D tri-axially braided composite, (c) warp knitted composite, (d) weft knitted composite.

composites is their high shear resistance. It is well noticed that weaver yarns in 2D woven composites, 3D angle interlock woven composites and 2D braided composites have a similar undulated shape; the description and modelling of this type of yarns will be discussed in the following section.

Knitted composites are textiles made from basic construction units called loops. They are divided into two types: warp knitted (Figure 1.4c) and weft knitted (Figure 1.4d) fabrics. They are characterised by the number of loops into horizontal direction, called course (C = number of loops/unit length), and by the number of loops into vertical direction, called wale (W = number of loops/unit length).

1.2.6 Hybrid Composites

Hybrid composites are made from several layers of composites and other materials (Figure 1.1). In general, for automotive applications, it is well noticed that the aluminium –composite combination is used as a hybrid composite. The composite layer could be any kind of composites described previously. The composite layer is used to reinforce the aluminium structure in order to enhance its mechanical properties. It is remarked that the use of hybrid composites is increasing, especially in tubular structures subjected to impact loading.

1.3 Application of Composite Materials in the Automotive Industry

It is well remarked that the application of composites in the automotive industries is increasing. In which concern, the focus of this study is on new research concerning advanced composites. These new kinds of composites consist of polymeric or metallic matrices reinforced with long fibres of carbon, glass and Kevlar materials.

The application of advanced composites will be shown in the next sections by reviewing the recent works that have been done in order to enhance the understanding of composites during impact, fatigue load and other complex loads. The review of studies dealing with

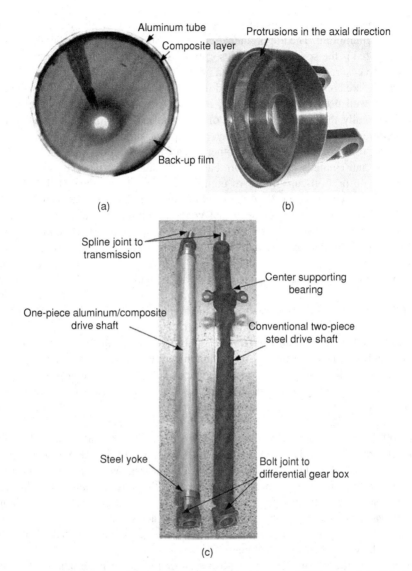

Figure 1.5   Hybrid composite driveshafts [35]. Reproduced from Ref. [35]. Copyright 2004 Elsevier.

advanced composites is divided into: crashworthiness studies, the development of automotive parts subjected to heavy static and dynamic loads, such as driveshafts (Figure 1.5), composite springs and gas turbines, and the use of natural fibre reinforced polymers as green composites replacing glass fibre reinforced polymer.

1.3.1   Crashworthiness

It is remarked that great attention is given to structural component material design related to car safety from impact and shock loading at different velocities. It is well known composites give the ability to design a lightweight structure that also can sustain much higher damage than steel and aluminium. The important parameter in crashworthiness studies of any structure to be investigated is the specific energy absorption (SEA; KJ/Kg) as well as the rate of work decay (KJ/s) to ensure a safe design that can protect the passengers [31].

Most studies are conducted to understand the behaviour of composite tubular parts. These structures are well used in the car chassis in order to enhance its crashworthiness during impact. Historically, the experimental work of Thornton and Edwards [32], in 1982, should be noticed. Different experimental tests subjected composite tubes made from glass, Kevlar and graphite fibres to impact loading. Later, other research was conducted as the work of Farley [33], in 1991, and Hamada [34], in 1996. Farley [33] studied the effects of crushing speed on the energy absorption capabilities of composite tubes. The investigated materials were graphite fibres/epoxy (Thronel 300/Fiberite 934) and Kevlar/epoxy (Kevlar fibres/Fiberite 934) composites. The main objective of that work was to determine the energy absorption capabilities as a function of crushing speed. However more recent work led by Hamada and Ramakrishna [34] concerning the optimisation of composite laminate fibre orientation for tubes subjected to impact loading was presented in 1996. The composite was a laminated carbon fibre reinforced poly-ether-ether-Kreton (PEEK) with 61% of fibre volume fraction. It was found that composite tubes with a fibre orientation of $\pm15°$ absorbed the most specific energy (225 KJ/Kg), and this was reported as the highest ever noticed in the literature for that time [34].

More advanced composites are used for energy absorption. Textile composites and especially carbon/epoxy braided composites are promising materials to be used, as shown by Chiu et al. [36]. In this study, the influence of braiding angle and axial yarn content was investigated. It was shown that the average width of the splaying fronds increased with increasing braiding angle, while it decreased with increasing axial yarn content. However, the highest specific energy absorption of 88 kJ/Kg was noticed for 20° of braiding angle. However, Bouchet et al. [37], in 2000, investigated the crushing behaviour of hybrid composite tubes. The composite was an UD carbon fibre (THR/300 from Hexcel)/epoxy polymer DGEBA with a fibre volume fraction of 34%. The three layered carbon/epoxy UD composite was wrapped on an aluminium alloy tube with the direction of fibres perpendicular to the longitudinal axis of the tube. The influence of surface treatments on an aluminium alloy before bonding with a carbon/epoxy composite was investigated. It was shown that the surface treatments increased the specific energy absorption capacities of around 30% between the chemically etched tube and the aluminium tube without surface treatment.

More recently, the advantage of composites over traditional materials as steel and aluminium in crush structures as tubes is well shown [38]. Quasi-static and intermediate rate axial crush tests were conducted on tubular specimens of carbon/epoxy (Toray T700/G83C) and glass/polypropylene (Twintex). The highest SEA measured (86 kJ/kg) was observed for carbon/epoxy tubes at quasi-static rates with a 45° chamfer initiator. Moreover, the highest energy absorption for Twintex tubes was observed to be 57.56 kJ/kg during 45° chamfer initiated tests at 0.25 m/s. However, with steel and aluminium, SEA values of 15 and 30 kJ/kg, respectively, were observed.

The study of crashworthiness has been also investigated with other types of automotive structural parts, such as bumpers or plates. In 2003, Corum et al. [39] studied experimentally the susceptibility of three candidate automotive structural composites subjected to low-energy impact damage. The reinforcements of composites that had the same urethane matrix consisted of random chopped-glass fibre and two stitch-bonded carbon-fibre mats, where one was in a cross-ply layup and the other in a quasi-isotropic layup. A pendulum device, representative of events such as tool drops, and a gas-gun projectile, representative of events such as kickups of roadway debris, were used to impact plate specimens. The glass-fibre composite was least vulnerable to damage, followed by the cross-ply carbon-fibre laminate, which had the same thickness. The quasi-isotropic carbon-fibre composite, which was thinner than the other two, sustained the most damage.

In addition, random chopped fibre reinforced composites were investigated by [40] as crash energy absorbers. According to [40], the Automotive Composite Consortium (ACC) is interested in investigating the potential use of these composites primarily because of the low costs involved in their manufacture, thus making them cost effective for automotive applications. The crashworthiness of composite plates subjected to quasi-static progressive crush tests was studied. The composite plates were made from three different kinds of materials: CCS100, HexMC and P4 composites. They were manufactured from Toray T700 chopped carbon fibre. The CCS100 composites were manufactured from chopped carbon fibre with YLA RS-35 epoxy resin using a compression moulding technique with a fibre volume fraction of 50% and a fibre length of 25.4 mm (1 inch). However, the random chopped carbon fibre epoxy resin HexMC composite plates, which had a fibre volume fraction of 57% and 50.8 mm (2 inch) fibre length, were compression moulded by Hexcel Composites LLC. The compression moulded P4 composite plates were manufactured from chopped carbon fibre having 50.8 mm (2 inch) fibre length and 36% fibre volume fraction with Hetron epoxy resin. It is remarked that all three materials have shown superior SEA as desired by the ACC, which could lead to their direct application in the automotive industries.

1.3.2 Composite Driveshaft and Spring

Composites have been recently used as structural materials of driveshafts and also springs, due to their light weight and high resistance to fatigue. In this section, some recent developments and research in this domain is investigated.

Lee *et al.* [41] investigated the torsional fatigue characteristics of a hybrid shaft of aluminium –composite co-cure joined shafts with axial compressive preload. It was observed that the fatigue strength of the hybrid shaft was much improved by the axial compressive preload, exceeding that of a pure aluminium shaft. Also, the degradation of the fatigue resistance of the hybrid shaft at sub-zero operating temperature was overcome by the axial compressive preload.

A finite element study of a composite driveshaft was done by [42]. The material consisted of hybrid carbon/glass fibre reinforced epoxy laminated composite. The layers were stacked in the following configuration [+45° glass/−45° glass/0° carbon/90° glass] which consisted of one layer of carbon/epoxy and three layers of glass/epoxy UD composites. It was shown that, with a change of carbon fibre orientation angle from 0 to 90°, the loss in the natural frequency of the shaft was 44.5%. Moreover, when shifting from the best to the worst stacking sequence, the drive shaft caused a loss of 46.07% in its buckling strength, which represented a major concern over shear strength in driveshaft design. In addition, the stacking sequence had an obvious effect on the fatigue resistance of the driveshaft.

More recently in 2011, Badie *et al.* [43] examined in their paper the effect of fibre orientation angles and stacking sequence on the torsional stiffness, natural frequency, buckling strength, fatigue life and failure modes of composite tubes. The studied composites consisted of carbon/epoxy and glass/epoxy laminates. The important remarks to be noticed form this study is that a carbon/epoxy driveshaft showed better torsional stiffens and fatigue life in comparison with a glass/epoxy driveshaft. Moreover, the stacking sequence with fibre orientation of $\pm45°$ showed a catastrophic sudden failure mode, while the stacking 90/0° experienced progressive and gradual failure. For carbon/epoxy tubes a higher fracture strain was shown than that of glass/epoxy tubes. In addition, in hybrid tubes, the severe difference in torsional stiffness of the layers led to initially suppressed twisting. Also with these tubes, the severe difference in torsional stiffness of the layers led to containing matrix cracks at the outer plies, not extending towards the tube ends. Concerning the natural frequency, the bending natural frequency increased by decreasing the fibre orientation angle. Decreasing the angle increased the modulus in the axial direction.

Recently, the use of elliptical springs made from E-glass/epoxy composite was studied [44]. A finite element model was used to investigate them, based on spring rate, log life and shear stress parameters. As a conclusion of this study, composite elliptic springs can be used for light and heavy trucks with a substantial weight reduction. Moreover the optimisation study of geometrical parameters of the cross-sectional area of the spring shows that a ratio of $a/b = 2$ yields the best mechanical properties.

1.3.3   Other Applications

The use of composites has also been observed with many other automotive parts, high-pressure full composite cylindrical vessels [45], brake discs [46], automotive door skins [47], car bumpers [48] and automotive radiator heads made from a nylon –glass fibre composite [49]. In the automotive industry, some applications of MMC can be found, such as brake rotors, pistons, connecting rods and integrally cast MMC engine blocks. The application of carbon ceramic brakes in automotive domains (beside motors sports) was started 10 years ago with the Enzo Ferrari F60 [50]. However nowadays, due to their high cost, the application of these advanced brake discs is still limited to supercars (e.g. Corvette ZR1, Ferrari 458 Italia, Ferrari California, Nissan GTR, Audi R8, Lamborghini Gallardo, Lexus LFA) and motor racing cars [50].

More advanced composites are used in manufacturing a Japanese 100 kW gas turbine for automotive applications (Figure 1.6). In order to achieve some requirements such as higher thermal efficiency over 40% at a turbine inlet temperature of 1350 °C, lower exhaust emissions to meet Japanese regulations and multi-fuel capabilities, the application of ceramic matrix composite was investigated by Kaya [51]. Parts made from different carbon fibre reinforced ceramic matrix showed higher mechanical properties, reliability against thermal shock, particle impact damage and creep resistance.

Other kinds of composites could also have a promising future in automotive applications as polymeric nanocomposites [52]. As reported; nanocomposites could afford a weight saving of about 80% in comparison with steel [52]. They could be produced by incorporating nanometre-size clay particles in polymeric matrices such as polypropylene (PP), polyethylene (PE), polyesters, or epoxies. Different methods could be used to produce nanoncomposites, as one of these methods is the in situ intercalation polymerisation method, pioneered by the Toyota

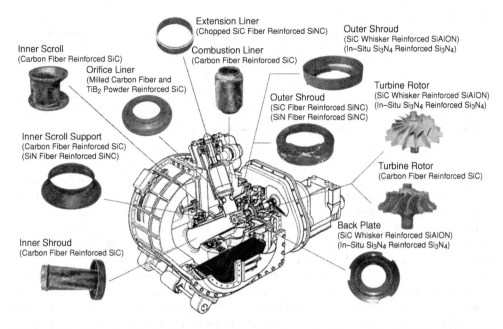

Inner Scroll
(Carbon Fiber Reinforced SiC)

Extension Liner
(Chopped SiC Fiber Reinforced SiNC)

Outer Shroud
(SiC Whisker Reinforced SiAlON)
(In–Situ Si_3N_4 Reinforced Si_3N_4)

Combustion Liner
(Carbon Fiber Reinforced SiC)

Orifice Liner
(Milled Carbon Fiber and
TiB_2 Powder Reinforced SiC)

Turbine Rotor
(SiC Whisker Reinforced SiAlON)
(In–Situ Si_3N_4 Reinforced Si_3N_4)

Outer Shroud
(SiC Fiber Reinforced SiNC)
(SiN Fiber Reinforced SiNC)

Inner Scroll Support
(Carbon Fiber Reinforced SiC)
(SiN Fiber Reinforced SiNC)

Turbine Rotor
(Carbon Fiber Reinforced SiC)

Inner Shroud
(Carbon Fiber Reinforced SiC)

Back Plate
(SiC Whisker Reinforced SiAlON)
(In–Situ Si_3N_4 Reinforced Si_3N_4)

Figure 1.6   Carbon ceramic 100 Kw gas turbine [51]. Reproduced from Ref. [51]. Copyright 1999 Elsevier.

Motor Company [53]. This method was used to create a nylon 6–clay hybrid (NCH), used to make a timing-belt cover, which could be considered as the first practical example of polymeric nanocomposites for automotive applications [54].

1.4   Green Composites for Automotive Applications

Due to the demanding needs for environmentally friendly composites, the automotive industry is seeking environmentally friendly biodegradable renewable composite materials and products. Over the last few years, a number of researchers have been involved in investigating the potential use of natural fibres as load bearing components in composite materials [55–58]. The use of such materials in composites has increased due to their relatively low cost, their ability to be recycled and their high strength to weight ratios.

Natural fibres show a potential as a replacement for inorganic fibres such as glass or aramid fibres in automotive components such as trim parts in dashboards, door panels, parcel shelves, seat cushions and cabin linings [59–61]. Mercedes-Benz used an epoxy matrix with the addition of jute in the door panels in its E-class vehicles [62].

Wambua *et al.* [63] studied the possibility of replacing glass fibres by natural fibres. They investigated the mechanical properties of sisal, hemp, coir, kenaf and jute reinforced polypropylene composites. The tensile strength and modulus increased with increasing fibre volume fraction. The mechanical properties of the natural fibre composites tested were found to compare favourably with the corresponding properties of glass mat polypropylene

Table 1.4 Mechanical properties of most used natural fibres compared to E-glass fibres [68, 69].

Mechanical properties	E-glass	Flax	Hemp	Jute	Ramie	Sisal
Density (g/cm³)	2.55	1.4	1.48	1.46	1.5	1.33
Young's modulus (Gpa)	73	60–80	70	10–30	44	38
Tensile strength (Mpa)	2400	800–1500	550–900	400–800	500	600–700
Elongation at failure (%)	3.0	1.2–1.6	1.6	1.8	2.0	2.0–3.0

composites. The specific properties of the natural fibre composites were in some cases better than those of glass. This suggested that natural fibre composites have a potential to replace glass in many applications that do not require very high load bearing capabilities.

Davoodi et al. [64] investigated the hybridisation of natural fibres with glass fibres for improving the mechanical properties over the natural fibres alone. A hybrid kenaf/glass fibre composite was investigated for a car bumper beam as a structural automotive component. They found that some mechanical properties such as tensile strength, Young's modulus and flexural modulus of the hybrid composite were similar to those of a typical glass mat thermoplastic bumper beam. However, the impact strength was lower, which showed the potential for the utilisation of the hybrid natural fibres by optimising some structural design parameters.

Recently the use of natural fibres in automotive domains has expended. It is reported that the growth of bio-fibres in automotive components is expected to increase by 54% per year [65]. Some kinds of strong, lightweight and low cost bio-fibres are introduced to replace glass fibre reinforced polymers in many interior applications. Fibres such as jute, kenaf, hemp, flax, banana, sisal and also wood fibre are making their way into the components of cars (Table 1.4). The bio-fibre reinforced polymers are used for door panels, seat backs, headliners, package trays, dashboards and trunk liners. Many review papers [60, 66, 67] have discussed the use of these fibres as reinforcement of a polymeric matrix, known as green composites. These papers reviewed the mechanical properties of fibres, matrices and composites. As well, they discussed the potential replacement of glass fibre reinforced polymer by the green composites, taking into account their ecological and economical influence.

Ashouri [61] presented in 2008 a review paper that discussed the use of wood plastic composite in the automotive industries. The main advantages shown by wood fibres compared to other kinds of natural fibres are to reinforce plastics due to their relative high strength and stiffness, low cost, low density, low CO_2 emission, biodegradability and being annually renewable. In the same prospect, in 2010, Alves et al. [70] pointed out the advantages of applying other kinds of natural fibre composites, jute fibre composites, in buggy enclosures. The study aimed at presenting the ability to replace glass fibre composites by jute fibre composites, by showing that the use of natural fibres will improve the overall environmental performance of the vehicles.

More recently, an important review study of the green composites properties was done by Koronis et al. [60]. The potential use of these kinds of composites for automotive application was discussed. It was shown that bio-fibres, also known as natural fibres, could have good mechanical properties in comparison with glass fibres. Databases were presented for the important mechanical properties of some natural fibres, polymeric matrices, as well as their composites.

1.5   Modelling the Mechanical Behaviour of Composite Materials

The modelling of the mechanical behaviour of composite materials has taken a major interest in recent years. Two main methods are used to determine the mechanical behaviour of composites: analytical models and numerical models (based on finite element analysis). Numerical models have taken most attention, due to the development in computation tools and software. It is noticed that these models are more trusted and widely used by industries. However, their implementation consumes a lot of time, beside the need to use super computers. In contrast, analytical models offer straightforward solutions and are more flexible and easier tools in terms of geometrical modelling and homogenisation methods. Moreover, they need less computation time without the necessity of super computers, and there is the ability of simple integration of analytical models in optimisation tools.

The modelling of the mechanical behaviour of composites must pass at first by the prediction of the stiffness matrix of the composite as a homogenised material. In contrast with metals, which have isotropic behaviour, composites with long fibres are treated as orthotropic materials. In addition, it should be noticed that some fibres, like carbon and graphite fibres, have anisotropic behaviour, in general considered transversely isotropic, while others, like glass fibres, are considered to have isotropic behaviour. Thus, the estimation of elastic properties and ultimate strengths presents a major challenge for researchers.

Analytical or numerical modelling of the mechanical behaviour of composites aids the optimisation of any structural component made of it. However, the variety of composites imposes a real challenge and intensive work in order to determine all the required mechanical properties. The detailed discussion of methods used to determine elastic properties, ultimate strengths, fatigue and impact behaviour, is considered to be out of the scope of this chapter. However a brief review of available analytical and numerical models used as homogenisation methods is presented.

1.5.1   Modelling the Elastic Properties of Unidirectional Composites

The prediction of the elastic properties of unidirectional composites using micro-mechanical models or FE methods is considered to be a basic stage in order to determine the mechanical properties of laminates or textile composites. The effective stiffness and compliance matrices of a transversely isotropic material are defined in the elastic regime by five independent engineering constants: longitudinal and transversal Young's moduli, longitudinal (also known as axial) and transversal shear moduli and major Poisson's ratio. Analytical models used to predict these engineering constants are known as micro-mechanical models. They can be categorised into four categories: phenomenological models, elasticity approach models, semi-empirical models and homogenisation models:

- Phenomenological models: rule of mixture (ROM); Reuss [71] model.
- Semi-empirical models: modified rule of mixture (MROM); Halpin–Tsai model [72]; Chamis model [73].
- Elasticity approach models: composite cylinder assemblage (CCA) model of Hashin and Rosen [74]; Christensen generalised self-consistent model [75].
- Homogenisation models: Mori–Tanaka (M-T) model [76]; self-consistent (S-C) model [77]; bridging model [78].

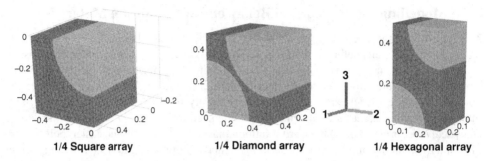

Figure 1.7   Meshing of square, diamond and hexagonal array unit cells.

Numerical FE modelling is widely used in predicting the mechanical properties of composites. Numerical modelling is a reliable tool, but the time consumed on the geometrical dimensions definition and the corresponding calculation time represent a major disadvantage against analytical models. Moreover there are many discussions and studies that deal with the appropriate boundary, symmetric and periodic conditions required to evaluate the elastic properties of UD composites. In this domain, a major work was done by Li [79]. It should be noticed that numerical FE modelling requires geometrical modelling or representation of the unit cell. For UD composites, there are three types of idealised fibre arrangements: square array, diamond array and hexagonal array (Figure 1.7).

1.5.2   Modelling of Laminated and Textile Composites

1.5.2.1   Analytical Modelling

Analytical models used for laminated and textile composites rely on two major factors for determining the stiffness of laminates and textiles: geometrical modelling and the homogenisation method. Geometrical modelling yields the required components for the homogenisation method, such as the volumes of the representative elementary volume (REV), layers, yarns and yarn sub-volumes. Furthermore, geometrical modelling gives the fibre volume fractions in the REV and in yarns as well as the orientations of yarns and yarn sub-volumes.

It should be noticed that the general scheme of homogenisation methods adopted for almost all analytical models are mainly based on a two-step homogenisation procedure: micro and macro levels. At micro level, the sub-volumes of subdivided REV are considered as unidirectional lamina with long fibres. Homogenisation at micro level is achieved, to predict the stiffness matrix of sub-volumes in a local coordinate system (123) where 1 is along the fibres. At this level, micro-mechanical models are employed.

At macro level, the transformation matrices are used to estimate the stiffness matrix of sub-volumes in a global coordinate system (XYZ) related to the REV. Then, a homogenisation method for the macro level is used.

For laminated composites, the famous classical lamination theory (CLT) method is the most reliable and used homogenisation method. However, for textiles, various homogenisation methods and approaches are used to assemble different sub-volumes, yarns and layers in order to predict the global stiffness matrix of the REV. These methods are based on: iso-strain assumptions, iso-stress assumption, mixed iso-strain/iso-stress assumptions, or inclusion methods (Mori–Tanaka or self-consistent methods). A brief review of recent works concerning the modelling of elastic properties of textiles will be presented, while detailed review of these methods can be found in [80–84].

Historically, Ichikawa and Chou [84–88] were the first who tried to find analytical models in order to predict the elastic properties of 2D woven composites. They proposed three basic analytical models: the mosaic model [84], the fibre undulation model [84] and the bridging model [85]. Later, many models were proposed as extensions of these models [89–91].

Recently, most works have been conducted towards the modelling of textiles with complex geometries such as 3D n-directional braided composites and 3D woven.

The iso-strain model was used by Li *et al.* [92] for modelling the elastic behaviour of 3D five-directional braided composites. The predicted results showed good agreement with experimental results for axial Young's modulus and tensile strength with an error not exceeding 10%. In addition, a 3D iso-strain model was presented by Shokrieh and Mazloomi [93]. The model was used to evaluate the stiffness of tri-axially braided composite. They compared their results to those experimental data presented by Quek *et al.* [94]. Results showed good predictions of the transverse Young's modulus E_y and the in-plane shear modulus G_{xy}.

Concerning 3D woven composites, Yanjun *et al.* [95] and El-Hage *et al.* [96] used the iso-strain model to predict the effective elastic properties of 3D angle interlock woven ceramic composite and orthogonal and angle interlock woven composites. Also, mixed iso-strain/iso-stress models have been introduced in order to improve the prediction of elastic properties of such complex composites. Pochiraju and Chou [97, 98] proposed a model for the prediction of anisotropic elastic stiffness of 3D textiles. The stiffness of the REV, decomposed into small subdivisions, is evaluated using an effective response comparison (ERC) technique. Concerning the angle interlock composites, the model yields good results in comparison to experimental data. Later, Tan *et al.* [99–102] developed two analytical models, the ZXY and ZYX models, to determine the mechanical properties and the thermal expansion coefficients of 3D orthogonal and angle interlock woven fabrics. The idealised REV is decomposed into sub-blocks which are assembled, under iso-strain and iso-stress conditions, using three models: the X model, Y model and Z model along the x, y and z directions, respectively. The comparison with experimental results has shown that the models predict well the effective elastic properties of 3D orthogonal woven composites while less agreement is obtained in estimating the longitudinal Young's modulus E_x of through-the-thickness 3D angle interlock woven composite [99]. However, Hallal *et al.* [103, 104] proposed analytical models in order to enhance the prediction of E_x of 3D angle interlock woven composite, especially for composites with a high content of undulated yarns. The models are also based on an assembling scheme of REV subdivisions in parallel, under the iso-strain condition, and in series, under the iso-stress condition. In [104], it is noticed that the iso-strain model have significant problems in predicting Young's modulus along a direction where the volume fraction, in the REV, of undulated yarns is high (x direction). However, the improved three-stage homogenisation method (3SHM) model has shown very good agreement with both experimental and numerical finite element (FE) results, especially for E_x.

1.5.2.2   Numerical FE Modelling

The prediction of the elastic properties and the stress –strain fields of laminated and textile composites have taken major interest with numerical FE modelling. The major problems remain in the time consumption and the meshing of complex geometries, as well as defining the contact between yarns and matrix.

A brief review of modelling composites with FE methods is presented. Cox *et al.* [105] and Xu *et al.* [106] developed a FE model, the binary model, to predict the mechanical properties of 3D interlock composites. Nie *et al.* [107] introduced the problem that represented the angle-interlock fabric in a more concise and efficient way, which is a major problem for the computer-aided design (CAD) of multi-layered angle-interlock woven fabric, as referred by Ko [108]. In the approach of Barbero *et al.* [109], a FE model of plain weave fabrics based on geometrical measurements from photographs was developed to determine the damage evolution using a meso-mechanical continuous damage formulation under tensile loading. Also a multi scale 3D mosaic model was applied for woven composites by Bogdanovich [110] where the stress and failure of 3D woven composites were studied. Moreover, Nehme *et al.* [111] used different meshing techniques, with tetrahedral mesh for the isotropic matrix and mapped hexaedral mesh for the transversely isotropic yarns, in order to achieve the FE discretisation.

1.6   Discussion

From the above sections, it is clearly shown from recent applications of composites in automotive industries that most attention is on using composites for crashworthiness applications in order to enhance car safety and to reduce the effect of impacts. In addition, there is an increasing consideration of a potential use of green composites in large masses by replacing glass fibre reinforced composites. Moreover, composites have been also used for other structural components, such as driveshaft, springs, brake discs, gas turbine and so on. However, the emergence of new types of composites has led to further development of new enhanced manufacturing processes for a large number of structures made from composites such as resin transfer moulding (RTM), performing and the liquid compression moulding manufacturing process.

Concerning the crashworthiness applications, the use of composites as the main structural component or as part of the hybrid material in tubular structures has taken the most consideration and most research. It is important to notice that these studies started more than 30 years ago. The important physical factors studied were the specific energy absorption and the work rate decay necessary to keep deceleration below a certain value during impact (<20 *g*). However, many parameters have been investigated in order to observe their effects on these factors, including the material kind, type of composite, stacking sequence of laminated composites and fibre orientation. Also, the review study reveals that for architecture, laminated composites are the most used; however, more recently braided composites started to receive more attention with their high shear strength. Moreover, glass fibres and carbon fibres are the most studied reinforcement with polymeric matrices. It is also clear the mechanical advantage that graphite and carbon fibres brought over glass fibre reinforced polymers and traditional materials. Concerning hybrids, the influence on the energy absorption of a surface treatment between the aluminium layer and carbon fibre reinforced polymer is studied. In addition, carbon short fibre reinforced polymers show great potential, with relatively low manufacturing cost and good mechanical properties concerning an impact crash.

Recent research also revealed that new directions in materials design tend to adopt composites as the first choice material for automotive structural parts that are subjected to severe static and dynamic loads. In these cases, the ultimate strength as well as the fatigue resistance of the material is crucial. One of these applications is the composite drive shaft. Finite element simulations have been conducted in order to find an optimised material that has high buckling strength, can resist the torsional fatigue loading and can increase the fundamental natural bending frequency. Hybridisation of studied driveshafts is done with carbon-epoxy UD/aluminium or carbon fibre-epoxy/glass fibre-epoxy laminates. All of the studied composites are made from long fibres, where the fibre orientation and the stacking sequence of laminate layers, as well as the material type (carbon fibre or glass fibre) are well investigated. It is well found these parameters have a good influence on the mechanical properties of driveshafts.

Moreover, other new applications of composites in automotive domains are shown, with E-glass/epoxy springs that could be used for heavy trucks and could save much weight, high performance gas turbine, advanced carbon ceramic brake discs and so on.

As recently observed, new natural fibres have emerged to substitute glass fibres for mostly interior automotive parts. Naturel fibre reinforced polymers are known as green composites due to their ecological and economic effects. Natural fibres such as woods, jute, Kenaf and flax have lower densities than most used glass fibres. It is found that these composites could have a promising future especially with automotive applications, while until now the dominant natural fibres are wood fibres. Much of the research observed recently just focused on some of the mechanical properties of these composites in comparison with glass fibre reinforced polymers, with more focus on their ecological and economical influences.

In general, it is remarked how composites gives many opportunities to offer solutions for many automotive applications. However, the large variety of fibres, matrices and architectures to form a composite material, led to much research work involving the optimisation and material design of automotive structures. In addition, it is noticed that unidirectional long-fibre laminated composites well dominate the research and world market concerning automotive applications. While some attention is found related with more advanced types of composites, such as braided composites and nanocomposites, there is almost an absence of 3D woven and knitted composites.

1.7 Conclusion

This chapter has presented an overview of recent applications of composites in automotive domains. The use of composites is investigated for saving weight and enhancing stiffness and strength against fatigue or crash, which leads to the design of safer vehicles. It is shown that the advantage of composites, especially by saving weight or the use of natural fibres has enormous ecological and economic effects.

It is found that most attention has been conducted to use composites for structural components to enhance crashworthiness. In addition, the use of composites will lead to saving much weight, when replacing traditional materials such as steel. Most of the composites used for tubular parts or plates consist of laminates. The influence of fibre types and laminate stacking sequence are the most investigated. However, the use of tri-axial braided composites as a more advanced composite has also been noticed in recent works. It is found that carbon and graphite fibres possess higher mechanical properties, while their manufacturing cost still presents a great disadvantage.

Other automotive parts could benefit from composites, such as driveshafts, springs, gas turbine and brake discs. Composite laminates made from carbon fibre or glass fibre reinforced polymers are the most used for driveshafts and springs. From numerical finite element simulations, it is found that composites enhance the fatigue life as well as the stiffness of these components subjected to heavy dynamic load. For gas turbine and brake discs, advanced carbon ceramic composites have been used in order to sustain both thermal stresses and mechanical loads. Due to their high cost, the use of carbon ceramic brake discs is still limited to racing cars and modern supercars.

It is also significant to notice the attention green composites have taken recently. These composites, formed from natural fibre reinforced polymers, have been used to replace glass fibre reinforced polymers for many automotive parts. The ecological and economic benefits that green composites could afford lead to them being considered as a promising material.

References

[1] McWilliams, A. (2007) Advanced Materials, Lightweight Materials in Transportation, report, Report Code: AVM056A.

[2] Ghassemieh, E. (2011) *Materials in Automotive Application, State of the Art and Prospects, New Trends and Developments in Automotive Industry*, InTech, Marcello Chiaberge, ISBN: 978-953-307-999-8.

[3] Taub, A., Krajewski, P., Luo, A. and Owens, J. (2007) The Evolution of Technology for Materials Processing over the Last 50 Years: The Automotive Example. JOM, February.

[4] Auto News (2012) www.compositestoday.com, Posted on 9th October, 2012 in Automotive News (Accessed 8th February 2013).

[5] Auto News (2012) www.compositestoday.com, Posted on 15th June, 2012 in Automotive News (Accessed 8th February 2013).

[6] Brooks, R. (2000) *Composites in Automotive Applications: Design, Comprehensive Composite Materials*, Volume 6: Design and Applications, Springer, New York, pp. 341–363.

[7] Schouwenaars, R., Cerrud, S. and Ortiz, A. (2002) Mechanical analysis of fracture in an automotive radiator head produced from a nylon-glass fibre composite. *Composites: Part A*, **33**, 551–558.

[8] Lee, S. and Lee, D. (2005) Composite hybrid valve lifter for automotive engines. *Composite Structures*, **71**, 26–33.

[9] Imihezri, S., Sapuan, S.M, Sulaiman, S. *et al.* (2006) Mould flow and component design analysis of polymeric based composite automotive clutch pedals. *Journal of Materials Processing Technology*, **171**, 358–365.

[10] Beetz, C.P., Schmueser, D.W. and Hansen, W. (1989) Summary of panel discussion, "Challenges to the Researchers of Carbon Fibers and Composites from the Automotive and Boatbuilding Industries". *Carbon*, **27**(5), 767–771.

[11] Chang, D.C. and Khetan, R.P. (1980) *Advances in Material Technology in the Americas*, Volume 1-Material Recovery and Utilization, ASME, New York, p. 51.

[12] Feraboli, P. and Masini, A. (2004) Development of carbon/epoxy structural components for a high performance vehicle. *Composites: Part B*, **35**, 323–330.

[13] Feuillade, V., Bergeret, A., Quantin, J.-C. and Crespy, A. (2006) Characterisation of glass fibres used in automotive industry for SMC body panels. *Composites: Part A*, **37**, 1536–1544.

[14] Ning, H., Pillay, S. and Vaidya, U. (2009) Design and development of thermoplastic composite roof door for mass transit bus. *Materials and Design*, **30**, 983–991.

[15] Palanivelu, S., Paepegem, W.V., Degrieck, J. *et al.* (2011) Crushing and energy absorption performance of different geometrical shapes of small-scale glass/polyester composite tubes under quasi-static loading conditions. *Composite Structures*, **93**, 992–1007.

[16] Zarei, H., Kroger, M. and Albertsen, H. (2008) An experimental and numerical crashworthiness investigation of thermoplastic composite crash boxes. *Composite Structures*, **85**, 245–257.

[17] Pitarresi, G., Carruthers, J.J., Robinson, A.M. *et al.* (2007) A comparative evaluation of crashworthy composite sandwich structures. *Composite Structures*, **78**, 34–44.

[18] Guedes, R.M. (2007) Durability of polymer matrix composites: viscoelastic effect on static and fatigue loading. *Composites Science and Technology*, **67**, 2574–2583.

[19] Hosseinzadeh, R., Shokrieh, M. and Lessard, L. (2006) Damage behavior of fiber reinforced composite plates subjected to drop weight impacts. *Composites Science and Technology*, **66**, 61–68.

[20] Bartus, S.D. and Vaidya, U.K. (2005) Performance of long fiber reinforced thermoplastics subjected to transverse intermediate velocity blunt object impact. *Composite Structures*, **67**, 263–277.

[21] Lanzi, L., Castelletti, L.M.L. and Anghileri, M. (2004) Multi-objective optimisation of composite absorber shape under crashworthiness requirements. *Composite Structures*, **65**, 433–441.

[22] Mamalis, A.G, Manolakos, D.E., Ioannidis, M.B. and Papapostolou, D.P. (2004) Crashworthy characteristics of axially statically compressed thin-walled square CFRP composite tubes: experimental. *Composite Structures*, **63**, 347–360.

[23] Lee, D.G., Lim, T.S. and Cheon, S.S. (2000) Impact energy absorption characteristics of composite structures. *Composite Structures*, **50**, 381–390.

[24] Findik, F. and Tarim, N. (2003) Ballistic impact efficiency of polymer composites. *Composite Structures*, **61**, 187–192.

[25] Davoodi, M.M., Sapuan, S.M. and Yunus, R. (2008) Conceptual design of a polymer composite automotive bumper energy absorber. *Materials and Design*, **29**, 1447–1452.

[26] Bisagni, C., Di Pietro, G., Fraschini, L. and Terletti, D. (2005) Progressive crushing of fiber-reinforced composite structural components of a formula one racing car. *Composite Structures*, **68**, 491–503.

[27] Launay, A., Marco, Y., Maitournam, M.H. *et al.* (2010) Cyclic behavior of short glass fiber reinforced polyamide for fatigue life prediction of automotive components. *Procedia Engineering*, **2**, 901–910.

[28] Ruggles-Wrenn, M.B., Corum, J.M. and Battiste, R.L. (2003) Short-term static and cyclic behavior of two automotive carbon-fiber composites. *Composites: Part A*, **34**, 731–741.

[29] Verpoest, I. (2009) *Micromechanics of Continuous and Short Fibre Composites*, Department of Metallurgy and Materials Engineering, Katholieke Universiteit Leuven, Belgium.

[30] Hexcel (2012) www.hexcel.com (Accessed 8th February 2013).

[31] Jacob, G., Fellers, J., Simunovic, S. and Starbuck, J. (2002) Energy absorption in polymer composites for automotive crashworthiness. *Journal of Composite Materials*, **36**(7), 813–850.

[32] Thornton, P.H. and Edwards, P.J. (1982) Energy absorption in composite tubes. *Journal of Composite Materials*, **16**(6), 521–545.

[33] Farley, G.L. (1991) The effects of crushing speed on the energy-absorption capability of composite tubes. *Journal of Composite Materials*, **25**(10), 1314–1329.

[34] Hamada, H., Ramakrishna, S. and Sato, H. (1996) Effect of fiber orientation on the energy absorption capability of carbon fiber/PEEK composite tubes. *Journal of Composite Materials*, **30**(8), 947–963.

[35] Lee, D.G., Kim, H.S., Kim, J.W. and Kim, J.K. (2004) Design and manufacture of an automotive hybrid aluminium/composite driveshaft. *Composite Structures*, **63**, 87–99.

[36] Chiu, C.H., Tsai, K.H. and Huang, W.G. (1998) Effects of braiding parameters on energy absorption capability of triaxially braided composite tubes. *Journal of Composite Materials*, **32**, 1964–1983.

[37] Bouchet, J., Jacquelin, E. and Hamelin, P. (2000) Static and dynamic behavior of combined composite aluminium tube for automotive applications. *Composites Science and Technology*, **60**, 1891–1900.

[38] Brighton, A., Forrest, M., Starbuck, M. *et al.* (2009) Strain rate effects on the energy absorption of rapidly manufactured composite tubes. *Journal of Composite Materials*, **43**(20), 2183–2200.

[39] Corum, J.M., Battiste, R.L. and Ruggles-Wrenn, M.B. (2003) Low-energy impact effects on candidate automotive structural composites. *Composites Science and Technology*, **63**, 755–769.

[40] Jacob, G., Starbuck, J., Fellers, J. *et al.* (2006) Crashworthiness of various random chopped carbon fiber reinforced epoxy composite materials and their strain rate dependence. *Journal of Applied Polymer Science*, **101**, 1477–1486.

[41] Lee, D.G., Kim, J.W. and Hwang, H.Y. (2004) Torsional fatigue characteristics of aluminium –composite co-cured shafts with axial compressive preload. *Journal of Composite Materials*, **38**(9), 737–756.

[42] Abu Talib, A., Ali, A., Badie, M. *et al.* (2010) Developing a hybrid, carbon/glass fiber-reinforced epoxy composite automotive driveshaft. *Materials and Design*, **31**, 514–521.

[43] Badie, M.A., Mahdi, E. and Hamouda, A.M.S. (2011) An investigation into hybrid carbon/glass fiber reinforced epoxy composite automotive driveshaft. *Materials and Design*, **32**, 1485–1500.

[44] Abu Talib, A.R., Ali, A., Goudah, G. *et al.* (2010) Developing a composite based elliptic spring for automotive applications. *Materials and Design*, **31**, 475–484.

[45] Ruban, S., Heudier, L., Jamois, D. *et al.* (2012) Fire risk on high-pressure full composite cylinders for automotive applications. *International Journal of Hydrogen Energy*, **37**, 17630–17638.

[46] Vasconcellos, M.A.Z., Hinrichs, R., da Cunha, J.B.M. and Soares, M.R.F (2010) Mossbauer spectroscopy characterization of automotive brake disc and polymer matrix composite (PMC) pad surfaces. *Wear*, **268**, 715–720.

[47] Puri, P., Compston, P. and Pantano, V. (2009) Life cycle assessment of australian automotive door skins. *The International Journal of Life Cycle Assessment*, **14**, 420–428.

[48] Marzbanrad, J., Alijanpour, M. and Saeid Kiasat, S. (2009) Design and analysis of an automotive bumper beam in low-speed frontal crashes. *Thin-Walled Structures*, **47**, 902–911.

[49] Schouwenaars, R., Cerrud, S. and Ortiz, S. (2002) Mechanical analysis of fracture in an automotive radiator head produced from a nylon-glass fibre composite. *Composites: Part A*, **33**, 551–558.

[50] Brembo (2012) www.brembo.com (Accessed 8th February 2013).

[51] Kaya, H. (1999) The application of ceramic-matrix composites to the automotive ceramic gas turbine. *Composites Science and Technology*, **59**, 861–872.

[52] Garcos, J.M., Moll, D.J., Bicerano, J. *et al.* (2000) Polymeric nanocomposites for automotive applications. *Advanced Materials*, **12**(23), 1835–1839.

[53] Giannelis, E.P (1996) Polymer layered silicates nanocomposites. *Advanced Materials*, **8**(1), 29–35.

[54] Kurauchi, T., Okada, A., Nomura, T. *et al.* (1991) Nylon 6-clay hybrid - synthesis, properties and application to automotive timing belt cover. SAE Technical Paper 910 584.

[55] Du, Y., Yan, N. and Kortschot, M. (2012) Light-weight honeycomb core sandwich panels containing biofiber-reinforced thermoset polymer composite skins: fabrication and evaluation. *Composites: Part B*, **43**, 2875–2882.

[56] Akil, H.M., Omar, M.F., Mazuki, A.A.M. *et al.* (2011) Kenaf fiber reinforced composites: a review. *Materials and Design*, **32**, 4107–4121.

[57] Summerscales, J., Dissanayake, N., Virk, A. and Hall, W. (2010) A review of bast fibres and their composites. Part 2 – Composites. *Composites: Part A*, **41**, 1336–1344.

[58] Joshi, S.V., Drzal, L.T., Mohanty, A.K. and Arora, S. (2004) Are natural fiber composites environmentally superior to glass fiber reinforced composites? *Composites: Part A*, **35**, 371–376.

[59] Jacob, M. and Thomas, S. (2008) Biofibers and biocomposites. *Carbohydrate Polymers*, **71**, 343–364.

[60] Koronis, G., Silva, A. and Fontul, M. (2013) Green composites: a review of adequate materials for automotive applications. *Composites: Part B*, **44**, 120–127.

[61] Ashori, A. (2008) Wood–plastic composites as promising green-composites for automotive industries! *Bioresource Technology*, **99**, 4661–4667.

[62] Suddell, B. and Evans, W. (2005) Natural fibers, biopolymers, and biocomposites, in *Natural Fiber Composites in Automotive Applications* (eds A.K. Mohanty, M. Misra and T.L. Drzal), CRC Press.

[63] Wambua, P., Ivens, J. and Verpoest, I. (2003) Natural fbres: can they replace glass in fibre reinforced plastics? *Composites Science and Technology*, **63**, 1259–1264.

[64] Davoodi, M., Sapuan, S.M., Ahmad, D. *et al.* (2010) Mechanical properties of hybrid kenaf/glass reinforced epoxy composite for passenger car bumper beam. *Materials and Design*, **31**, 4927–4932.

[65] EU (2002) Annual Report of the Government–Industry Forum on Non-Food Uses of Crops, Department of Environment, Food and Rural Affairs Publications, EU, August.

[66] La Mantia, F.P. and Morreale, M. (2011) Green composites: a brief review. *Composites: Part A*, **42**, 579–588.

[67] Zini, E. and Scandola, M. (2011) Green composites: an overview. *Polymer Composites*, **32**(12), 1905–1915.

[68] Bismarck, A., Baltazar-Y-Jimenez, A. and Sarlkakis, K. (2006) Green composites as Panacea? Socio-economic aspects of green materials. *Environment, Development and Sustainability*, **8**(3), 445–463.

[69] Gassan, J. and Bledzki, A.K., (2000) Possibilities to improve the properties of natural fiber reinforced plastics by fiber modification – jute polypropylene composites. *Applied Composite Materials*, **7**(5–6), 373–385.

[70] Alves, C., Ferrao, P.M.C., Silva, A.J. *et al.* (2010) Ecodesign of automotive components making use of natural. Jute fiber composites. *Journal of Cleaner Production*, **18**, 313–327.

[71] Reuss, A. (1929) Berechnung der Fliessgrense von Mischkristallen auf Grund der Plastizitätsbedingung für Einkristalle. *Zeitschrift Angewandte Mathematik und Mechanik*, **9**, 49–58.

[72] Halpin, J.C. and Kardos, J.L. (1976) The Halpin-Tsai equations: a review. *Polymer Engineering and Science*, **16**(5), 344–352.

[73] Chamis, C.C. (1989) Mechanics of composite materials: past, present, and future. *Journal of Composite Technology and Research*, **11**, 3–14.

[74] Hashin, Z. and Rosen, B.W. (1964) The elastic moduli of fiber reinforced materials. *Journal of Applied Mechanics*, **31**, 223–232.
[75] Christensen, R.M. (1990) A critical evaluation for a class of micromechanics models. *Journal of Mechanics and Physics of Solids*, **38**(3), 379–404.
[76] Mori, T. and Tanaka, K. (1973) Average stress in matrix and average elastic energy of materials with misfitting inclusions. *Acta Metallurgica*, **21**, 571–574.
[77] Hill, R. (1965) Theory of mechanical properties of fibre-strengthen materials-III. Self-consistent model. *Journal of Mechanics and Physics of Solids*, **13**, 189–198.
[78] Huang, Z.M. (2001) Micromechanical prediction of ultimate strength of transversely isotropic fibrous composites. *International Journal of Solids and Structures*, **38**, 4147–4172.
[79] Li, S. (2008) Boundary conditions for unit cells from periodic microstructures and their implications. *Composites Science and Technology*, **68**, 1962–1974.
[80] Tan, P., Tong, L. and Steven, G.P. (1997) Modelling for predicting the mechanical properties of textile composites – a review. *Composites: Part A*, **28**, 903–922.
[81] Crookston, J.J., Long, A.C. and Jones, I.A. (2005) A summary review of mechanical properties prediction methods for textile reinforced polymer composites. *Proceeding of IMechE, Part L: Journal of Materials: Design and Applications*, **219**, 91–109.
[82] Ansar, M., Xinwei, W. and Chouwei, Z. (2011) Modeling strategies of 3D woven composites: a review. *Composite Structures*, **93**, 1947–1963.
[83] Hallal, A., Younes, R. and Fardoun, F. (2013) Review and comparative study of analytical modeling for the elastic properties of textile composites. *Composites Part B: Engineering*, Available online 8 February 2013, ISSN 1359-8368. doi: 10.1016/j.compositesb.2013.01.024
[84] Ishikawa, T. and Chou, T.W. (1982) Elastic behaviour of woven hybrid composites. *Journal of Composite Materials*, **16**(1), 2–19.
[85] Ishikawa, T. and Chou, T.W. (1982) Stiffness and strength behaviour of woven fabric composites. *Journal of Material Sciences*, **17**, 3211–3220.
[86] Ishikawa, T. and Chou, T.W. (1983) One-dimensional micromechanical analysis of woven fabric composites. *AIAA Journal*, **21**(12), 1714–1721.
[87] Ishikawa, T. and Chou, T.W. (1983) In-plane thermal expansion and thermal bending coefficients of fabric composites. *Journal of Composite Materials*, **17**(2), 92–104.
[88] Ishikawa, T. and Chou, T.W. (1983) Nonlinear behaviour of woven fabric composites. *Journal of Composite Materials*, **17**(5), 399–413.
[89] Yang, J.M., Ma, C.L. and Chou, T.W. (1986) Fiber inclination model of three-dimensional textile structural composites. *Journal of Composite Materials*, **20**(5), 472–484.
[90] Whitney, T.J. and Chou, T.W. (1989) Modeling of 3-D angle-interlock textile structural composites. *Journal of Composite Materials*, **23**(9), 890–911.
[91] Byun, J.H., Whitney, T.J., Du, G.W. and Chou, T.W. (1991) Analytical characterization of two-step braided composites. *Journal of Composite Materials*, **25**(12), 1599–1618.
[92] Li, D.S., Lu, Z.X., Chen, L. and Li, J.L. (2009) Microstructure and mechanical properties of three-dimensional five-directional braided composites. *International Journal of Solids and Structures*, **46**, 3422–3432.
[93] Shokrieh, M. and Mazloomi, M. (2010) An analytical method for calculating stiffness of two-dimensional tri-axial braided composites. *Composite Structures*, **92**, 2901–2905.
[94] Quek, S.C., Waas, A.M., Shahwan, K.W. and Agaram, V. (2003) Analysis of 2-D triaxial flat braided textile composites. *International Journal of Mechanical Sciences*, **45**, 1077–96.
[95] Yanjun, C., Guiqiong, J., Bo, W. and Wei, L. (2006) Elastic behavior analysis of 3D angle-interlock woven ceramic composites. *Acta Mechanica Solida Sinica*, **19**(2), 152–159.
[96] El-Hage, C., Younes, R., Aboura, Z. *et al.* (2009) Analytical and numerical modeling of mechanical properties of orthogonal 3D CFRP. *Composites Science and Technology*, **69**, 111–116.
[97] Pochiraju, K. and Chou, T.W. (1999) Three-dimensionally woven and braided composites. I: a model for anisotropic stiffness prediction. *Polymer Composites*, **20**(4), 565–580.
[98] Pochiraju, K. and Chou, T.W. (1999) Three-dimensionally woven and braided composites. II: an experimental characterization. *Polymer Composites*, **20**(6), 737–747.
[99] Tan, P., Tong, L. and Steven, G.P. (1999) Micromechanics models for mechanical and thermomechanical properties of 3D through-the-thickness angle interlock woven composites. *Composites: Part A*, **30**, 637–648.

[100] Tan, P., Tong, L., Steven, G.P. and Ishikawa, T. (2000) Behavior of 3D orthogonal woven CFRP composites, part I. Experimental investigation. *Composites: Part A*, **31**, 259–271.

[101] Tan, P., Tong, L. and Steven, G.P. (2000) Behavior of 3D orthogonal woven CFRP composites. part II, FEA and analytical modeling approaches. *Composites: Part A*, **31**, 273–281.

[102] Tan, P., Tong, L. and Steven, G.P. (2001) Mechanical behavior for 3-d orthogonal woven E-Glass/epoxy composites. *Journal of Reinforced Plastics and Composites*, **20**(4), 274–303.

[103] Hallal, A., Younes, R., Nehme, S. and Fardoun, F. (2011) A corrective function for the estimation of the longitudinal young's modulus in a developed analytical model for 2.5D woven composites. *Journal of Composite Materials*, **45**(17), 1793–1804.

[104] Hallal, A., Younes, R. and Fardoun, F. (2012) Improved analytical model to predict the effective elastic properties of a 2.5D interlock woven fabric composite. *Composite Structures*, **94**, 3009–3028.

[105] Cox, B.N., Carter, W.C. and Fleck, N.A. (1994) A binary model of textile composites-I. Formulation. *Acta Metallurgica et Materialia*, **42**, 3463–3479.

[106] Xu, J., Cox, B.N., McGlockton, M.A. and Carter, W.C., (1995) A binary model of textile composites-II. The elastic regime. *Acta Metallurgica et Materialia*, **43**, 3511–3524.

[107] Nie, J., Lu, S. and Gu, B., (2006) Fractional formula description of angle-interlock woven fabric construction. *Journal of Industrial Textiles*, **36**(2), 125–132.

[108] Ko, F.K. (1999) *3D Textile Reinforced in Composite Materials, 3-D Textile Reinforcements in Composite Materials*, Woodhead Publishing, CRC Press, England, pp. 9–40, ISBN 1 85573 376 5.

[109] Barbero, E.J., Trovillion, J., Mayugo, J.A. and Sikkil, K.K. (2006) Finite element modelling of plain weave fabrics from photomicrograph measurements. *Composite Structure*, **73**, 41–52.

[110] Bogdanovich, A.E., (2006) Multi-scale modeling, stress and failure analyses of 3-D woven composites. *Journal of Materials Science*, **41**, 6547–6590.

[111] Nehme, S., Hallal, A., Fardoun, F. *et al.* (2011) Numerical/analytical methods to evaluate the mechanical behavior of interlock composites. *Journal of Composite Materials*, **45**(16), 1699–1716.

2

High-Volume Thermoplastic Composite Technology for Automotive Structures

Neil Reynolds and Arun Balan Ramamohan
WMG, University of Warwick, Warwick, UK

2.1 Introduction – Opportunities for Thermoplastic Composites

Targets set in order to dramatically reduce the environmental impact resulting from the usage of vehicles are leading to a research focus on alternative propulsion technologies and aggressive vehicle lightweighting solutions. New cars to be registered in Europe must meet fleet averaged emissions targets of 130 g CO_2/km in 2012 and 95 g CO_2/km by 2020, or vehicle manufacturers will face substantial financial penalties for each car sold that exceeds the emission limits [1].

Light-weighting to promote vehicle efficiency is a key theme in the UK's New Automotive Innovation and Growth Team (NAIGT) strategy. The research and development roadmap defines an industry-led short- and medium-term research agenda to deliver lightweight structures and functional integration of parts. Additionally, a portfolio of university focused research being carried out at lower technology readiness levels aims to deliver 50% weight savings beyond 2020 [2].

Reducing the weight of a vehicle will obviously reduce the fuel consumption and hence the CO_2 emissions per kilometre travelled. An average fuel reduction factor has been determined for a range of conventional light-duty internal combustion engine (ICE) platforms as 0.69 l/100 km for every 100 kg of weight save [3]; another study has found that a total vehicle mass reduction of approximately 15% could result in a 10% decrease in fuel consumption and hence CO_2 emissions [4]. The fuel reduction factor is strongly dependent upon vehicle segment and powertrain selection.

The vehicle body in white (BIW) and continuous load-bearing structural elements present an attractive target for mass-reduction engineering. Class leading manufacturers' BIW structures

Advanced Composite Materials for Automotive Applications: Structural Integrity and Crashworthiness,
First Edition. Edited by Ahmed Elmarakbi.
© 2014 John Wiley & Sons, Ltd. Published 2014 by John Wiley & Sons, Ltd.

constitute approximately 15–25% of a vehicle's total mass when manufactured in steel and approximately 10–15% for aluminium intensive vehicles [5]. Therefore, the ability to achieve weight savings across the vehicle primary structure readily translates into substantial reductions of total vehicle weight.

Automotive components manufactured using composite materials can make a significant contribution to meeting the required weight reduction targets, particularly in the case of vehicle structures. One study investigated the alternative materials and process technologies currently offered by suppliers to the automotive market that could deliver substantial mass reductions across all areas of a current production vehicle [6]. Different research and development levels were considered in line with materials and process maturity and the potential weight savings achievable. The "high development" scenario for the vehicle body structure would deliver a mass reduction of >40% as compared to the current production solution; some of this significant weight reduction was achieved via the use of composite material and process technologies that accounted for 21% of the total vehicle body structure by weight.

Figure 2.1 compares the performance of a range of composite material types against structural automotive steel and aluminium grades using the structural efficiency metrics of specific bending stiffness and specific buckling strength (mass normalised using each material's specific gravity). A range of composite materials are included in the comparison – random reinforced glass fibre materials (e.g. injection/compression-moulded) and aligned glass fibre and aligned carbon fibre reinforced laminate materials (stamp-formed and resin transfer moulded). Specific bending stiffness allows easy comparisons between the performance of different materials in panel type applications (e.g. roof, load-floor), reflecting the ability to

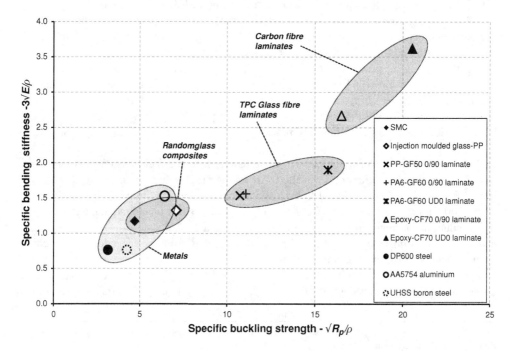

Figure 2.1   Material performance comparison using structural efficiency metrics.

create panel components more resistant to deflections in bending modes through increased part thickness with no increase in part weight. Specific buckling strength enables the comparison of material performance in beam applications (e.g. longitudinal elements in crash), accounting for relative material densities and the ability to obtain geometric advantage through the scaling of beam section properties.

In addition to exhibiting high specific mechanical properties such as stiffness and strength, a key benefit of using composite materials in automotive primary structures is superior crash energy management through the initiation and propagation of highly mass-efficient failure modes as compared to structures made from steel and aluminium.

A critical issue when considering material selection and substitution for an automotive component or assembly is the embodied energy within the material and the available end of life treatments of the part. European legislation currently stipulates that 85% of end of life vehicles must be reused or recovered and 80% must reused or recycled (by total vehicle weight); in 2015 these targets rise to 95% reuse/recovery and 85% reuse/recycling [7]. The defined reuse/recycling targets preclude the use of recovery techniques such as incineration for energy recovery, meaning only a small fraction of end of life vehicles can be recovered in this manner (currently 5% of vehicle by mass, rising to 10% in 2015). These stringent recycling targets do not present a problem for readily recycled steel and aluminium, but can be an issue for alternative materials such as some polymeric composites, particularly those based on thermosetting polymer matrices.

However, composites based on thermoplastic polymer matrices can be easily recycled through the process of mechanical chopping/grinding and then subsequent re-moulding, albeit with limited loss of mechanical properties. Despite the attendant loss of mechanical properties through mechanisms such as shortening of fibre reinforcement length and polymer matrix degradation (e.g. polymer chain scission), recent research shows that thermoplastic composite (TPC) recyclate materials still possess sufficiently high mechanical properties to be used in automotive applications and hence retain commercial value [8]. Due to this inherent recyclability, TPCs are therefore an attractive proposition for intensive use in automotive applications, as their usage provides no barrier to meeting recycling targets.

Furthermore, although thermoplastic polymer matrices exhibit generally lower mechanical properties than most thermosetting matrices in terms of stiffness, strength and creep performance, TPCs can offer significantly improved ductility over relatively brittle thermosetting polymers, providing improved damage tolerance, particularly in the case of low energy impact loading. These potential issues of stiffness, strength and creep can be readily overcome by selecting the optimal thermoplastic polymer, focusing on engineering polymers (e.g. polyamide instead of polypropylene) and by specifying a suitable fibre weight fraction and fibre architecture according to the desired application. The TPC glass fibre laminates included on the structural efficiency performance comparison in Figure 2.1 are based on polypropylene and polyamide matrices and can be seen to offer attractive relative properties.

2.2 Recent Developments in Automotive TPCs

The use of TPC materials within the automotive industry has generally been restricted to random glass-reinforced commodity polymeric materials for high-volume non-structural or semi-structural applications. Examples of typical non-structural/semi-structural automotive applications are exterior components, such as rear bumpers, under-body pans (e.g. PP-glass compression-moulded glass mat thermoplastic, GMT), and interior components, such as injection moulded door modules, centre consoles, dashboard mounts.

However, driven by material and processing innovations and enhanced computer-aided engineering (CAE) techniques, TPCs are now finding applications further into the primary vehicle structure. One important trend in TPC material developments includes the shift away from commodity polymers towards engineering polymers that have higher levels of fibre reinforcement whilst still retaining a high degree of mouldability. Composites suitable for injection moulding such as glass-filled polyamide (PA6, PA66, etc.), poly butyl terephthalate (PBT) and polyethylene terephthalate (PET) are available with a variety of fibre lengths and fibre weight fractions of up to 60%. Fibre lengths can vary from >10 mm (so-called long fibre technology, LFT) to highly formable <1 mm short fibre composites. Material suppliers are also offering carbon fibre and carbon/glass fibre hybrid compounds for even higher performance.

The structural potential of this family of flow-formed materials can be further enhanced through the use of "integrative" CAE techniques, whereby the fibre orientation that occurs due to flow alignment during the moulding process is modelled and optimised for a given component design and corresponding loading condition. The fibre distribution and orientation in so-called randomly reinforced flow-formed TPCs (e.g. injection-moulded) is rarely truly random, and the resultant reinforcement fibre architecture will be determined through key parameters such as mould gating [the location(s) at which the polymer–composite melt enters the mould] and overall part design that determines how the part fills during moulding. Once this moulding process can be reliably modelled, predictive design studies can be executed that take output data from the manufacturing process simulation that incorporates the fibre distribution and orientation and feeds it into the performance simulation. In this manner, the design of the part can be optimised for minimum weight and maximum mechanical performance by making the changes in part geometry and tool design to favourably influence fibre architecture.

This integrative design approach to structural automotive components has been used by BMW for the transmission support crossbeam of the Gran Turismo 550i [9]. The material used was a BASF Ultramid® A3WG10CR (PA66-GF50%), and the part was designed using BASF's proprietary ULTRASIMTM software process to gain increased performance through design optimisation and fibre flow alignment during the moulding process. The material was selected due to the requirement of good mechanical properties at elevated temperatures, maintaining a Young's modulus of approximately 10 GPa and a strength of 225 MPa at 100 °C. The production part is 50% lighter than the original aluminium casting and offers improved crash performance.

Another emergent technique used to increase the structural capability of high-volume TPCs is the augmentation of injection-moulded TPC semi-structural parts with discrete metal (typically aluminium) or aligned fibre thermoplastic composite (so-called organo-sheet) inserts. These flat or preformed inserts can be placed robotically into the injection mould tool prior to mould closure and are then over-moulded and incorporated into the component. This creates a structurally improved hybrid part, having enhanced performance in critical regions that experience high loading. Due to the high level of automation coupled with an already high-volume injection-moulding route, cycle times are in line with mass production expectations.

The new Audi A8 features several components that use this hybridisation technique. The first is the injection-moulded A8 front-end module that has been developed jointly with Lanxess. The component uses a combination of three pre-formed aluminium inserts and a single TPC laminate insert which are all over-moulded with Lanxess Durethan® PA6-GF30 to create the final part. The aligned fibre TPC insert is made from 1 mm thick continuous fibre laminate named Tepex® from Bond-Laminates and is thermo-formed in a secondary step prior to insertion into the injection mould tool [10]. The new front end module provides increased rigidity and strength to the front of the vehicle in comparison with typical front ends, whilst still being 50% lighter than the original design that used a steel insert.

The new A8 spare wheel is the second component to use hybridised composite mouldings and an integrative design methodology to provide improved properties whilst enabling mass reduction. The component is injection moulded using Durethan® PA6-GF60, includes an aluminium insert and has been developed by Audi in conjunction with Lanxess and Polynorm [11]. The component is of particular interest as it is a structural compartment designed to carry 70 kg payload (spare wheel, compressor, etc.) and also makes a contribution to the global stiffness and crash performance (rear impact) of the vehicle. The injection-moulding process also employs gas injection technology (GIT) to create rigid hollow sections. In the process, a critically timed pulse of high-pressure nitrogen (20 MPa) is injected into the mould through needles or nozzles at specific points; the gas initiates and enlarges a bubble in the chosen section, creating a hollow cross-section and displacing the polymer–composite melt into a designated tool overflow. It is the largest series production automotive part to use the GIT moulding process, particularly in such radical manner with a 2 kg material displacement overflow. A 2 kg overflow constitutes 30% of the total material injected, with approximately 4 kg of TPC in the finished part; the material displaced into the tool overflow is then recycled and re-used in subsequent moulding processes. The materials and techniques used to manufacture the part confer a 30% weight reduction over a functionally identical steel part.

Aside from these commercially available TPC production solutions for structural automotive applications, there are key research developments emerging from material and component suppliers. Lotus Engineering Ltd and Jacob Composite GmbH carried out a research project (Ecolite) to investigate the use of aligned fibre TPCs to manufacture front crash structures [12]. The front crash structure concept combined recent developments for low-speed crash energy management such as TPC bumper /crush can assemblies and expertise in thermosetting composite full crash structures (e.g. Lotus Elise, Aston Martin Vanquish), resulting in a component made using thermoformed PA6-GF45 laminate. Dynamic testing revealed a specific energy absorption rate close to resin transfer moulded (RTM) carbon fibre-epoxy (35–55 J/g for PA6-GF45 versus 35–80 J/g for epoxy-carbon). Accurate simulation models were created and run, and the thermoformed crash structure was predicted to be more economical than steel for volumes lower than 80 000 parts per year due to reduced investment costs in tooling and other capital. The thermoformed assembly was significantly lighter than a traditional steel crash structure at 16 kg versus 30 kg.

The carbon fibre manufacturer Teijin Ltd has developed a novel TPC material framework and applied it in the development and production of a concept demonstrator vehicle. The concept relies on a range of three intermediate (pre-impregnated or so-called pre-preg) carbon fibre reinforced TPC materials, used in different areas of the vehicle according to particular performance requirements [13]. The three materials comprise a uni-directionally reinforced prepreg for the highest stiffness and strength performance, a more isotropic material with multi-directional properties and finally an injection moulding grade for manufacturing parts with high geometric complexity. The composite material formats can be made using different thermoplastic polymeric matrices, such as polyamide and polypropylene. This portfolio of materials has been used to make a full vehicle structure that weighs 47 kg, using mass production processes that can deliver moulded parts in less than a minute (e.g. injection moulding).

2.3   Case Study: Rapid Stamp-Formed Thermoplastic Composites

The following practical case study into the use of TPCs for automotive primary structures was carried out at WMG as part of the ERDF-AWM funded Low Carbon Vehicle Technology Programme (LCVTP). Partners in the project were Jaguar–Land Rover, Tata Motors European Technical Centre (TMETC), WMG, Coventry University, Ricardo, Zytek and MIRA.

The remit of LCVTP was to develop and deploy technology and skills across the automotive supply base in the UK West Midlands region, significantly accelerating the introduction of low carbon technologies into the automotive industry, in line with the previously discussed NAIGT strategy. The programme of research and development was broken down into 15 discrete work streams of activity. These streams spanned hybrid/electric vehicle technology (battery performance, drive motors, power electronics, energy storage/recovery, auxiliary power units), vehicle systems (vehicle supervisory control, HVAC/system cooling), vehicle dynamics, parasitic losses, aerodynamic performance, lightweight structures and the application of the research on vehicle integration platforms.

The objective of the lightweight structures work stream was to research and develop materials and process technologies suitable for application in vehicle structures to enable a predicted weight save of at least 20% over conventional technology. The purpose of the weight saving was to facilitate reductions in vehicle usage phase related CO_2 emissions. The research objectives stipulated that the weight saving had to be achieved without incurring any detrimental environmental impacts when considering the entire product life cycle, such as manufacturing and end of life phases. Any materials and process technologies that were developed needed proof to a concept-level, and a range of individual tasks were formulated within the work stream to deliver the information and data to achieve this. Consideration was given to material selection, process development, CAE simulation tool development, component performance and the economic aspects and environmental impact associated with material substitution. Areas considered as being outside this concept-level research study included issues such as durability (fatigue performance) and reparability.

2.3.1   Materials Selection: Exploring the Potential of Aligned Fibre TPCs

Material choice was influenced by several key criteria – performance, manufacturing route (including production volume applicability), cost and environmental performance. The candidate material type clearly had to be suitable for structural applications and feasible for short- to medium-term application in a medium production volume automotive manufacturing environment. It was agreed with the project partners that the target production volumes under consideration would be circa 30 000 to 50 000 parts/year, with a clear route demonstrable to higher production volumes beyond this number. To comfortably achieve these volumes with some remaining overhead to allow future increases in production, a target cycle time of 60–90 s was defined. Another key factor driving the material selection process was high-temperature stability, as it was decided that the preferred production concept required any lightweight structures to be able to pass through a standard automotive paint line without suffering any dimensional issues such as distortion due to heat deflection. The maximum temperature recorded from an experiment by project partners whereby a full BIW was instrumented with thermocouples and put through the paint shop cycle was identified as being around 180 °C. The typical maximum in-service design temperature for automotive applications would be 100 °C.

Following a procedure of supplier engagement and industry review, where current and emergent materials/process technology was evaluated against these selection criteria, the candidate material was selected as continuous aligned glass-fi bre reinforced PA6. As already identified in Section 2.2, polyamide is emerging as the polymer matrix of choice for high-volume automotive applications, due to an attractive combination of price, performance, ease of processing and recyclability. Although the PA6 is also available with carbon fibre reinforcement, the high cost of carbon fibre means that the PA6-CF60 raw material is approximately five times the cost of the E-glass reinforced PA6-GF60 and hence not feasible for immediate mass-market application.

Aligned continuous fibre reinforced PA6-GF60 is available in several formats – extruded section, woven fibre (either PA-GF commingled fabric or pre-calendered sheets) and thin tape. Thin extruded tape was the preferred option as it confers the best mechanical properties, as laminate stacks have no crimp, as compared to woven fabrics, and the ability to infinitely vary the ply stacking sequence, whereas with bi-axial woven fabrics the smallest laminate building block is typically 0°/90°. Stitch-bonded unidirectional (UD) woven reinforcement fibres are commercially available, but TPC pre-pregs based on this architecture are not. An additional benefit of the continuous fibre reinforced tape over some of the woven laminates (supplied as either dry fabrics woven from commingled polymer and glass fibres or as calendered sheets) is that, in the tapes, the fibres are already fully impregnated with resin due to the high-pressure tape extrusion process, which leads to decreased void content and hence reduced strength in rapidly manufactured parts.

Figure 2.2 shows the dynamic mechanical thermal analysis (DMTA) data from an experiment on some PA6-GF60 raw tape material. It can be seen from this data that the material retains over 50% of its dynamic stiffness up to 200 °C, with an obvious drop in stiffness beyond 210 °C. The Tan(δ) data, which provides a measure of the ratio between the elastic (in-phase) and visco-elastic (out of phase) material response to the DMTA's mechanical perturbation reveals the melting temperature of the PA6 polymer matrix as between 215 and 225 °C, corresponding to the Tan(δ) reaching a maximum.

In addition to the high-temperature properties of the PA6 polymer matrix, the mechanical properties of laminates manufactured from the PA6-GF60 tape were studied and proven to be applicable to the structural applications, and the relevant CAE simulation model input data was generated. Figure 2.3 shows a plot of the static tensile properties of a variety of ply stacking sequences and the compression properties of the UD ply. The UD ply tested in 0° direction had an average Young's modulus and a tensile strength of 35 GPa and 730 MPa and an average compressive strength of 430 MPa. When the performance of the tape was evaluated in a laminate constructed in a bi-axial configuration, the on-axis average Young's modulus was 19 GPa and tensile strength was 340 Mpa; and tested off-axis at 45° to ascertain the in-plane shear performance, the strain to failure was over 17%. This portfolio of mechanical performance according to ply configuration demonstrates the way in which the polyamide-based laminate can be constructed for high strength and stiffness, high ductility, or a combination of the two.

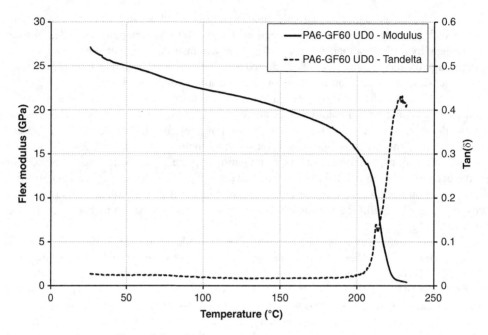

Figure 2.2   DMTA trace for PA6-GF60 extruded tape.

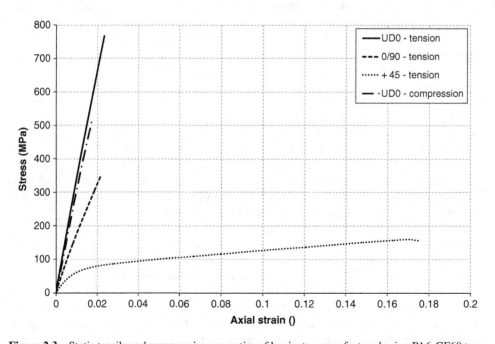

Figure 2.3   Static tensile and compressive properties of laminates manufactured using PA6-GF60 tape.

Figure 2.4   LCVTP longitudinal demonstrator beam CAD model (approx. 40 × 70 × 450 mm).

2.3.2   Demonstrator Beam Component

The selection and manufacture of a case-study demonstrator component enabled the practical performance evaluation of the candidate PA6-GF60 laminate material against typical automotive grade metallic benchmarks, such as steel and aluminium. Performing the practical build and test activities allowed key aspects of manufacturing (including joining) and mechanical performance to be investigated and established to the required concept-level.

A selection process was carried out that resulted in the decision to adopt a scaled front longitudinal section (so-called top-hat) as the demonstrator component (Figure 2.4). The component design was based on a stamped steel cross-section from an OEM partner's vehicle development programme, with a closure plate that would be joined as a secondary operation. The component cross-sectional dimensions (overall size) were downscaled to maintain compatibility with the test equipment available within the project consortium (e.g. maximum load), resulting in a beam section size of approximately 40 mm in height × 70 mm in width × 450 mm in length. Additionally, the part was simplified by removing the original section's slight longitudinal taper to promote ease of testing and manufacture.

Three beam variants were considered for the manufacture and test study (Table 2.1). Beam A was a structural steel beam and the baseline in this study for current automotive structural materials technology. The material for beam A was chosen as a cold-rolled dual-phase high-strength steel with an ultimate strength of 600 MPa, having a generic material name of DP600 [14].

Table 2.1   Case study longitudinal demonstrator beam variants.

Beam type	Top hat section	Closure plate	Beam weight (g)
A – baseline	DP600 1.5 mm	DP600 1.5 mm	1880
B	5754 aluminium 2.5 mm	5754 aluminium 2.5 mm	1020
C	PA6-GF60 3 mm, 11 layers	PA6-GF60 3 mm, 11 layers	790

Figure 2.5   TPC laminate concept supply chain.

Beam B was the lightweight metallic technology demonstrator made using high-performance aluminium, selected as cold-rolled AA5754 [15]. Beam C was the all-TPC beam manufactured from laminates assembled from extruded PA6-GF60 tape [16].

2.3.3   TPC Process Development

The candidate manufacturing process was developed with several key criteria in mind. The maximum target cycle time of the process was around 60 s, in order that the highest target production volumes could be met. A clear objective throughout the process development was to adopt a production approach that could be easily implemented in the existing automotive supply chain. The chosen route also had to be capable of delivering acceptable part quality and making parts greater than 1 m^2 using aligned fibre laminates. Finally, whilst meeting the above conditions, the candidate process had to still be economical at the lower end of the applicable annual production volumes of 30 000, particularly in terms of tooling investment costs.

A hot stamp-forming process was selected in order to fulfil all of the process selection requirements. Previous research projects undertaken by the research team at WMG had demonstrated that a 60 s cycle time could be met using a rapid stamp-forming process with polypropylene-glass based laminates, for material blank thicknesses below 5 mm. The production approach would match a higher tier supplier of pre-laminated, semi-finished sheet (made using the PA6-GF60 extruded tape) with a lower tier supplier press shop, as already supplies stamped sheet metal parts to the automotive industry (Figure 2.5). The only difference in the process at the lower tier for component-level production would be the necessity to install inexpensive continuous heating equipment (e.g. conveyor belt fed infra-red ovens, as already used for GMT processing) that could provide a regular supply of pre-heated blanks to the press station every 60–90 s. This concept allows that the existing press infrastructure would remain unchanged and could even remain interoperable with existing sheet metal processing activities. In terms of press capacity requirements, a 1 m^2 part stamp formed at 10 MPa (100 bar) would require a 1000 t press station.

Currently, the biggest barrier to the proposed manufacturing concept is the gap that exists in the supply chain regarding semi-finished (sheet) material blanks, at sufficiently high production volumes to meet the agreed annual production targets. Here, a high-volume process needs to be established and suppliers set up. It is considered that a semi-finished sheet production model would be based on a continuous automated tape laying (ATL) process; research and market development is ongoing in the area of ATL, some in the area of TPCs [17], but most research is applied to thermosetting prepregs as used for aerospace applications and some learning could be transferred from this work.

The hot-stamping process was implemented at WMG on a 100 t vertical clamp compression moulding press via the procurement of a bespoke oil-heated steel matched tooling set

Figure 2.6   Rapid TPC hot stamp-forming tool for LCVTP longitudinal demonstrator beam.

(Figure 2.6). A large (>2 m^2) contact heating lamination machine capable of reaching 250 °C was procured for the rapid pre-heating of the PA6-GF60 laminate material blanks.

2.3.4   Beam Manufacture

The metallic beams (beams A and B) were manufactured using a standard automotive industry sheet metal stamp forming process on purpose-built tooling at a supplier to one of the OEM project partners.

The composite beams (beam C) were manufactured using the candidate TPC hot stamp-forming process developed at WMG. Laminate pre-forms (11-ply, approx. 3 mm thick) were assembled by hand using a generic non-optimised cross-ply stacking sequence, where the 0° datum was in line with beam length (0°/90°/90°/0°/0°/90°/0°/0°/90°/90°/0°). The pre-forms were then heated to 235 °C and manually transferred to the heated tool for stamp-forming at approximately 10 MPa (100 bar). The time between loading the pre-heated blank and de-moulding the part was 60 s.

All beam variants were trimmed and assembled in the same way, mimicking standard automotive assembly processes. After stamping, the beams were trimmed to size using an industrial water-jet cutting process. They were then cleaned and joined via a hybrid joining process using adhesive and mechanical fixings (rivets and bonding, or riv-bonding). Single-part structural epoxy was applied to the stamped beam section flange and the closure plate was brought together with the beam using bespoke assembly tooling. The uncured assembly was then mechanically joined using a self-piercing rivet (SPR) process that was controlled by the assembly tooling. Finally, the assembled uncured parts were put in a hot-air oven that was programmed to replicate the automotive BIW paint-bake cycle, whereby the single part epoxy adhesive would heat-cure.

2.3.5 Demonstrator Beam Structural Performance

The mechanical performance of the TPC longitudinal demonstrator beams was evaluated against the steel benchmark and lightweight aluminium variant in static three-point flexure and dynamic axial crush.

The three-point bend test allowed the assessment of static structural capability of each material system, using an identical outer package (section) geometry whilst only varying the material gauge inwards from the fixed outer surface. It could be considered that the three-point test is analogous to automotive load cases such as side impact scenarios, where stiffness and energy absorption properties have to be balanced in order to minimise intrusion whilst also minimising occupant accelerations. The quasi-static flexure tests were conducted at 20 mm/min (0.333 mm/s) up to a maximum deflection of 50 mm. Critical performance parameters such as beam stiffness, load at ultimate failure, energy absorption rates and also failure mode (s) were assessed and compared for each material system.

Representative specific (mass-normalised) three-point load versus deflection curves are shown in Figure 2.7. It can be seen that the TPC beam performs favourably, with the highest specific peak load. Figure 2.8 shows the summarised specific performance data, specific beam stiffness, and specific energy absorbed to peak load. The beams exhibited a failure mode of progressive localised crush under the central loading point, and the local crush started at moderately low loads; this means that the beam stiffness data is a measure of both the resistance to flexure and resistance to local crush combined. It is obvious that the beams have similar specific flexural performance in terms of this mixed-mode flexural/crush deflection. The PA6-GF60 has a combination of toughness from the polyamide matrix and stiffness/strength from the high aligned glass content and this leads to enhanced specific energy absorption up to peak (failure) load. The peak load represents the point in the test when local crush starts to dominate the system response, after which the beam then typically starts to fold about the load introduction point.

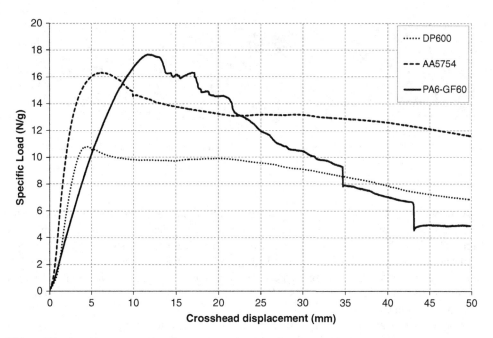

Figure 2.7 Specific (mass normalised) three-point flexure curves for steel, aluminium and PA6-GF60 demonstrator components.

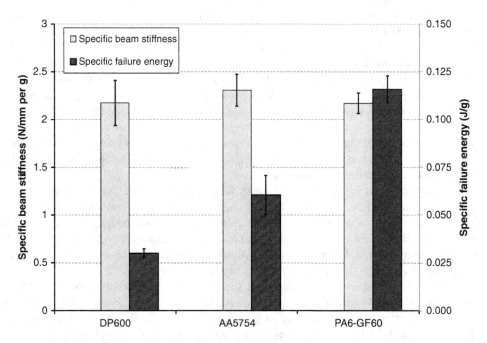

Figure 2.8 Comparison of specific three-point flexure performance of demonstrator beam variants.

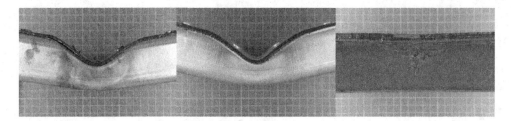

Figure 2.9 Failed central sections of (a) steel, (b) aluminium and (c) PA6-GF60 beams tested in three-point flexure.

A montage image comprising representative photographs of the central sections of tested DP600, AA6754 and PA6-GF60 beams to illustrate the typical failure modes is shown in Figure 2.9. The characteristic "V" shape resulting from the local crush of the highly ductile steel and aluminium sections is clearly evident, whereas the composite beam exhibits longitudinal and transverse cracking resulting from compressive side-wall loading and tensile loading on the upper beam surface as it has conformed to the central load introduction point (cylindrical roller).

In addition to static flexure, the beam variants were also tested in dynamic axial crush, simulating an automotive front longitudinal component in a frontal crash scenario. The beams were tested on an instrumented spring-assisted drop-weight tower at two different energy levels. The equipment allowed the measurement of the speed at the time of impact and the dynamic load throughout the event. High-speed digital video was also captured to enable failure mode evaluation. Energy level 1 (4 kJ of applied kinetic energy) corresponded to an impactor mass of 133 kg moving at a speed of approximately 8 m/s; energy level 2 (8 kJ) used a total mass of 74 kg accelerated to 14.6 m/s.

For the dynamic axial crush tests, the 450 mm beams were shortened to 375 mm to stabilise the crush column, therefore preventing global buckling. Failure initiators were also introduced onto the beam ends to ensure that a stable and progressive crush mode occurred in each test. For the metallic crush test beams, the initiating mechanism consisted of a swage that was introduced into each of the four beam faces at the edge of first contact using machined guides. The composite beams had a 1 mm wide notch hand-cut into each of the upper "top-hat" radii, 10 mm up from the contact end of the beam.

The metallic beams had similar axial crush failure modes, with a characteristic mode of accumulating uniform ductile folds. The TPC laminate failure mode consisted of progressive failure along the joint between the riv-bonded closure plate and the stamped "top-hat" section and tearing failure of the upper "top-hat" radii leading to the formation of large-scale fronds and some fragmentation. Figure 2.10 demonstrates a comparison between the failure modes of the aluminium beam versus that of the TPC laminate beam.

In terms of performance comparison, the specific (mass-normalised) energy absorption (SEA) of each of the beam variants was calculated. The extent of the damaged region of each beam was accurately measured and used to calculate the mass of material destroyed/deformed in each test. The SEA for each beam variant and test energy is shown in Figure 2.11. At 4 kJ (8 m/s), the TPC laminate beam has an SEA of >30 J/g, over three times that of the DP600 steel and over twice that of the aluminium 5754. However, at the higher impact energy of 8 kJ

Figure 2.10 Comparison of dynamic axial beam crush: aluminium and PA6-GF60 progressive failure modes.

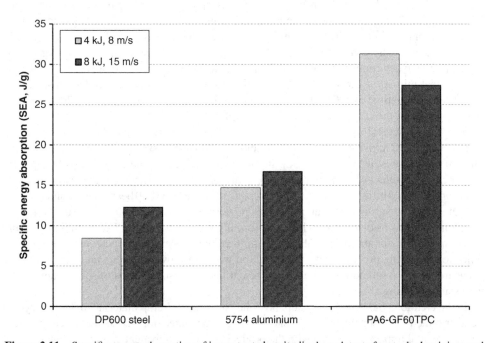

Figure 2.11 Specific energy absorption of impact rate longitudinal crush tests for steel, aluminium and PA6-GF60 TPC beams.

(14.6 m/s), the strain-rate stiffening in the response of the metallic beams and a small decrease in global crush stability for the TPC component meant that the SEA of the TPC decreased to only slightly over two times that of the steel and under two times that of the aluminium.

2.3.6   Environmental Impact Assessment

A life cycle assessment (LCA) study was conducted to gain an understanding of the effect of material substitution on the overall environmental impact of the LCVTP demonstrator beam component. The LCA study goal was to compare the complete life cycle impact of the three beam variants in terms of climate change due to atmospheric emissions.

The scope of the study covered the evaluation and comparison of the environmental impact arising from the cradle to grave life cycle of the beam variants, where a functional unit of the study was one LCVTP demonstrator beam. The specific environmental impact category assessed was climate change and the characterisation factor was global warming potential 100 years (GWP_{100}) in kg CO_2 equivalent [18]. The GWP_{100} factor allows the climate change-related environmental impact of each of the materials/process solutions to be measured by calculating the integrated radiative forcing (over 100 years) caused by the cradle to grave atmospheric emissions associated with each and comparing the result to the integrated radiative forcing caused by the instantaneous atmospheric emission of 1 kg CO_2 [19].

The industry standard LCA software GaBi 5 from PE International was used to build, execute, and interpret the LCA process models. The life cycle inventory input data was based on the PE International GaBi Professional Database with extension databases. The LCA models were broken down into three discrete phases: creation, usage and end of life (Figure 2.12). The creation phase covered raw material extraction, secondary and primary processes and final component assembly. The usage phase consisted of a process that uses an average fuel reduction factor (0.14 l/100 km reduction in consumption per 100 kg vehicle weight save) to determine the emissions created or saved by increasing/reducing the total mass of a typical mid-sized European vehicle by the mass of the component under consideration over a vehicle lifetime of 200 000 km. The end of life phase considered the ELV 2015 scenario whereby 85% of the component by mass is recycled [7].

There were several assumptions made to simplify the models and allow the creation of the necessary LCA process plans. The principal simplifying assumption was that, across all three beam variants, the manufacturing supply chain would be identical in terms of suppliers and their geographical locations and hence transport steps for raw materials and secondary components could be ignored. In the creation phase of the PA6-GF60 beam, the secondary sheet manufacture and subsequent stamp-forming process was adapted from an existing injection moulding process as there was no laminate forming process available. The moulding pressure and temperature for the thermoplastic stamp-forming process is equivalent to the existing injection moulding process, and so the solution was to use two times the input energy of the standard injection moulding process to represent the two discrete steps of sheet manufacture and stamp-forming.

In the creation phase, some valuable post-industrial waste/scrap material could be recycled (e.g. trimmed sheet metal waste). This would be accounted for via the subtraction of the environmental impact associated with the production of an equivalent amount of raw material to that of the waste material (steel, aluminium, or PA6-GF60) from the overall environmental

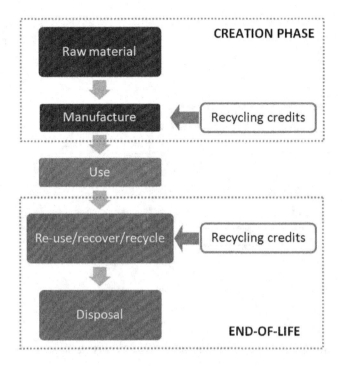

Figure 2.12   LCA modelling flow chart.

impact of the creation process, thus gaining recycling credits via the avoidance of raw material production (illustrated in Figure 2.12).

The same approach of gaining environmental impact credits via material production avoidance was used at the end of life phase. For aluminium and steel variants, credit was gained for the production of ingot mass equivalent to the component mass multiplied by the recycling rate (85%). In considering the PA6-GF60 laminate beam, the same calculation of 85% of the beam total weight was used to calculate the applicable recycling credit for the production of short-fibre injection moulding grade PA6-GF60, reflecting the recycling technique of grinding and re-moulding [8].

The comparison of the selected environmental impact (GWP$_{100}$, CML2001 normalised) arising from the different LCVTP beam variant life-cycles is shown in Figure 2.13 [20]. The environmental impact associated with each variant's total life cycle and the individual phases are displayed, with the end of life recycling credits shown as negative numbers.

It is clear that the PA6-GF60 laminate variant has the lowest overall environmental impact, at about one-third that of the steel part and one-half that of the aluminium. The steel variant's largest impact is in the usage phase due its significantly higher mass (1.9 kg, see Table 2.1 for list of beam variants), has a creation phase impact in line with the composite part and has a modest recycling credit associated with it as a result of steel having a comparatively low embodied energy as compared to the aluminium and TPC. Conversely, the aluminium part has a low impact associated with the usage phase (lower mass at around 1 kg), but much higher creation phase and recycling credits due to the energy-intensive aluminium raw

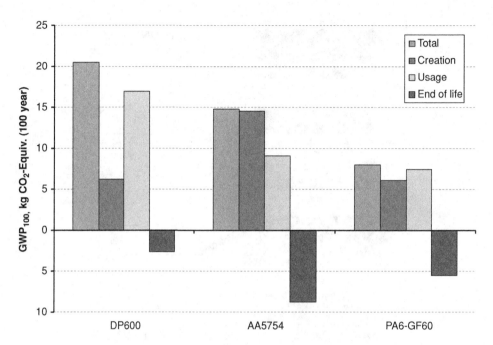

Figure 2.13   Life cycle global warming potential for LCVTP demonstrator beam variants.

material production process. The TPC variant has the lowest usage phase impact (lowest mass at approximately 0.8 kg) and an intermediate credit associated with the end of life recycling process.

2.3.7   Economic Analysis

The final part of the case study considered the economic aspects of material substitution for vehicle lightweighting. A process-based cost modelling (PBCM) approach was used to calculate and compare the manufacturing costs associated with the three different scenarios.

In PBCM the manufacturing process required for the component are split into the individual processes [21]. Each step is then allocated the corresponding process cost elements associated with it such as labour costs, equipment and tooling. These parameters are then combined with the operational parameters such as operating hours and number of shifts to calculate the unit cost of the component under consideration [22]. PBCM has advantages over traditional costing systems. The predictive nature of the PBCM approach enables the user to compare alternate technology and material choices even before the actual implementation. A transparent system of steps followed by the modeller improves the credibility of a model thereby acting as a platform of discussion amongst the designers, management and developers [23]. One possible disadvantage to PBCM is the high levels of input data required to create the model. As the model functions by breaking the manufacturing operation down into individual processes, the data should be available for each input parameters of the particular step. The accuracy of the results obtained therefore depends upon the quality of data set available for each of the operations.

Table 2.2   PBCM cost elements.

Cost element
Material costs
Tooling costs
Equipment costs
Labour costs
Building costs
Energy costs

In this PBCM study, six cost elements were considered (Table 2.2). The data required for these cost elements were obtained from secondary sources of data such as journals, books and publications. Apart from the cost elements mentioned above, data is required regarding critical process parameters such as cycle time, annual working hours, working days and so on.

In order to represent the sources of the data used in the modelling, a summary of the references and the key data that are available in the reference is summarised in Table 2.3. Material costs were obtained from private communications with materials suppliers in the automotive industry.

Once the models had been constructed, the PBCM was carried out to predict annual production volumes ranging from one-offs to 250 000 parts per annum (ppa), and the piece cost (in GBP) versus annual production volume was calculated. The predicted piece cost versus production volume for all three beam types is shown in Figure 2.14.

It is obvious that the PA6-GF60 part is economically competitive for production volumes below 50 000 ppa. This arises from the lower investment costs required in the TPC stamp-forming process as compared to metal press-forming for capital items such as press stations and also press tooling. Above approximately 50 000 ppa, the TPC variant becomes the more expensive option due to much higher raw material costs, which means that any potential weight saving would be achieved at a cost premium. This may be acceptable to some vehicle segments (e.g. premium), as emissions legislation is driving the market to make weight savings, and the notion of OEMs allocating allowable costs per kg weight save is becoming common. This is reflected in the slight premium across all production volumes for the aluminium variant,

Table 2.3   PBCM cost element data sources.

Cost element input data	Data source
Interest rate, power requirements per line, building area, equipment investment, tooling investment	Nadeau *et al.* [24]
Equipment cost for stamping press and tools, power usage, machine area requirements, cycle time	Turner *et al.* [21]
Annual working days, shift durations, equipment life, tooling life, building life, building area costs, labour rates, energy rates, stamping loss rates, material rejection rates	Johnson and Kirchain [25]
Steel and aluminium stamping cycle times	Davies [26]
Energy rates	International Energy Agency [27]
Labour rates	Eurostat European commission [28]

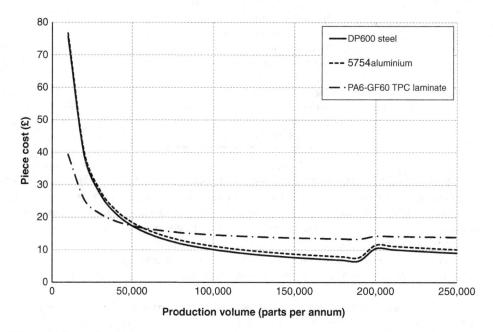

Figure 2.14 Predicted piece cost (GBP) versus annual production volumes for the LCVTP beam variants.

arising from the higher raw material costs. At production volumes of around 200 000 ppa, there is a discrete increase in piece cost; this is due to the requirement at this production volume to make additional capital investment to set up a parallel line due to reaching a process utilisation factor greater than 1. At this point, the lower capital costs associated with the production of the TPC variant significantly lowers the piece cost premium over the metallic parts.

2.4 Conclusion

TPCs are currently experiencing an increase in applications for the production of automotive structures due to a combination of price, performance, design freedom and inherent recyclability. Development in materials and process technology and CAE capabilities has facilitated the use of various thermoplastic composite technologies in the manufacture of semi-structural and structural automotive parts, conferring significant weight savings and/or performance gains. These developments include production techniques such as hybridisation, where the performance of high-volume injection moulded TPC parts is significantly improved via the inclusion of aligned fibre TPC or metallic inserts, providing a useful and unique combination of structure and high part complexity, whilst maintaining recyclability.

It is considered that the next stage of industrial application will be towards the development of supply chains and processes that enable the high-volume manufacture of aligned fibre TPC parts. Recent collaborative research carried out in LCVTP has demonstrated that high-performance aligned fibre TPCs can be manufactured in volumes that are relevant to the automotive industry using the newly developed rapid hot stamp-forming process. Additionally within this research, the outline structural performance of aligned fibre TPC components has been confirmed, with the potential environmental benefits and economic implications of switching to TPCs for automotive structures established.

Acknowledgements

The work carried out in this chapter's case study constituted a part of the research in Work Stream 7 of the Low Carbon Vehicle Technology Programme, which was funded by AWM-ERDF. The authors would like to acknowledge the LCVTP Work Stream 7 team members who have all made significant contributions to the work described herein – Dr Darren Hughes, Dr Mark Pharoah, Dr Geraint Williams, Darren Stewardson, Dave Mossop, Nicholas Blundell, Prof Richard Dashwood, Dr Kerry Kirwan and the WMG LCVTP administrative team and our colleagues at Tata Motors European Technical Centre (TMETC, including Terry Wheeldon, Tony King and Mike Cromarty) and Jaguar-Land Rover.

References

[1] EU (2009) Regulation (EC) No 443/2009 of 23 April 2009 setting emission performance standards for new passenger cars as part of the Community's integrated approach to reduce CO2 emissions from light-duty vehicles. Official Journal of the European Union.

[2] NAIGT (2009) An Independent Report on the Future of the Automotive Industry in the UK. Department for Business, Enterprise and Regulatory Reform.

[3] Bandivadekar, A.E.A. (2008) *On The Road In 2035: Reducing Transportation's Petroleum Consumption And GHG Emissions*, MIT Laboratory for Energy and the Environment, Cambridge, Massachusetts.

[4] Wohlecker, R., Johannaber, M. and Espig, M. (2007) Determination of Weight Elasticity of Fuel Economy for ICE, Hybrid and Fuel Cell Vehicles. SAE World Congress and Exhibition, April 2007. Detroit, MI, USA: Society of Automotive Engineers INC.

[5] WorldAutoSteel (2008) Progress in Weight Loss Steel: Body Structures Keep the Slimming Trend Going [Online]. http://www.worldautosteel.org/why-steel/mass-reduction/progress_vehicle_weight_loss/:WorldAuto Steel (Accessed 28th May 2012).

[6] Lotus Engineering Inc. (2010) *An Assessment of Mass Reduction Opportunities for a 2017–2020 Model Year Vehicle Program*, The International Council on Clean Transportation (ICCT).

[7] EU (2000) Directive 2000/53/EC of 18 September 2000 on end-of life vehicles. Official Journal of the European Communities.

[8] Bernasconi, A., Davoli, P. and Armanni, C. (2010) Fatigue strength of a clutch pedal made of reprocessed short glass fibre reinforced polyamide. *International Journal of Fatigue*, **32**, 100–107.

[9] Stewart, R. (2010) Automotive composites offer lighter solutions. *Reinforced Plastics*, **54**, 22–28.

[10] Stewart, R. (2011) Rebounding automotive industry welcome news for FRP. *Reinforced Plastics*, **55**, 38–44.

[11] Risch, H., Ries, T., Voge, F. and Broos, L. (2010) Use of Plastic–Metal Hybrid Add-on Components in the Body Structure of the New Audi A8, in *Plastics in Automotive Engineering* (ed. F. Matejiček), VDI, Mannheim.

[12] Marler, D., Rowe, J., Wacker, M. and Russ, J. (2005) Moulding the Future of Composite Crash Structures. 5th Annual Automotive Composites Conference & Exhibition (ACCE), Troy, MI, USA: Society of Plastics Engineers (SPE).

[13] Whitfield, M. (2011) *Teijin accelerates CFRP composites production* [Online]. ICIS.COM. Available: http://www.icis.com/Articles/2011/10/17/9500366/innovation-awards-teijin-accelerates-cfrp-composites-production.html (Accessed 31st May 2012).

[14] TATA Steel Europe (2009) *Cold-rolled DP600 (HCT600X) Advanced high-strength dual phase steel* [Online]. Available: http://www.tatasteeleurope.com/file_source/StaticFiles/Business_Units/CSPUK/DP600%20Cold%20Datasheet.pdf (Accessed 3rd June 2012).

[15] Novelis (2012) *Automotive Bodies* [Online]. Available: http://www.novelis.com/en-us/Pages/Automotive-Bodies.aspx (Accessed 3rd June 2012).

[16] TICONA (2012) *Celstran® CFRT/LFRT Composites* [Online]. Available: http://www.ticona.com/home_page/homepage/beta_composites /beta_composites-celstran.htm (Accessed 3rd June 2012).

[17] Fiberforge (2012) *Fiberforge – RELAY® Station Technology* [Online]. Available: http://www.fiberforge.com/thermoplastic-composites/relay-technology.php (Accessed 31st May 2012).

[18] EU (2006) Environmental Management – Life Cycle Assessment – Requirements and Guidelines (ISO 14044:2006). European Committee for Standardization.

[19] Solomon, S. (ed.) (2007) Climate Change 2007: The Physical Science Basis. Contribution of Working Group I to the Fourth Assessment Report of the Intergovernmental Panel on Climate Change, *TS.2.5 Net Global Radiative Forcing, Global Warming Potentials and Patterns of Forcing.* Intergovernmental Panel on Climate Change.

[20] PE-International (2012) *Description of the CML 2001 Method* [Online]. Available: http://documentation.gabi-software.com/1_LCIA.html#CML_2001 (Accessed 6th June 2012).

[21] Turner, T.A., Harper, L.T., Warrior, N.A. and Rudd, C.D. (2008) Low-cost carbon-fibre-based automotive body panel systems: a performance and manufacturing cost comparison. *Proceedings of the Institution of Mechanical engineers, Part D: Journal of Automobile Engineering,* **222**(1), 53–63.

[22] Montalbo, T., Lee, T.M., Roth, R. and Kirchain, R. (2008) Modeling Costs and Fuel Economy Benifits of Lightweighting Vehicle Closure Panels. SAE World Congress and Exhibition, Detroit, 2008. SAE International.

[23] Field, F., Kirchain, R. and Roth, R. (2007) Process cost modeling: strategic engineering and economic evaluation of material technologies. *Journal of the Minerals, Metals and Materials Society,* **59**(10), 21–32.

[24] Nadeau, M.-C., Kar, A., Roth, R. and Kirchain, R. (2010) A dynamic process-based cost modeling approach to understand learning effects in manufacturing. *International Journal of Production Economics,* **128**, 223–234.

[25] Johnson, M. and Kirchain, R., (2009) Quantifying the effects of parts consolidation and development costs on material selection decisions: a process-based costing approach. *International Journal of Production Economics,* **119**, 174–186.

[26] Davies, G. (2003) *Materials for Automobile Bodies,* Butterworth-Heinemann, Oxford.

[27] International Energy Agency (2010) Key World Energy Statistics, Energy statistics report. International Energy Agency Energy Statistics Division, IEA, Paris.

[28] Eurostat European Commission (2009) *Labour Market Statistics,* Eurostat Pocketbooks, Publications office of the European Union, Luxembourg.

3

Development of Low-Cost Carbon Fibre for Automotive Applications

Alan Wheatley[1], David Warren[2], and Sujit Das[2]

[1] *Department of Computing, Engineering, and Technology, University of Sunderland, Sunderland, SR6 0DD, UK*
[2] *Oak Ridge National Laboratory, Oak Ridge, USA*

3.1 Introduction

In pursuit of the goal to produce ultra-lightweight fuel efficient vehicles, there has been great excitement during the last few years about the potential for using carbon fibre reinforced composites in high volume applications. Currently, the greatest hurdle that inhibits wider implementation of carbon fibre composites in transportation is the high cost of carbon fibre when compared to other candidate materials. However, significant research is being conducted to develop lower cost, high volume technologies for producing carbon fibre. This chapter will highlight ongoing research in this area.

Through United States Department of Energy (US DoE) sponsorship, Oak Ridge National Laboratory (ORNL) and its partners have been working with the United States Automotive Composites Consortium (ACC) to develop technologies that would enable the production of a carbon fibre at between US$5 and US$7 per pound (0.454 kg). That cost goal would allow the introduction of carbon fibre-based composites into a greater number of applications for future vehicles. The approach has necessitated the development of both (i) alternative precursors and (ii) alternative production methods.

Alternative precursors under investigation include textile grade polyacrylonitrile (PAN) fibres and fibres from lignin-based feedstocks. Previously, as part of the research programme, Hexcel Corporation developed the science necessary to allow textile grade PAN to be used as a precursor rather than typical carbon fibre grade precursors. Efforts have also been continuing to develop carbon fibre precursors from lignin-based feedstocks. ORNL and its partners are working on this effort with domestic wood and paper producers.

Advanced Composite Materials for Automotive Applications: Structural Integrity and Crashworthiness,
First Edition. Edited by Ahmed Elmarakbi.
© 2014 John Wiley & Sons, Ltd. Published 2014 by John Wiley & Sons, Ltd.

In terms of alternative production methods, a microwave-based carbonisation unit has been developed that can process pre-oxidised fibre at over 6 m/min. In addition, ORNL have developed a new method of high speed oxidation, and are defining new methods for precursor stabilisation. Additionally, novel methods of activating carbon fibre surfaces have been developed which allow atomic oxygen concentrations as high as 25–30% to be achieved rather than the more typical 4–8% achieved by ozone treatment.

3.2 Research Drivers: Energy Efficiency

The need to improve vehicle fuel efficiency is a major research priority on both sides of the Atlantic Ocean. The United States' transportation systems are 95% dependent on petroleum-based fuels. To supply this need, the United States imports roughly 55% of its petroleum requirements from abroad. To lessen dependence upon foreign sources, the National Energy Policy was developed and is administered by the United States Department of Energy (DOE). The National Energy Policy calls for an aggressive agenda to reduce petroleum demand through energy efficiency while simultaneously increasing energy supply and diversifying the sources of energy used.

The European Union (EU) is even more heavily dependent on imported oil, currently importing around 10 million barrels per day – around 70% of its needs [1]. These figures make the EU the second largest oil consumer in the world, after the United States. The United Kingdom is, by far, the greatest oil-producing country within the EU. Even so, since 1999, United Kingdom oil production has decreased steadily and from 2005, the United Kingdom became a net importer of oil [2], hence contributing to the EU oil problem rather than helping it. When environmental concerns, such as global warming from CO_2 emissions, are added to the equation, the problems of fuel efficiency (or lack thereof) become even more pressing.

Motorised transportation systems consume about 19% of the world's total energy supplies, with 95% of this amount being petroleum, accounting for about 60% of the total world petroleum production [3]. In the United States, about 80.5% of the motorised transportation energy is consumed by road vehicles [4]. The recent increase in petroleum prices, expanding world economic prosperity, the probable peaking of conventional petroleum production in the coming decades, concerns about global climate changes and the recent release of significant quantities of oil as a result of the failure of the deep sea well in the Gulf of Mexico all suggest the need to focus efforts to increase the efficiency of the use of, and develop alternatives for, petroleum-based fuels used in road transportation. Efforts to increase the energy efficiency of a vehicle will require improvements in materials and processes for propulsion systems and structures, new advanced propulsion systems, batteries and alternative fuels. In many industrial countries, road transportation accounts for a significant portion of the country's energy consumption. In developing countries, the use of energy for transportation is on the rise. Most studies indicate that 70–80% of the energy usage in the life cycle of a road transportation vehicle is in the use phase. The remainder is energy usage is in the production of the vehicles, including the production of the materials, supply of the fuel and disposing of the vehicles. Thus, advances in many materials and processes will be required in efforts to increase the energy efficiency of motorised vehicles for road transportation.

Automotive technology has been a concern of the governments worldwide since the environmental movement of the 1960s and the oil crisis of the 1970s. During those decades, in the

United States, the Clean Air Acts established standards for emissions [5]. The Energy Policy and Conservation Act of 1975 [6, 7] established corporate average fuel economy (CAFÉ) standards for light duty vehicles. Fuel prices are increasing as are CAFÉ standards for model years 2017–2025 along with tightening emission regulations. Collectively, in the United States, these will require car and light truck performance equivalent to 20.7 km/l (54.5 mpg) and reductions in greenhouse gas emissions to 101 g/km (163 g/mile) in 2025.

United States Environmental Protection Agency (EPA) data for light vehicles from 1975 to 2011 [8] suggest that, in this period, there were dramatic improvements in automobile technology. However, the problem has been that this was focused on making the cars larger and better performing rather than in reducing fuel consumption. Original equipment manufacturers' (OEMs') objectives for safety, coupled with customer demands for space, performance, extra features and minimal noise, vibration and harshness (NVH) caused the average light-duty vehicle weight to increase each year between 1987 and 2004. Since 2004 the average light vehicle weight has levelled off. Nevertheless, gains from technological advances during this period have been largely negated by the increased fuel consumption of heavier cars. This is not a sustainable situation and current research is directed towards a longer-term solution in which significant vehicle mass reductions become a reality.

As a result, the pressure to produce lighter, more fuel efficient vehicles is stronger than ever before. It is generally accepted that a reduction in vehicle mass of 10% results in an improvement in fuel economy of 6–8%. This represents the area of focus of this chapter – specifically the development of low-cost carbon fibre materials destined for automotive applications and the associated potential mass reductions achievable.

3.3   Lightweight Automotive Materials

If the potential mass savings associated with migration from one automotive material to another are well understood, then the goals for any low-cost carbon fibre programme (properties vs cost vs mass saving) become clearer. These issues are dealt with in more detail in Chapter 17. For the moment, let us consider how materials science and technology have affected the make-up of a typical family car.

The trend here has been in migration from conventional steels to: (i) high strength steel, (ii) lighter alloys (e.g. aluminium and magnesium) and (iii) polymer composite materials.

In particular, the migration from conventional steel to higher strength steel is clear. Figure 3.1 shows details of how the materials' make-up of light vehicles has changed between 1995 and 2011 [9].

Table 3.1 [10] shows the typical mass reduction potential of a range of other structural materials when migrating from mild steel. The potential mass savings illustrated are typical but, of course, depend on a multitude of factors which will alter the achievable weight saving. Again, these issues are explored further in Chapter 17.

Vehicle lightweighting represents one of several design approaches automakers are currently evaluating to improve fuel economy and lower emissions. The next few years will likely see considerable lightweighting across the automotive industry. Weight reduction can be accomplished in three main ways: (i) fleet downsizing by emphasising smaller models which requires a shift in consumer purchasing decisions, (ii) design changes, such as the shift from body on frame to unibody construction, which results in mass reduction, and (iii) the use of lighter weight or higher strength materials.

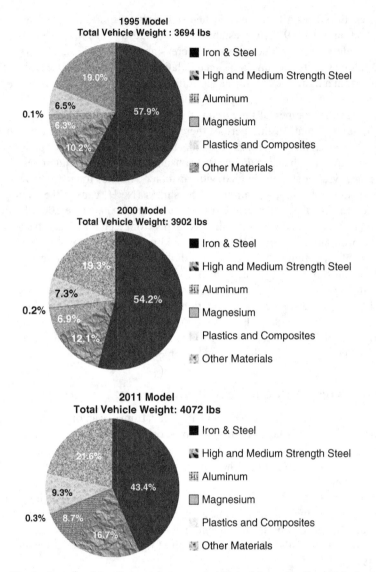

Figure 3.1   Distribution of weight of materials in a typical United States family vehicle for model years 1995, 2002 and 2009 [9].

Manufacturers generally favour lightweighting through material substitution, design optimisation and adopting other advanced manufacturing technologies, rather than compromising a vehicle's attributes and functionalities, such as occupant or cargo space, vehicle safety, comfort, acceleration performance and so on. It is projected that the lightweight materials' share in the automotive sector will increase from 30 to 70% by 2030 (with high strength steel considered as a lightweighting material) [11]. Several technical barriers prohibit most lightweight materials from being commercialised widely, with material cost being one of those significant

Table 3.1 Typical mass reduction potentials of various candidate lightweighting materials [10].

Material	Density (g/cm³)	Strength/ density	Modulus/ density	Cost	Mass reduction potential (%)
			Relative to steel		
Mild steel	7.87	1	1	1×	0
High strength steel	7.87	1.86×	1×	0.9–1.2×	0–10
Advanced high strength steel	7.87	2.25×	1×	0.8–1.5×	0–30
Aluminium	2.7	3.95×	1.02×	1.3–2.0×	30–60
Magnesium	1.74	3.66×	1.02×	1.5–2.5×	30–70
Titanium	4.51	4.73×	0.98×	1.5–10.0×	40–55
Metal matrix composites	1.9–2.7	3.81×	1.45×	1.5–3.0×	30–70
Ceramics	3.9	0.7×	3.05×	1.5–3.0×	10–30
Plastics	0.9–1.5	0.82×	0.08×	0.7–3.0×	20–50
Sheet moulding compound	1.1–1.9	4.39×	1.16×	0.5–1.5×	20–30
Glass fibre composites	1.4–2.0	4.74×	0.5×	0.9–1.5×	25–35
Carbon fibre composites	1.4–1.6	18.3×	1.93×	1.5–5.0×	50–90

barriers. Even after taking into account the reduced mass of material needed and parts consolidation, which reduces manufacturing steps and reduces joining after parts consolidation, raw material costs can still be a major inhibitor to advanced materials implementation. The cost of carbon fibre composites is the obstacle to their implementation that is most commonly voiced by automotive engineers and executives.

3.4 Barriers to Carbon Fibre Adoption in the Automotive Industry

Many obstacles stand in the way of successful incorporation of significant quantities of carbon fibre into the cars of the future. Some of those obstacles, such as those listed below, are technical in nature and may be overcome with enough research and development. However, many are perhaps more practical than technical.

The high cost of the fibre remains the largest hurdle to production implementation of carbon fibre composites as a structural vehicle material. The primary technical challenges are huge and are generally divided into five broad categories [12] with research efforts by both government and industry in each of these areas:

- **Cost** – of raw materials and manufacturing.
- **Manufacturability** – rapid, repeatable low-cost methods.
- **Design data/test methodologies** – how to design and test components and subsystems to assure long term durability and reliability.
- **Joining** – especially to dissimilar materials but also rapidly in an assembly plant.
- **Recycling and repair** – how to repair without replacing and avoiding landfill of post-consumer parts.

In March of 2011, the United States Department of Energy conducted a workshop to determine the most pressing needs for materials research related to potential lightweighting materials. The workshop received input from automotive original equipment manufacturers (OEMs), part suppliers (tier I), materials suppliers (tier II) and researchers from government, industry and academia. During that workshop all material systems were considered, and outlined below are the most pressing research and development needs for the incorporation of carbon fibre and carbon fibre composites [13]. These challenges are essentially an expansion of the above list:

- **Carbon fibre cost**: Technologies for producing carbon fibre from precursor materials are not optimised for automotive grade materials.
- **Carbon fibre conversion**: Current processes for converting available carbon fibre precursors to usable fibre are too energy intensive and slow to produce carbon fibre at the cost and in the volumes necessary to support significant use for light duty vehicles.
- **Carbon fibre precursors**: A route employing non-petroleum-based precursors, that would remove dependence on oil, is not yet available. Likewise, significantly lower cost precursors are not available.
- **Carbon fibre supply infrastructure**: A fourth obstacle to the incorporation of carbon fibre composites in the automotive industry is past experience of the car companies with the carbon fibre community. The carbon fibre industry over the past three decades has been something of a "boom or bust" business. When business was good, there was rarely enough capacity to keep up with demand. As a result, prices rocketed and lead times lengthened. Unstable pricing and uncertain supplies are avoided at all costs in the automotive industry. The automotive industry has a two to five year development cycle and must have development partners that are consistent during those times. *The automotive industry needs stable long term pricing and stable long term partners.*
- **Fibre/resin compatibility**: For many resins of interest to the automotive industry, the inability to achieve strong bonding of the fibre to the resin is inadequate to take full advantage of the inherent properties of the fibre. Existing surface treatments such as coupling chemistries are not optimised for most thermoplastic resin systems that are used to incorporate carbon fibres. Current sizings are not optimised for wetting characteristics, and for the ability to transfer load to enable carbon fibre to be used with polyester resins, for example.
- **Joining technologies**: Joining technologies for carbon fibre composites to each other or within a multi-material system are inadequate. Joining technologies to incorporate CFCs in suspension and body applications are either not developed or are not compatible with the fast processes necessary in a production environment. Joining composites to other materials such as magnesium, steel, aluminium, plastic or glass fibre composites to produce multi-material components has not advanced far enough to be used extensively in high volumes.
- **Design data, standardisation and predictive modelling**: To take advantage of the large mass reduction afforded by carbon-based materials requires replacing steel in large sections of the body and chassis with carbon fibre composites. The body and chassis are the skeletal and exterior portions of the vehicle that protect occupants during a crash. Designers need to become comfortable with the crash response of alternative materials and the methods used to join them. Errors in judgement for these subsystems can result in tens of thousands of lawsuits and a poor corporate reputation that could last for decades. *Full vehicle and subsystem prototyping and demonstration could do much to alleviate these concerns.*

Automotive-grade composites are (currently) rarely laminate-based and therefore many failure criteria used in the composites industry are not directly transferable. Historically, there is a wealth of information regarding epoxy and epoxy-like systems. However, automotive composites are equally likely to be based on vinyl ester, isocyanurate, polyester or similar systems. Research is being conducted to develop test methods, failure criteria and material models that are applicable to more ductile, lower glass transition (T_g) automotive resin systems and less directed fibre architectures. Reliable materials design data and predictive modelling tools are vital to engender confidence in the material.

In a similar manner, standard uniform test procedures must be employed to generate such design data – otherwise diverse sets of data will not be directly comparable. The carbon fibre family includes traditional carbon/graphite fibres, but also incorporates developments in low-cost and recycled carbon fibres. The development and acceptance of appropriate standard test techniques for production, supply and fabrication are vital steps in promoting widespread adoption. Capabilities for accurate predictive models supporting design, processing and crash energy management are either inadequate or insufficiently validated to avoid "over design" that incorporates the additional safety factors:

- **Lack of standard fibre/resin systems**: Few fibre/resin systems are inexpensive enough and provide rapid cycle time to be compatible with high volume demands and performance needs. Short cycle time in component manufacturing and joining processes are required for the production of inexpensive complex-shaped carbon fibre composite components. The fifth, and often overlooked, obstacle to carbon fibre market penetration relates to fibre sizing. Most PAN-based carbon fibres are chemically and microstructurally very similar, with major differences being in the purity of the initial precursor, the amount of tensioning during production and the final graphitisation step (if employed). What vary greatly are the surface activation techniques, the sizing chemistry and the quantity of sizing. To use carbon fibres with automotive grade resin systems, the fibre surface chemistry must be matched to the resin chemistry and downstream production methods. Fibre manufacturers and sizing manufacturers tend to consider surface chemistry highly proprietary and therefore have been extremely reluctant to truly work with the automotive companies. They are willing to try different existing technologies and then send them to the automotive suppliers to try, but most have not been very willing to work collaboratively with resin suppliers to develop optimal systems targeted specifically at automotive grade resin systems and production methods. Additionally, industry standard sizings that are interchangeable with similar resins are needed. If a company is using a commercial grade carbon fibre with vinyl ester, they need a standard surface chemistry that allows the use of any resin from any of the major resin suppliers and carbon fibre from any commercial grade carbon fibre supplier, without having to re-engineer the fibre–matrix interface.
- **Long term durability**: Understanding of carbon fibre behaviour under service conditions is insufficient. There is a need to better understand how carbon fibre materials behave under a variety of service conditions. Comprehensive testing/measurement of carbon fibre composite components will enable the development of predictive tools for short/discontinuous/chopped fibre in matrix, for example.
- **Damage detection and repair**: The technology to detect damage in carbon fibre composites is immature, as is the technology to repair components. The community needs the development of tools for detecting damage based on non-destructive evaluation such as

ultrasonic, thermography or computer aided tomography (CAT), for example. The ability to detect damage and repair it needs to be as easy and reliable with composites as it is now with metal structures.

- **Manufacturability**: Body and chassis subsystems require an extensive amount of capital equipment to manufacture. Before investing in new equipment, manufacturers are going to have to be assured that production techniques are fully developed. Low volume aerospace techniques are rarely optimised for cost effectiveness to sufficient levels and are most often far too slow for high volume manufacturing. Even if production methods are developed, the automotive companies have already sunk hundreds of millions into steel processing equipment that cannot simply be scrapped for the next generation of material. For this reason, penetration into the automotive market will be a slow process occurring only when the metal forming tooling needs to be replaced and then only if the production and design technologies are ready. Methods for the high-volume production of automotive components from composite materials are being developed that yield the required component shape and properties in a cost-effective, rapid, repeatable and environmentally conscious manner. New processing technologies must be compatible with automotive manufacturing plants and methodologies. Much current manufacturing research has centred on using fibre reinforced thermosets. More work needs to be carried out on thermoplastic composites and sheet moulding compound (SMC) solutions within the context of carbon fibres.

Removing such barriers would make widespread carbon fibre adoption by the automotive industry a much more attractive option. Of course, immediate migration is not possible. Established fabrication technologies cannot be dismantled overnight. In reality the migration from steel to carbon fibre composites will be an evolving process. This is unavoidable for another very good reason. While growing rapidly, the current global carbon fibre production capacity is simply not yet of the magnitude required to allow incorporation of anything but a trivial mass of carbon fibre per new vehicle produced. This topic is considered in more detail below.

3.5   Global Production and the Market for Carbon Fibre

The instantaneous global capacity for carbon fibre supply is somewhat difficult to pin down. First, nameplate capacity never translates into what is achieved or achievable (typical yields may be 60–70% of nameplate capacity). Second, figures for small tow (1 to 24 k) fibres and large tow (>24 k) fibres are often quoted separately. However, to a first approximation, the current supply of small tow higher performance carbon fibres is similar to or slightly greater than the supply capacity for larger tow commercial grade fibre. This balance between small tow and large tow is likely to shift towards large tow as industrial applications and associated demand increase. Within the "industrial" category, the largest contributor (and fastest growing) by far is wind power applications.

Figure 3.2 shows how carbon fibre supply and demand have grown since 1998. There are three distinct regions on this graph. From 1998 to 2005 demand rose at the rate of around 1000 t (2 million lb)/year. From 2005 to 2012 demand surged such that the annual increase was 4000 t (8 million lb)/year. The acceleration in growth rate is predicted to continue through to 2020 with demand rising by 7000 t (15 million lb)/year. Major reasons for the periodic

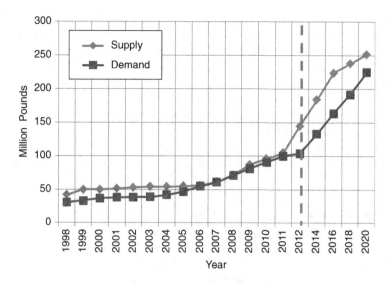

Figure 3.2 Carbon fibre supply and demand trends: 1998 to 2020.

upturns are: (i) the evolution (ca. 2005) of aircraft such as the Boeing 787 and Airbus A350 and A380 which incorporate over 50% by weight of carbon fibre in their make-up and (ii) the huge increase in carbon fibre wind turbine blade deployment, particularly following the economic downturn of 2009–2010.

Current global carbon fibre nameplate capacity for all PAN-based fibre is around 110 000 t (240 million lb)/year [14–16]. Allowing for a 60% yield, this approximates to a current (2012/2013) supply capacity of 66 000 t (145 million lb)/year. Around 30 000 t (66 million lb) of this capacity relates to commercial grade fibre appropriate to the automotive market.

Current North American vehicle production is >14 million units per year [17]. Incorporating just 4.5 kg (10 lb) of carbon fibre in each North American vehicle would create a demand for >140 million lb (>64 000 t) of carbon fibre. That is equal to the global capacity for all PAN-based fibre and over twice the current world supply capacity of commercial grade PAN-based fibre. If global vehicle production (>80 million vehicles in 2012) [17] is considered then using 10 lb (4.5 kg) of commercial grade carbon fibre per vehicle would require a 12× increase in supply capacity.

Of course, using 4.5 kg of carbon fibre per vehicle is not going to produce any significant weight reduction. The migration from steel to CFRP composites needs to be on a much larger scale to achieve this. In terms of the potential applications of carbon fibre composites in an automobile, our estimate is that 300 kg of such composite material could be used as steel replacement. Even if this material contained only 40% w/w carbon fibre, this equates to 120 kg (264 lb) of carbon fibre per vehicle. Global supply capacity would need to increase by a factor of 56 to satisfy the needs of North American vehicle production or by a factor of 320 to satisfy the needs of global vehicle production. That increase in capacity would only meet the needs of the automotive industry, not the many other industries having similar requirements that would grow as a result of automotive industry penetration and experience. To achieve such drastic industry scale-up will require changes to carbon fibre production technologies, production facilities, packaging and emission control procedures and changes in the precursors used to manufacture carbon fibre. That increase in capacity would only meet the needs of the automotive industry and not the many other industries having similar requirements that would grow as a result of automotive industry penetration and experience. However, even if capacity were massively increased, there would still remain the paramount issue of COST [18].

The current single greatest interest for some automotive companies seems to be carbon fibre sheet moulding compound (CFSMC). CFSMC does have the disadvantage of not maximising the material performance (and thus the weight savings available) because practical considerations limit the fibre content to around 50% w/w (40% v/v). However, CFSMC offers several advantages, namely:

- SMC is commonly used in non-crash critical components and therefore can be incorporated without adding the risk of changing energy management structures of the vehicle.
- SMC is a familiar material to designers and therefore working with it will afford them the opportunity to become more familiar with the material.
- Since most automotive plants are already tooled for manufacturing components from SMC, a minimum of capital investment will be necessary.
- Because of the improved material properties that are theoretically available by replacing glass in SMC with carbon fibre, a familiar material with familiar processing techniques may become available for more demanding applications.
- Carbon fibre SMC can be a "drop-in" replacement for conventional SMC, yielding the opportunity to get field experience with real parts, without the need to redesign entire systems in order to exploit the benefits of composites. One composite would simply be replaced with another, so the reluctance to use composites is not relevant.
- About 70% of current automotive composite demand is for SMC. It is a rather large opportunity.

3.6   Low-Cost Carbon Fibre Programme

As more aggressive vehicle weight becomes needed, it has become apparent that carbon fibre-based composites are serious candidates for enabling those goals to be met. Carbon fibre-based composites are restricted in automotive industry use due to the high current cost of carbon fibre in comparison to other potential vehicle structural materials. The greatest cost factors in carbon fibre production are the high cost of precursors (45–60% of production costs) and the high capital equipment costs (20–35% of production costs) [19].

Carbon fibre is typically made by highly controlled thermal pyrolysis of only a few types of precursors (typically pitch, polyacrylonitrile or rayon). This naturally begs the question: "Are these the only materials, energy methods and production technologies capable of producing carbon fibre?" In response to this, ORNL has developed a research portfolio aimed at reducing the cost of carbon fibres. The current research portfolio includes the development of non-traditional precursors, novel fibre production methods and a means for fully developing these advances in a demonstration user facility which will be available to potential suppliers. The synergistic effects of improving both precursor and manufacturing costs aims to bring carbon fibre costs within acceptable ranges for the automotive industry.

In 2012, ORNL completed a new cost model for the cost of making carbon fibre. The cost model consisted of a baseline cost model to obtain an estimation of the manufacturing cost for an industrial grade carbon fibre and then looked at the effects of increasing production volume and the impact of new technologies and materials that are being proposed to lower the production costs of carbon fibre. While the cost model aims to get as close to production costs as possible, the main purpose of the cost model is to qualitatively determine the relative importance of various proposed new technologies and precursor materials.

3.6.1 Project Aims

Cost goals have been coordinated with the automotive industry. Specifically, these goals are to reduce carbon fibre costs to a range of US$11.00–15.40/kg (US$5–US$7/lb). Costs within or approaching that range would also have a significant impact on wind power, oil and gas, construction, power transmission and a variety of other industries.

This must be achieved while meeting minimum property specifications of stiffness (modulus, $E_f \geq 25$ Msi: 172 GPa), strength (ultimate tensile strength, $\sigma_{uf} \geq 250$ ksi: 1.72 GPa) and corresponding failure strain ($\varepsilon_{uf} \geq 1\%$).

3.6.2 Precursor Materials

Earlier work on this project involved an investigation into a range of potential precursor fibres. These included polyolefins (HDPE, PP and LLDPE), PVC, cellulosics and polystyrene (e.g. see Ref. [12]). However, the results of this work were that future efforts were focused on two particular precursors, namely (i) textile commodity-grade PAN and (ii) lignin-based fibres.

3.6.2.1 Commodity PAN-Based Precursors

Early work in this field was conducted by Hexcel Carbon Fibres under DOE contract. As a result of this work, the base technologies necessary to use commodity grade PAN textile as a precursor for making large tow carbon fibre were developed [20]. Initially the project started as a "clean sheet" look at what possible precursors were available and what their advantages and disadvantages were. Those candidate materials were then reduced to a suitable short list based on a variety of criteria and evaluated in laboratory-scale trials.

The results of the laboratory-scale evaluation of the candidate systems led to the down-selection of a very promising technology for reducing carbon fibre price which is based on commodity textile acrylic tow with a single chemical modification added during textile manufacture or possibly a radiation pre-treatment prior to oxidation. Around 50% of the cost of traditional carbon fibre derives from the PAN precursor fibre. At around US$2–3/lb (US$4.40–6.60/kg), commodity textile-grade acrylic fibre (see Figure 3.3) is less than half of the cost of traditional PAN-based precursors and therefore presents very significant cost saving benefits. These cost benefits are discussed in more detail in Chapter 17.

In the course of down-selecting the technologies, Hexcel performed the following: (i) processed and evaluated various state-of-the-art and developmental PAN precursors, (ii) developed the continuous in-line chemical modification and demonstrated a reduction in stabilisation

Figure 3.3   Commodity PAN precursor.

times of ~50%, (iii) evaluated and selected E-beam radiation technology for pre-stabilisation and (iv) developed cost models for the selected technologies.

Hexcel's project focused on technologies that could be implemented into commercialisation plants using available industrial facilities and infrastructures with the least capital investments. The end product of the project was the definition of the technologies for the production of low-cost carbon fibre. This included the required materials and facilities, and is supported by detailed manufacturing cost analysis and processing cost models. Mechanical data that resulted from this work is shown in Table 3.2 below. It can be seen that, even at this relatively early stage of the research, the program goals were being achieved.

Today, the science and technology behind the use of textile-grade PAN precursors has developed to full commercial production. For example, Zoltek's Panex 35 fibre is manufactured from textile-grade PAN fibre and has quoted modulus and strength figures of 35 Msi (242 GPa) and 600 ksi (4140 MPa) respectively. The use of textile grade PAN has allowed significant cost reductions to be achieved. Further cost reductions are still needed, however, in order to make carbon fibre more attractive to the automotive industry. Such cost savings will need to be made in respect of either (i) even lower cost precursors or (ii) lower cost conversion techniques. These factors are considered below.

Table 3.2   Mechanical properties of textile PAN-based carbon fibres.

Property	Textile spool 1	Textile spool 2	Program goal
Young's modulus (Msi/GPa)	30.2/208	30.8/212	25.0/172
Ultimate strength (ksi/MPa)	393.4/2712	398.6/2748	250.0/1720
Elongation at break (%)	1.30	1.29	1.0

3.6.2.2 Lignin-Based Precursors

There is currently great interest in exploiting lignin as a cheap precursor for carbon fibres [21–25]. Lignin is a natural material found in plant cell walls and, importantly, is available as a byproduct from wood pulping and bio-refinery processes. As such, it is cheap, abundant and represents a sustainable, renewable source of precursor fibre.

Research at ORNL in this field has been directed at the development of methods for the production of carbon fibre feedstock from high-volume, low-cost materials based on renewable resources – and lignin fits this requirement well. The goal has been the development of technologies which reduce precursor and processing costs, and which could be implemented to produce low-cost carbon fibre on a scale sufficient to support its large-scale use in passenger vehicles. Use of renewable materials also decreases sensitivity of carbon fibre cost to changes in petroleum production and energy costs. Lignin is currently produced in quantities of millions of tons per year with the majority of that being burnt as a fuel in the paper making process. More recently, lignins have also been isolated from biomass conversion operations, notably ethanol production [24].

Lignin presents many research challenges. Its chemistry is rather complex and this has a great bearing on successful conversion. For example, hardwood lignin is easier to melt spin than softwood lignin but is harder to crosslink – rendering stabilisation more difficult. Such issues dictate that research is carried out into identifying methods of achieving a balance between melt spinnability and stabilisation speed – while attaining target goals in terms of mechanical performance [24].

Initial research work [22, 23] involved proof of concept production of single fibres from a variety of natural, renewable, materials available in high volume. Single fibre, melt spun from blends of Kraft hardwood lignin (an inexpensive, high-volume wood pulping byproduct) could be processed using conventional stabilisation (oxidising atmosphere), carbonisation and graphitisation (inert atmosphere) furnace technologies to yield carbon fibres. Graphite content, as measured by X-ray diffraction, increased with increasing temperature of treatment. Carbon fibre yield from the lignin-based feedstocks was approximately 50%. Physically, the fibres were dense, smooth-surfaced and uniformly round. Figure 3.4 shows carbon fibres produced from lignin precursors.

Later, larger quantities of a lignin –polyester (recycled) feedstock were melt spun as a multifilament tow (28 fibres) using a 27 mm diameter twin-screw Leistritz extruder. This showed that: (i) lignin-polyester blend can be melt-extruded as a small tow using near-commercial scale melt spinning equipment, (ii) resin-fibre composites made using the graphitised lignin/polyester-based carbon fibres exhibited normal fracture patterns and (iii) although very smooth surfaced, lignin/polyester-based carbon fibres can be plasma treated and silanated to provide good fibre/resin adhesion. In summary, research and development progress [24] in this area has resulted in:

- Sustained melt spinning of 10 μm fibre, from both Kraft and Organosolve pulped lignins (hardwoods), at speeds of 1500 m/min demonstrated, but with speeds of 5000 m/min possible with the necessary winding equipment.
- These figures are almost three times those achieved with commercial mesophase pitch-based fibres and four times commercial wet-spinning speed with PAN-based fibres.
- Low-cost modification of lignin chemistry (to increase molecular weight and glass transition temperature) has a profound, beneficial effect on carbon fibre yield and stabilisation.

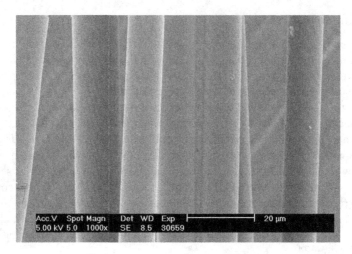

Figure 3.4   Lignin-based carbon fibre.

- The time required for stabilisation of lignin-based fibres has been reduced from days to the order of minutes by means of chemical modifications. Indeed, on the fly stabilisation (immediately after the spinneret) is proposed.

Parallel developments by Zoltek and Weyerhaeuser [25] relate to wet-spun (rather than melt-spun) lignin-based fibres. This work involves lignin/PAN blends as precursor raw materials, with lignin present at up to 45% concentration. The reduced reliance on PAN produces cost savings of 11% on carbon fibre cost. With improvements in conversion speed and efficiency, cost reductions of up to 25% are predicted.

3.6.2.3   Polyolefin-Based Precursors

With textile-grade PAN fibres costing US$2–3/lb (US$4.40–6.60/kg) and polyolefin fibre precursors costing US$1–2/lb (US$2.20–4.40/kg), the ability of polyolefin precursors to take yet more cost out of carbon fibre production is clear. In addition, polyolefin fibres contain 86% w/w of carbon compared to 68% in PAN. This represents a useful starting advantage in the journey toward carbon fibre. Polyolefin fibres offer good handling capability and are very amenable to melt processing. Given these factors they offer a high potential performance to cost ratio – indeed better than just about any rival precursor.

Initial work [26] involved the use of 20 µm diameter polyolefin (LLDPE) fibres. Following thermochemical stabilisation and carbonisation, these produced carbon fibres with a modulus of 15 Msi (103 GPa) and a tensile strength of 150 ksi (1030 MPa). These low values have been attributed to insufficient sulfonation at the core of the fibre, leading to hollow-cored carbon fibres. Subsequent work on 10 µm diameter polyolefin fibres showed that crosslinking was consistent across the fibre diameter and carbon fibres of 21 Msi (145 GPa) modulus and 200 ksi (1380 MPa) strength were produced. More recently, carbon fibres with strengths > 230 ksi (1590 MPa) have been produced and work is currently underway to reach program goals of 25 Msi (172 GPa) modulus and 250 ksi (1720 MPa) strength.

Figure 3.5 ORNL MAP reactor during operation.

3.6.3   Advanced Processing Techniques

3.6.3.1   Microwave Assisted Plasma Processing

A rapidly maturing project for reducing the processing costs of converting precursors into carbon fibres is based on microwave assisted plasma (MAP) technology [16]. Typically pitch, PAN or rayon precursors are processed by highly controlled thermal pyrolysis of the precursors under mechanical stressing leaving only the carbon in a preferred structure and orientation of that structure. The ORNL processing work includes the use of microwave energy generated plasmas to carbonise and graphitise precursors. The initial continuous pilot unit was designed to achieve a line speed of 6 in/min (15.24 cm/min) to demonstrate technical feasibility. After many iterations, a carbon fibre production line speed of >200 in/min (508 cm/min) was achieved. Figure 3.5 shows the MAP reactor in operation.

Program activities then refined the process parameters for running full spools of continuous 50 K tow precursors to obtain desired properties. The economic cost advantage is a 12–20% reduction in overall fibre cost. Mechanical data indicates the microwave assisted plasma processed fibre surpasses the values targeted for the automotive industry. ORNL-produced fibres have achieved moduli up to 32 Msi (221 GPa) and ultimate tensile strengths of up to 424 ksi (2920 MPa). Correspondingly conventionally manufactured (with the same PAN precursor) carbon fibre featured a modulus of 31 Msi (214 GPa) and an ultimate strength of 485 ksi (3340 MPa). Electrical and morphological properties of these fibres are comparable to those in current commercial 50 K tows. Next the project developed the ability to run multiple tows simultaneously through the processing unit. Some results from those trials are shown in Table 3.3 below.

Table 3.3   Results for microwave assisted plasma (MAP) carbonisation.

Variable	3 Tow MAP south (red)	3 Tow MAP middle (blue)	3 Tow MAP north (yellow)	Program goal
Production line speed (in/min)	12.3	12.3	12.3	—
Modulus (GPa)	222	207	199	172
Ultimate strength (GPa)	2.96	3.14	2.94	1.72
Elongation at break (%)	1.32	1.47	1.44	1.0

Table 3.4 Physical and mechanical properties of fibres provided to ACC.

Variable	Sample 1	Sample 2	Sample 4	Program goal
Production line speed (in/min)	36	36	36	—
Calculated filament diameter (μm)	7.77	7.72	7.69	
Modulus (GPa)	154	178	194	172
Ultimate strength (MPa)	3010	2830	2630	1720
Elongation at break (%)	1.95	1.59	1.36	1.0

More recently [27], at the request of the Automotive Composites Consortium (ACC), two each of 1-lb (0.454-kg) spools of pre-oxidised, 50 k tow were single-tow MAP carbonised and test coupons and plaques were delivered to ACC researchers for characterisation. ORNL also characterised samples of this fibre. Physical and mechanical properties are shown in Table 3.4 Wide angle X-ray scattering morphology data are shown in Figure 3.6, and indicated that MAP-carbonised fibre morphology can compare favourably with that of contemporary commercial grade fibres.

3.6.3.2   Advanced Stabilisation/Crosslinking

The purpose of this aspect of the project was to investigate and develop a technique to rapidly and inexpensively stabilise a polyacrylonitrile (PAN) precursor [28]. Such new processing techniques are being developed for the purpose of reducing the cost of carbon fibre conversion. Previous and ongoing research at ORNL has demonstrated that plasma processing shows great

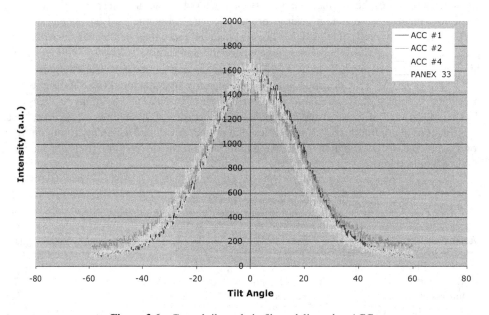

Figure 3.6   Crystal tilt angle in fibres delivered to ACC.

promise for inexpensively and rapidly carbonising, and graphitising polymer precursors to convert them to carbon fibres. However, the precursor needs to be lightly stabilised, or cross-linked, before it is exposed to plasma-generated oxidative species. Stabilisation and oxidation together are estimated to represent $\sim 18\%$ of the cost of commercial grade carbon fibre. A rapid, inexpensive and robust stabilisation technique is needed to complement the aforementioned advanced process modules and enable the development of an integrated advanced technology conversion line that converts polymer precursor fibres into carbon fibres at significantly lower cost than conventional conversion technology.

This project therefore intends to develop an advanced stabilisation module that integrates with other advanced fibre processing modules to produce inexpensive carbon fibre with properties suitable for use by the automotive industry. Critical technical criteria include:

(As previously) ≥ 25 Msi (172 GPa) tensile modulus, ≥ 250 ksi (1720 MPa) tensile strength and $\geq 1.0\%$ ultimate strain in the finished fibre.

Uniform properties along the length of the fibre tow.

Repeatable and controllable processing.

Significant unit cost reduction compared with conventional processing.

3.6.3.3 Plasma Oxidation

The purpose of this part of the project was to investigate and develop a plasma processing technique to rapidly and inexpensively oxidise polyacrylonitrile (PAN) precursor fibres [29]. Oxidation is a slow thermal process that typically consumes over two-thirds of the processing time in a conventional carbon fibre conversion line. A rapid oxidation process could dramatically increase the conversion line throughput and appreciably lower the fibre cost. A related project (see Section 3.6.3.1) has already demonstrated the potential for greatly increasing line speed in the carbonisation and graphitisation stages, and rapid stabilisation techniques have been developed (see Section 3.6.3.2), but the oxidation time must be greatly reduced to effect fast conversion.

Researchers are investigating PAN precursor-fibre oxidation using non-equilibrium, non-thermal plasma at atmospheric pressure. Plasma processing is believed to enhance oxygen diffusion and chemistry in the PAN oxidation process. Atmospheric-pressure plasma provides better control over the thermal environment and reaction rates than evacuated plasma, in addition to eliminating the sealing problems accompanying evacuated plasma processing. Various fibre characterisation tools and instruments are used to conduct parametric studies and physical, mechanical and morphological evaluations of the fibres to optimise the process.

Exposure to plasma products at or near atmospheric pressure provides superior thermal control because the gas flow should convectively heat or cool the fibres. This is deemed particularly important to avoid fibre melting from exothermic reactions. However, the short mean free path and life span of the chemically reactive species at atmospheric pressure presents another set of challenges, principally associated with finding a process recipe that delivers high process stability and short residence times.

A potentially very important discovery was that the cores of plasma-oxidised fibres are more chemically stable than are those of conventionally processed fibres. Figure 3.7 illustrates the

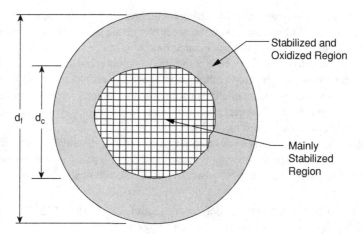

Figure 3.7   Illustration of advancing oxidation and stabilisation through filament cross-section.

advancement of oxidation and stabilisation through a filament cross-section. Most of the fibre must be oxidised, as represented by the outer region, before it can withstand carbonisation. In thermal oxidation, the outer stabilised and oxidised region grows inward slowly, purely by diffusion. However in plasma oxidation, it has been observed that oxidation over the filament cross-section is much more uniform, and the entire fibre becomes highly oxidised at a lower fibre density (fibre density is normally used as an indicator of degree of oxidation advancement). This suggests that plasma oxidation may allow onset of carbonisation at a lower fibre density. If carbonisation can indeed commence at a lower fibre density, this could significantly reduce the residence time required for oxidation.

In ORNL's conventional pilot line, which represents the baseline process, PAN stabilisation and oxidation occur in four successive furnaces in air, at temperatures increasing from about 200 to 250 °C. Although there is no precise transition from stabilisation to oxidation, in general, one can consider stabilisation to occur in the first furnace and (chemical) oxidation in the last three, so in this project the researchers are working to reproduce the conversion advancement from the last three furnaces. The advancement from the first furnace is being addressed in the parallel advanced stabilisation project.

Interfacing the oxidation module to other modules has received significant attention. As illustrated in Figure 3.8, the module interfaces may not occur at the same degree of advancement (marked by increasing density in the conventional process) for conventional and advanced-technology production lines. As a result of the lack of "hollow core" structure that is common in conventionally oxidised fibres, earlier onset of carbonisation may be possible with plasma oxidised fibres (i.e. at lower fibre density), as illustrated by the uncertain location of the oxidation-carbonisation interface in Figure 3.8. This would likely reduce the overall conversion residence time. Furthermore, the processing protocols in every module are sensitive to the prior processing history. For example, the oxidation protocols developed for conventionally stabilised fibres did not work for some of the alternative process stabilised fibres, and it was necessary to modify the oxidation process parameters significantly when oxidising these fibres. Acceptable oxidation process recipes have been developed for all stabilisation routes conducted via plasma oxidation thus far.

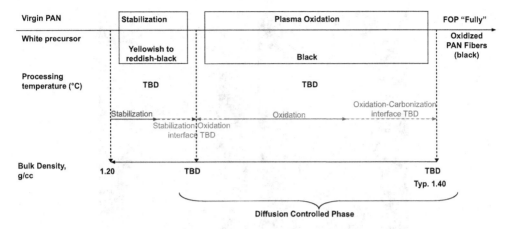

Figure 3.8   A schematic representation of stabilisation and oxidation modules.

Conventional module interfaces are represented by the vertical dotted lines in Figure 3.8. As shown by the horizontal arrows, the interface locations may change with advanced technology.

In summary, plasma oxidation shows great promise. A number of principal metrics have been developed to measure forward progress, and these will no doubt be refined as the researchers continue to grow their understanding of technology development and deployment. The combined technology in this effort could replace all four stages of oxidative stabilisation in the conventional process. Using a precursor that conventionally requires 90–120 min of residence time in thermal oxidative stabilisation ovens, the advanced modules have accomplished the same job in less than 30 min. Later carbonisation of those oxidised samples demonstrated strengths approaching 600 ksi (4140 MPa).

3.6.3.4   Advanced Surface Treatment and Sizing

In terms of surface activation, the fibre surface chemistry is very important because of the need to ensure that there are sufficient reactive sites at the fibre surface to promote fibre–matrix adhesion and to allow the most efficient use of suitably tailored coupling agents and sizes. This will allow for greatest flexibility in matching fibres to resin systems beyond the conventional epoxies and including polyesters, vinyl esters and even thermoplastics. In this respect, oxygen concentration at fibre surface is very important because the reactive groups present allow chemical bonding to that surface. Such reactive groups vary, but may commonly include carbonyl, carboxyl or hydroxyl groups. Preliminary ORNL plasma surface treatment results appear far superior to those achieved by conventional treatment results in terms of residual oxygen concentration at the surface. Researchers have investigated the effects of plasma surface oxidation with various gas compositions. It was confirmed that the chemistry can be quite sensitive to feed gas composition. A visual depiction is shown in Figure 3.9.

A wide range of feed gas compositions were investigated and the results recorded. The results remain confidential at the time of writing. However, it can be reported that atomic

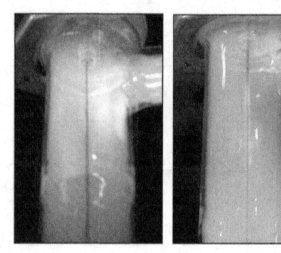

Figure 3.9   Plasma with different feed gas compositions in evacuated plasma reactor.

oxygen concentrations as high as 25–30% have been achieved. This contrasts favourably with the more typical 4–8% achieved by ozone treatment. Table 3.5 summarises the results in this particular study. Oxygen concentration data is from XPS analysis.

3.6.4   Integration: Low-Cost Carbon Fibre Pilot Line

Several processes and alternate materials developed to significantly reduce the cost of carbon fibre have been described. Alternative precursors being pursued are textile-based PAN, polyolefin and lignin-based materials. Alternative production methods include microwave plasma-assisted carbonisation, advanced stabilisation and advanced oxidation methods. Chemical pretreatment of precursors has also been developed. The technologies that have been developed follow the production process from raw material development through final product forms ready for delivery. However, in order for the technologies developed to be ready for commercialisation, they must be integrated [30].

Table 3.5   A comparison of atomic oxygen concentrations for carbon fibre samples oxidised by various routes.

Condition	Oxygen concentration (%)
Untreated commercial fibre	4.7
O_3 treated commercial fibre	6.2
AP-A	24
AP-B	30
AP-C	29
AP-D	21
AP-E	28

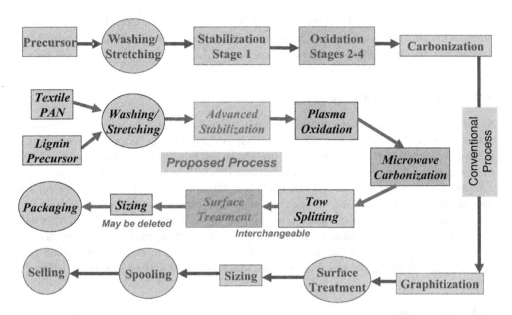

Figure 3.10 A comparison of the conventional carbon fibre processing route with the proposed ORNL low-cost carbon fibre route.

The goal of the final project in the low-cost carbon fibre portfolio is to develop an integrated, pilot-scale production line which will work out the interface issues associated with dramatically changing the materials and methods by which carbon fibre is produced. While each of the technologies offers significant cost savings, the synergistic effects of savings from many areas of the production process offer the best opportunity to reach the program cost goals. The pilot line will incorporate the advances made in textile precursors, lignin-based precursors, advanced oxidation, microwave carbonisation, advanced tow splitting and carbon fibre optimisation for industry use projects. The prototype line has been constructed at Oak Ridge National Laboratory with support from the United States Department of Energy. It is shown schematically in Figure 3.10.

As the various other strands of work come to fruition, the integration process remains the final key role in realising overall project objectives.

This project, specifically the development of a "carbon fibre systems integration" facility for testing and demonstrating new carbon fibre manufacturing technology, is underway. A conventional carbon fibre pilot line has been installed and tested. Modifications are underway to enable sustained, automated operation of the conventional pilot line. An advanced-technology pilot line will be constructed next to the conventional pilot line, with the capability to utilise any combination of conventional and advanced-conversion processes provided by the two adjacent lines. The MAP carbonisation module is the first advanced-technology module and has been installed adjacent to the conventional pilot line. MAP carbonisation scaling is underway to increase line speed and bandwidth. Advanced stabilisation and oxidation projects are developing modules that will be installed in the advanced-technology pilot line in future years. A precursor evaluation system is being developed to evaluate conversion protocols for lignin and other alternative precursor fibres. Metrics have been established against which progress toward system integration and commercialisation can be evaluated. More details on the commercialisation issues can be found in Chapter 17.

3.7   International Cooperation

Significant advances have been made in the United States through joint government–industry research toward reducing our dependence on foreign oil, minimising the environmental impact of the automobile and improving our economy. PNGV and FreedomCAR are successful models of how to do this, however much more remains to be done. The automotive industry is no longer comprised of a few domestic companies, but instead by many multi-national corporations. The need for scientific research and the talented people who can conduct that research extend far beyond the shores of the United States. It now may be time for cooperation in research not only between government and industry but also between the governments of different countries.

To this end, in 2006, the United Kingdom became involved in the low-cost carbon fibre programme, with the University of Sunderland joining in the collaborative research effort.

Acknowledgements

The authors wish to thank the DOE program management team and the board and staff of the Automotive Composites Consortium. This work was sponsored by the United States Department of Energy, Office of FreedomCAR and Vehicle Technologies, Automotive Lightweighting Materials Program. Oak Ridge National Laboratory is operated by UT-Battelle, LLC, under contract DE-AC05-00OR22725.

References

[1] European Commission (2011) DG for Energy, *"Registration of Crude Oil Imports and Deliveries in the EU"*. See: http://ec.europa.eu/energy/observatory/oil/import_export_en.htm (Accessed 28th May 2012).

[2] DUKES (Digest of United Kingdom Energy Statistics) (2012) *Petroleum* (Chapter 12). See: https://www.gov.uk/government/organisations/department-of-energy-climate-change/series/oil-statistics (Accessed 28th May 2012).

[3] Key World Energy Statistics (2007) (International Energy Agency, Paris 2007). See: www.iea.org/Textbase/nppdf/free/2007/key_stats_2007.pdf (Accessed 28th May 2012).

[4] Davis, S.C. and Diegel, S.W. (2007) Transportation Energy Data Book (Report ORNL-6978, edn 26; Tables 2.6.3: 3.1 and 3.2) Oak Ridge National Laboratory, Oak Ridge, TN.

[5] Rogers, P.G. (1990) The clean air act of 1970. *EPA Journal*, **1990**, 21–23. See: http://www.epa.gov/aboutepa/history/topics/caa70/11.html (Accessed 28th May 2012).

[6] Carpenter, J.A. Jr (1999) U.S. Government-Supported R&D on Materials for Highly Fuel-Efficient Automotive Vehicles. Proceedings of the 4th International Conference on Eco- Materials, Gifu, Japan, Nov 9-12, pp. 463–466.

[7] Carpenter, J.A. (2003) The Freedom Car Partnership. Proceedings of Ecomaterials and Ecoprocesses Symposium, Vancouver, British Columbia, Canada, 25 August.

[8] EPA (2011) Light Duty Automotive Technology, CO_2 Emissions and Fuel Economy Trends, 1975 through 2011. See: http://www.epa.gov/otaq/cert/mpg/fetrends/2012/420s12001a.pdf (Accessed 28th May 2012).

[9] Transportation Energy Data Book, 32nd Edition, 2013, Table 4.16, *http://cta.ornl.gov/data/download32.shtml*, Recopied from Ward's Communications, *Ward's Motor Vehicle Facts and Figures, 2012*, Detroit, MI, 2012, p. 62 and updates.

[10] Warren, C.D. (2012) High Volume Vehicle Materials. US Low Carbon Vehicles Workshop, Georgia Techno-logical University, Atlanta, Georgia.

[11] McKinsey and Company (2012). *"Lightweight, Heavy Impact,"* Feb. See: http://autoassembly.mckinsey.com/html/resources/publications.asp (Accessed 28th May 2012).

[12] Office of Transportation Technologies, Program Staff (1998) Energy Efficiency and Renewable Energy, Office of Advanced Automotive Technologies R&D Plan- Energy Efficient Vehicles for a Cleaner Environment, DOE/ORO/2065, March (1998), pp. 75–88.

[13] Sklad, P.S. (ed.) (2012) Vehicle Technologies Program Workshop Report (2012). Light-Duty Vehicles Material Technical Requirements and Gaps, Draft Report, October (2012).

[14] Black, S. (2012) *"Carbon Fibre Market: Gathering Momentum"*, High Performance Composites, Composites World, March 2012. See: http://www.compositesworld.com/articles/carbon-fiber-market-gathering-momentum (Accessed 28th May 2012).

[15] Sloan, J. (2013) *"Waiting for $5/lb Carbon Fibre?"* High Performance Composites, Composites World, January 2013. See: http://www.compositesworld.com/columns/waiting-for-5lb-carbon-fiber%284%29 (Accessed 8th February 2013).

[16] Red, C (2012) Composites World; *Carbon Fiber 2012: Supply and demand forecast.* See: http://www.compositesworld.com/news/carbon-fiber-2012-supply-and-demand-forecast (Accessed 28th May 2012).

[17] Sousanis, J. (2013) Wards Auto Data Center, *World Vehicle Sales.* See: http://wardsauto.com/sales-amp-marketing/world-vehicle-sales-surpass-80-million-2012 (Accessed 8th February 2013).

[18] Warren, C.D. and Wheatley, A.R. (2007) Novel Materials and Approaches for Producing Carbon Fibre. 11th European Automotive Congress, "Automobile for the Future", Budapest, Hungary, 2007.

[19] Das, S. and Cohn, S.M. (1998) A Cost Assessment of Conventional PAN Carbon Fibre Production Technology. Internal Report, Oak Ridge National Laboratory, Oak Ridge, TN.

[20] Abdallah, M.G. (ed.) (2005) Low Cost Carbon Fibre Development Program. FY2004 Progress Report for Automotive Lightweight Materials, US Department of Energy, Office of Energy Efficiency and Renewable Energy, April (2005), pp. 133–141.

[21] Leitten, C.F. Jr (ed.) (2005) Low Cost Carbon Fibres from Renewable Resources, FY2004 Progress Report for Automotive Lightweight Materials. US Department of Energy, Office of Energy Efficiency and Renewable Energy, April (2005), pp. 125–132.

[22] Baker, F.S. (ed.) (2006) Low Cost Carbon Fibres from Renewable Resources. FY2005 Progress Report for Automotive Lightweight Materials, US Department of Energy, Office of Energy Efficiency and Renewable Energy, April (2006), pp. 187–196.

[23] Baker, F.S. (ed.) (2007) Low Cost Carbon Fibres from Renewable Resources. FY2006 Progress Report for Automotive Lightweight Materials, US Department of Energy, Office of Energy Efficiency and Renewable Energy.

[24] Baker, F.S. (ed.) (2010) Low Cost Carbon Fibres from Renewable Resources. US Department of Energy, Office of Energy Efficiency and Renewable Energy.

[25] Husman, G. (2012) *Development and Commercialization of a Novel Low-Cost Carbon Fibre*, 2012 DOE Hydrogen and Fuel Cells Program and Vehicle Technologies Program Annual Merit Review and Peer Evaluation Meeting, Washington, DC. See: http://www1.eere.energy.gov/vehiclesandfuels/pdfs/merit_review_2012/lightweight _materials/lm048_husman_2012_o.pdf (Accessed 28th May 2012).

[26] Warren, C.D. and Naskar, A.K. (2012) *Low Cost Carbon Fibre Precursors*, DoE Report May 2012. See: http://www1.eere.energy.gov/vehiclesandfuels/pdfs/merit_review_2012/lightweight_materials/lm004_warren _2012_o.pdf (Accessed 28th May 2012).

[27] Paulauskas, F.L. (ed.) (2007) Low Cost Carbon Fibre Manufacturing Using Microwave Energy. Progress Report for Automotive Lightweight Materials, US Department of Energy, Office of Energy Efficiency and Renewable Energy.

[28] Paulauskas, F.L. (ed.) (2007) Advanced Stabilisation of PAN Fibre Precursor. Progress Report for Automotive Lightweight Materials, US Department of Energy, Office of Energy Efficiency and Renewable Energy, April (2007).

[29] Paulauskas, F.L. (ed.) (2007) Advanced Oxidation of PAN Fibre Precursor. Progress Report for Automotive Lightweight Materials, US Department of Energy, Office of Energy Efficiency and Renewable Energy.

[30] Paulauskas, F.L. (ed.) (2007) Carbon Fibre Systems Integration. Progress Report for Automotive Lightweight Materials, US Department of Energy, Office of Energy Efficiency and Renewable Energy.

Part Two

Impact and Crash Analysis

4

Mechanical Properties of Advanced Pore Morphology Foam Composites

Matej Vesenjak[1], Lovre Krstulović-Opara[2] and Zoran Ren[1]

[1] *Faculty of Mechanical Engineering, University of Maribor, Maribor, Slovenia*
[2] *Faculty of Electrical Engineering, Mechanical Engineering and Naval Architecture, University of Split, Split, Croatia*

4.1 Introduction

Porous materials and their composites have been increasingly used in modern engineering and medical applications over the past decade due to their advantageous mechanical and thermal properties. However, their structure in terms of shape, size and distribution of cellular pores cannot be fully controlled with existing industrial mass production processes. This results in a certain scatter of physical and other characteristics of these materials.

A new type of hybrid aluminium cellular structure material, named advanced pore morphology (APM) foam was developed recently. APM foam, consisting of sphere-like metallic foam elements, proves to have advantageous mechanical properties and unique application adjustability.

This chapter first presents the cellular materials and their composites in general and then mostly concentrates on the particularities of APM foam. Single metallic APM spheres and their mechanical properties are characterised with experimental testing and computational simulations, providing the basic properties and knowledge for an efficient design of composite APM foam structures. The final part of the chapter presents APM foam elements moulded with epoxy matrix, resulting in composite structures of different morphology types (partial and syntactic). The mechanical characterisation of composite APM foam structures based on experimental testing with free and confined boundaries is also addressed.

Advanced Composite Materials for Automotive Applications: Structural Integrity and Crashworthiness,
First Edition. Edited by Ahmed Elmarakbi.
© 2014 John Wiley & Sons, Ltd. Published 2014 by John Wiley & Sons, Ltd.

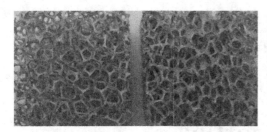

Figure 4.1   Open-cell cellular structure (left) and closed-cell cellular structure (right) [4]. Reproduced from Ref. [4]. Copyright 2000 Wiley.

4.2   Cellular Materials

Cellular materials are structured as interconnected network of solid/void struts or walls which form the edges and faces of cells/pores of an open-cell/closed-cell cellular structure (Figure 4.1). Cellular materials have some advantageous mechanical and thermal properties in comparison to solid materials, for example, low density (light-weight structures), high acoustic isolation and damping, high energy absorption capabilities, durability at dynamic loadings and fatigue [1].

Their micro- and macroscopic properties make them very attractive for use in automotive, rail, naval and aerospace industry as well in general engineering as heat exchangers, filters, bearings, acoustic dampers, core material in sandwich structures, bio-medical implants and elements for energy absorption [1–3]. One of the most important areas for the future application of cellular materials is in the automotive industry, where their high impact energy absorption through deformation is of crucial importance for increasing the passive safety of vehicles [1, 2].

4.2.1   Mechanical Behaviour of Cellular Materials

The mechanical behaviour of cellular materials mainly depends on relative density (porosity) and base material which can vary between metals, plastics, glass and ceramics. The other important parameters of the cellular structures are morphology (open or closed cell, Figure 4.1), geometry and topology (regular or irregular structure, cell size) and a possible pore filler. Regardless of their different base materials, topologies and morphologies, cellular materials have a characteristic stress–strain relationship in compression, which can be divided into four main areas (Figure 4.2).

After an initial quasi-linear elastic response [Figure 4.2, (i)], the cellular materials first experience buckling, plastic deformation and collapse of intercellular walls in the transition zone [Figure 4.2, (ii)]. Under further loading the mechanism of buckling and collapse becomes even more pronounced, which is manifested in large strains at almost constant stress [stress plateau, σ_P; Figure 4.2, (iii)] until the cells completely collapse [densification at ε_D; Figure 4.2, (iv)]. At this point, the cellular material stiffness increases and consequently converges towards the stiffness of the base material. The benefit of such a material response is the high efficiency of energy absorption that occurs within the stress plateau region. An ideal energy-absorbing material dissipates the kinetic energy of the impact while keeping the reacting force in certain limits in order to avoid any serious or even lethal injuries to the passengers [5]. Yu *et al.* [6] studied the influence of cell diameter on energy absorption properties of cellular structures. They observed that, while increasing the cell size, the energy absorption capability decreases.

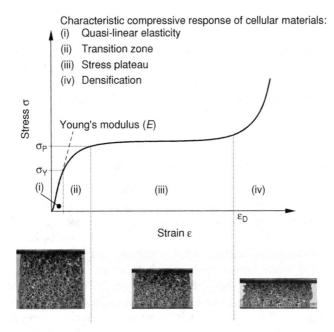

Figure 4.2 Characteristic behaviour of cellular materials under compressive loading.

Despite the many advantageous properties of cellular materials, there are still some technological problems related to control of the pore morphology and topology in industrial production processes, since the majority of existing technologies do not allow for precise control of the shape, size and distribution of the cellular pores [1, 4, 7, 8]. This results in a certain scatter of their mechanical and thermal characteristics. It is thus understandable that the development of cellular materials and their manufacturing techniques is orientated towards the production of cellular materials where the topology and morphology could be controlled, resulting in more regular and homogeneous structures. Such developments led to metallic hollow sphere structures, advanced pore morphology foam, lotus-type porous metals and wire-woven bulk Kagome structures [4, 9–13].

To further expand their application potentials, cellular materials are often used as parts or cores in composite structures, for example, sandwich structures (Figure 4.3). Sandwich

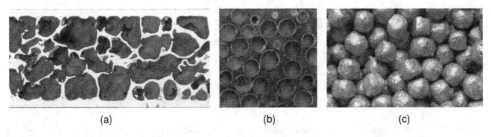

Figure 4.3 Closed-cell cellular structure as the core material in sandwich structure (a), metallic hollow spheres (b) and advanced pore morphology foam elements (c) adhered in composite structures.

structures are usually comprised of two face sheets with cellular material as a core in between. Such sandwich structures have an increasing relevance in various engineering applications due to their high stiffness, strength and reduced mass (light-weight). In some cases single cellular material elements (metallic hollow sphere structures, advanced pore morphology foam) are joined together using different technologies such as sintering, soldering and adhering [9, 11, 14]. Adhering provides the most economical way of joining and allows for further cost reduction and therefore the expansion of potential applications (Figure 4.3).

4.2.2   Energy Absorption Capabilities of Cellular Materials

The cellular materials in general are excellent elements for mechanical energy absorption through their deformation due to their prolonged plateau stress during plastic deformation. In the case of being used as filler materials in hollow structures (e.g. hollow pillars in a vehicle) their energy absorption capabilities can be increased even further (Figure 4.4). The cellular material under loading can absorb a certain amount of energy [Figure 4.4 (a)]. Its stress–strain curve is usually very smooth. The hollow part can also absorb energy during its deformation [Figure 4.4 (b)]. However, its stress–strain curve is much more oscillating due to the deformation mechanism (buckling). If such hollow part is filled with the cellular material energy [Figure 4.4 (c)] and loaded, its cumulative amount of energy absorption would represent not only the sum of the separate amounts of energy absorbed by cellular materials (W_{foam}) and hollow part (W_{tube}), but would be larger by ($W_{additional}$) due to interaction between the cellular material and the walls of the hollow part, as illustrated in Figure 4.4 [15]. Since the cellular material is confined by the hollow part, it can absorb more energy which can additionally be increased by friction between the cellular material and the hollow part. Such characteristics give cellular materials great applicative potential in the automotive industry,

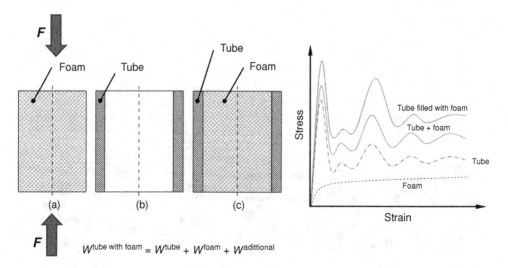

Figure 4.4   Stress–strain diagrams for different cases: (a) cellular material, (b) hollow structure – tube, (c) hollow structure filled with cellular material.

where hollow automotive parts or structures can be filled with cellular material, significantly increasing their energy absorption capabilities for a minimal added mass.

4.2.3 Influence of Pore Fillers

Additional influence on behaviour and energy absorption capabilities of cellular materials with open and closed cell morphology can be attributed to a possible pore filler in the cellular material [16–18]. After solidification and cooling down during the manufacturing process of closed-cell cellular materials with gas injection, it can be assumed that the gas is trapped in the base material. During loading of such materials the gas inside the pores influences the macroscopic cellular material behaviour [19, 20]. It was reported by Shkolnikov [21] that, even for open-cell honeycomb structures, the gas inside the pores contributed to 10% higher stiffness of the honeycomb, although this effect was observed only for larger specimens. Zhang [22] investigated the effect of gas pressure on the nonlinear macroscopic constitutive relationship of the porous materials. Under compressive loading the pore pressure increases due to cell deformation, thus contributing to the structure stiffness increase up to the point of base material yielding. This results in the collapse of intercellular walls and an increase of the material structure porosity [23]. The influence of pore filler can also be observed in open-cell cellular structures, where the filler can flow through the open network of the cellular structure and dissipate additional energy. At increased loading velocities (high strain rates during impact) the compressed gas pressure already flowing through the cellular structure increases and results in additional energy dissipation, according to Shim and Yap [24]. A logical solution to further increase the energy absorption in open-cell cellular materials is filling the cellular structure with a second phase filler material. Such a filler offers an additional level of energy dissipation during the deformation process. Filling the cellular structure with liquid or soft fillers results in an increase of energy absorption [18, 25, 26] which has been confirmed by experimental testing and computational simulations [26, 27].

4.2.4 Strain Rate Sensitivity of Cellular Materials

Another parameter that influences the cellular material energy absorption capability is the loading velocity and the strain rate sensitivity. The strain rate sensitivity and energy absorption capacity of commercially available cellular materials and the correlation between the mechanical response and the physical and geometrical properties have been studied by Montanini [28]. There are four influential parameters contributing to the strain rate sensitivity of the cellular structures: (i) cell morphology, (ii) cell topology, (iii) the micro-inertial effect and (iv) the material strain rate sensitivity [3, 14, 29, 30]. Each contributes to dynamic strengthening of the global behaviour of the foam structure [31, 32]. Usually the yield stress is also strain rate and temperature dependent [33]. The micro-inertia of the individual cell walls can affect the deformation of cellular materials as discussed in [34–36]. According to [37] the inertia effects associated with the dynamic localisation of crushing are responsible for the enhancement of the dynamic strength properties in high velocity regimes. In [36] the authors state that the inertia effects significantly contribute to the rate effect, even when the strain rate sensitivity of material properties is ignored. Vesenjak *et al.* [38], however, have shown that the micro-inertia has a minor effect in contributing to the stiffness of the cellular material. The influence of the material strain rate sensitivity is of much greater importance. Additionally, the strain rate sensitivity of the cellular material might also be affected by the gas (filler) flow through the structure [17, 39].

The strain rate sensitivity effect is more apparent for the higher density foams [39]. Mukai *et al.* [40, 41] discovered that the plateau stress of aluminium foam exhibits significant strain rate sensitivity. As a consequence, the energy absorption at dynamic strain rates is much higher compared to the quasi-static strain rate [41].

For many materials the strain rate sensitivity at a constant temperature and certain strain can be described with the following power law expression

$$\sigma = C \cdot \dot{\varepsilon}^m \tag{4.1}$$

where the exponent m represents the strain rate sensitivity [33]. The relative levels of stress at two strain rates, evaluated at the same strain, are given by

$$\frac{\sigma_2}{\sigma_1} = \left(\frac{\dot{\varepsilon}_2}{\dot{\varepsilon}_1}\right)^m \tag{4.2}$$

The strength of the material under dynamic loading (Figure 4.5) increases if $m > 0$ and decreases if $m < 0$, while it remains the same if $m = 0$. The values of the strain rate sensitivity parameter are listed in [33].

The most commonly used constitutive model in dynamic simulations accounting for strain rate sensitivity is the Cowper–Symonds model, which is based on scaling the quasi-static material stress–strain response

$$\sigma = \sigma_{stat} \cdot \left[1 + \left(\frac{\dot{\varepsilon}}{C}\right)^{1/p}\right] \tag{4.3}$$

where σ corresponds to the stress at strain rate $\dot{\varepsilon}$ and σ_{stat} is the quasi-static response of the material. Parameters C and p define the strain rate sensitivity [42–46]. An advanced constitutive model accounting for strain rate sensitivity is the Johnson–Cook model, which also considers

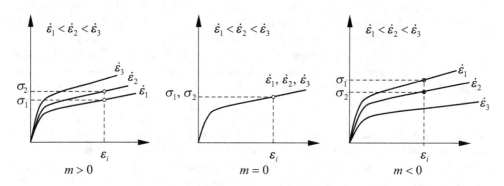

Figure 4.5   Material strain rate sensitivity.

the temperature influence [26, 42, 43, 46]. The comparison of strain rate sensitivity for the mentioned constitutive models can be expressed as

$$\frac{\sigma_2}{\sigma_1} = \underbrace{\left(\frac{\dot{\varepsilon}_2}{\dot{\varepsilon}_1}\right)^m}_{\text{power law}} \equiv \underbrace{1 + \left(\frac{\dot{\varepsilon}}{C}\right)^{1/p}}_{\text{Cowper–Symonds}} \equiv \underbrace{1 + c \cdot \ln\left(\frac{\dot{\varepsilon}_2}{\dot{\varepsilon}_1}\right)}_{\text{Johnson–Cook}} \qquad (4.4)$$

From Equation 4.4 it can be concluded that strain rate sensitivity is described in both power law and Johnson–Cook model with a single parameter. In contrast, the Cowper–Symonds model uses two parameters, which allows for better calibration of the model. However, it has to be taken into account that the response of the analysed model can be simulated with the highest accuracy if stress–strain curves for different strain rates obtained experimentally are directly used in a constitutive model.

The number of studies performed considering the effect of strain rate on densification strain is very limited due to difficulty characterisation of the high strain rate response. However, it was observed that, by increasing the strain rate, the densification strain decreases which results in lower capability of energy absorption [19, 26, 29, 30, 47].

4.3   Advanced Pore Morphology Foam

One of the new fabrication methods which was developed for a more convenient and flexible use of cellular materials in different applications and to overcome the technological problems related to the control of structure irregularity, also resulting in a more homogeneous and regular pore distribution, allows for the production of a new type of cellular materials – the advanced pore morphology (APM) foam [9, 10, 48, 49]. The APM foam consists of sphere-like metallic foam elements which have some advantageous mechanical properties and unique applicability [50]. The APM foam was developed at Fraunhofer IFAM (Bremen, Germany) and represents a new method for the production of composite cellular structures. The APM foam elements (spheres) consist of interconnected a closed-cell cellular structure (foam), as shown in Figure 4.6 [51, 52].

The manufacturing procedure consists of powder compaction (by the CONFORM® process) and rolling to obtain an expandable precursor material. The precursor material is then cut into

Figure 4.6   Advance pore morphology foam elements and their cross-section.

small volumes (granules) which are expanded (due to TiH_2 foaming agent) into spherical foam elements in a continuous belt furnace [53]. The matrix alloy of the APM foam is $AlSi_7$. Generally, three sizes of APM foam are manufactured: 5, 10 and 15 mm in diameter. Their density can vary from 500 to 1000 kg/m^3 [52]. Their detailed technology concept, production and properties are described in [51, 54]. The pore sizes in single APM foam elements vary between 1.5 and 2.5 mm [12]. The APM foam parts exhibit two types of porosity: (i) the inner porosity in single APM foam elements and (ii) the outer porosity between many APM foam elements [55].

The APM foam elements have a characteristic compressive stress–strain relationship similar to other cellular materials (see Figure 4.2). The APM foam elements have a wide potential application spectrum as an energy absorbing structure [56], stiffening elements (in shell structures), core layers, damping elements or bonded with a matrix in composite materials. One of their main advantages is their simple use as filler elements of hollow parts, for example, a hollow automotive part can be filled with APM foam elements covered with adhesive, then the filled part is heated up to join the APM foam elements. Such an APM foam filled part has much higher energy absorption capabilities at minimal mass increase. It should also be noted that this procedure is insensitive to hollow part shape and geometry, which can be a challenging problem for conventional aluminium foams. They are additionally able to undergo large deformations when subjected to compressive loading.

Since the technology concept and manufacturing procedure of APM foam elements was developed only recently, the characterisation of its mechanical properties is still under investigation [50]. Reference [51] along with the detailed description of the APM technology concepts explains the fundamental deformation behaviour of single and bonded APM foam elements under quasi-static compressive loading conditions. Results of uni-axial compression tests of APM foam are evaluated in [53], where the authors focus on variation and influence of adhesives and adhesive coating thicknesses used for bonding the APM foam elements with partial morphology. Additionally, they highlight some distinctive differences between APM and conventional aluminium foams. A practical example of filling a hollow profile with APM elements and comparing the stress–strain diagrams between the hollow profile, a hollow profile filled with APM elements and a hollow profile filled with bonded APM elements is given in [51, 55, 57]. The authors demonstrate that filling the profile with APM elements increases the capability of energy absorption which can be further improved by bonding the APM elements.

The mechanical characterisation of these materials is still very limited. The properties of a single APM foam element subjected to compressive loading are described in the next section, providing the basic properties and knowledge for an efficient composition of composite APM foam structures. Then, several APM foam composites with different morphologies and element sizes are moulded, where the APM elements are bonded with an epoxy resin. The deformation mechanism and the strain energy absorption are closely examined.

4.4   Mechanical Properties of Single APM Foam Elements

The experimental testing of single APM foam elements has been performed for quasi-static loading conditions [11, 50, 51, 53, 54] and for dynamic loading conditions [11, 50]. The following section comprises the mechanical properties of single APM foam elements given in

$\varepsilon = 0$ $\varepsilon = 0.15$ $\varepsilon = 0.3$ $\varepsilon = 0.45$

Figure 4.7 Deformation states of a single 10 mm APM foam element.

[11, 50]. The experiments were performed on the servo-hydraulic testing machine INSTRON 8801 under quasi-static and dynamic uni-axial compressive loading conditions. Two sizes of APM foam elements were tested: Ø5 and Ø10 mm. The matrix alloy of all specimens was AlSi$_7$ [58]. The APM foam elements ($\rho \approx 800$ kg/m^3) were not covered with any adhesive coating after expanding in the heating zone.

Figure 4.7 illustrates the deformation state of a single 10 mm APM foam element at 15% compressive engineering strain increment. The specimens provided very similar deformation behaviour. After an initial quasi-elastic region, the irregularities on the outer surface of the APM foam elements became visible due to the inner pores collapsing, which was also observed by Stöbener [51]. At further deformation those irregularities led to cracks and finally rupture of the outer surface. The deformation and collapse of inner pores was very even and gradually distributed through the specimen during the compressive loading.

Figure 4.8 shows the average relationship between force and engineering strain under quasi-static and dynamic compressive loading. The compressive engineering strain is represented as

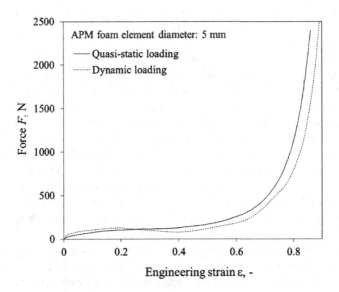

Figure 4.8 Average response of single APM foam elements (Ø5 mm) under quasi-static and dynamic compressive loading.

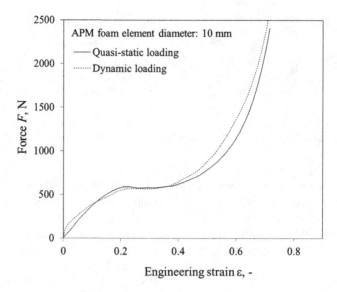

Figure 4.9 Average response of single APM foam elements (Ø10 mm) under quasi-static and dynamic compressive loading.

the ratio between diameter change due to deformation (displacement) and the initial diameter of the APM element. The experimental results show characteristic cellular material behaviour, that is, after the initial elastic response onset of yielding which is then manifested in typical stress plateau followed by the final densification [17]. The maximum strain rate of 33/s was achieved during dynamic testing. It can be observed from Figure 4.8 that, in the case of quasi-static loading, the stress increases more gradually compared to the dynamically loaded elements. However, this difference can only be observed up to 15% of deformation. For larger deformation, the response of quasi-statically and dynamically loaded APM elements is similar.

Figure 4.9 illustrates the quasi-static and dynamic compressive behaviour of 10 mm APM foam elements. The maximum strain rate of 20/s was achieved during dynamic testing. Comparing the results in Figures 4.8 and 4.9, it can be concluded that the (force) plateau region for larger specimens (Ø10 mm) is approximately 4.6× higher in comparison with the Ø5 mm APM foam elements. This difference is foremost due to a different inner porosity of the specimens. From weight measurements it was evaluated that the smaller APM elements (Ø5 mm) have approximately 10% higher inner porosity in comparison to Ø10 mm APM elements. Additionally, it was observed that larger foam elements experience lower densification strain (difference approx. $\Delta\varepsilon = 0.15$) which also confirms the above assumption. The results of the presented experimental tests are also in a very good agreement with results reported by Stöbener et al. [53].

The elastic–plastic deformation behaviour of APM foam elements was also studied by using computational modelling and computer simulations which offer an effective and efficient way to conduct further characterisation of different APM foam element variations. They are usually cheaper and faster in comparison to experimental testing, bearing in mind appropriate verification of the computational model and validation of computational results. Based on

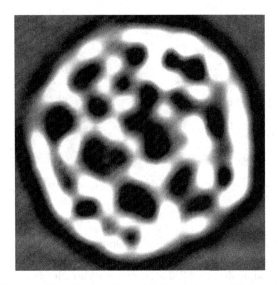

Figure 4.10   CT scan of an APM foam element (Ø10 mm).

CT scans of APM spheres (Figure 4.10), the structure was virtually reconstructed and three-dimensional finite element models were generated for dynamic computational simulations with the finite element code LS-DYNA with an explicit time integration scheme.

First, the space and time discretisation of the APM models were evaluated and verified. Then they were used to determine deformation behaviour at different deformation strain rates of APM foams elements.

The results of computational simulations give an extensive overview of the stresses and strains in every single point of the APM sphere (Figure 4.11). In this way the interior deformation mechanism can be closely observed.

Another computational approach is the use of homogenised constitutive material data of APM foam elements, whereby they can be modelled as solid (nonporous) elements with the same material properties as manufactured porous elements. Such a computational approach

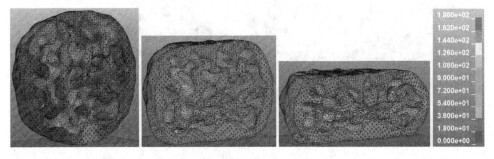

Figure 4.11   Von Mises stresses (MPa) and deformation pattern of the reconstructed APM foam during dynamic computational simulation model.

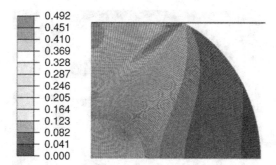

Figure 4.12 Plastic deformation of a single APM foam element (Ø10 mm) under compressive loading at deformation $\varepsilon = 0.20$.

allows for much easier and faster pre-processing, shorter computational times and more effective computational simulations of this material in large-scale structural engineering simulations. The homogenised models of APM foam elements are clearly showing the plastification zone (Figure 4.12) and their evaluation is described by Vesenjak *et al.* [11, 50].

The elastic–plastic deformation behaviour of APM foam elements was also examined by using thermography which directly shows the development of plastification zones in dynamically loaded structures [50]. Figure 4.13 shows the thermal distribution of dynamically compressed APM sphere taken at a 500 Hz frame rate.

The strain rate during dynamic testing was 20/s. The recorded heat generation, which is directly correlated to the plastic deformation, shows the plastification development of the APM sphere surface. It can be observed that the generated heat is significant and clearly indicates regions of the base material yielding with a high temperature increase. The yielding starts at the upper part of the APM element and propagates in a shear band to the lower part of the

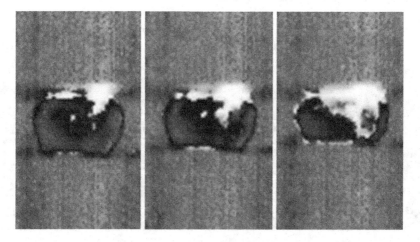

Figure 4.13 Thermography of an APM element (Ø10 mm) during dynamic loading.

Figure 4.14   Partially bonded APM structures (APM foam element: Ø10 mm).

sphere, finally resulting in a fully plastically compressed APM foam element. The results are in excellent agreement with the homogenised computational model (Figure 4.12).

4.5   Behaviour of Composite APM Foam

The characterisation of bonded APM foam elements in composite structures has been analysed for different loading conditions [11,12, 53]. This section provides the mechanical properties of adhered APM foam elements using epoxy resin. The experimental tests were performed on the servo-hydraulic testing machine INSTRON 8801 under quasi-static and dynamic compressive loading conditions with free and confined radial boundary surfaces. The experimental plan comprised of APM composites with different topologies and morphologies. The composites topologies varied by using APM foam elements with different diameters (Ø5 and Ø10 mm). The morphology of the composites was defined by bonding the APM sphere elements as: (i) syntactic APM composite (APM foam elements are completely embedded within the adhesive matrix, Figure 4.14 [59]) and (ii) partial APM composite (adhesive is applied only at the contact points of the APM foam elements, Figure 4.15).

The APM foam elements were bonded together with epoxy resin EPOCON ($\rho = 1080$ kg/m^3 and $\sigma_y = 90.4$ MPa). The testing of composite APM foam specimens was performed using two different boundary set-ups (Figure 4.16): (i) classical uni-axial compressive loading without radial constraints and (ii) compressive loading with confined boundaries – radial constraints. In order to reduce the friction between the specimen and the radial constraint tube (made of steel St37-2), the tube walls and loading plates were lubricated.

Figure 4.15   Syntactic APM structures (APM foam element: Ø10 mm).

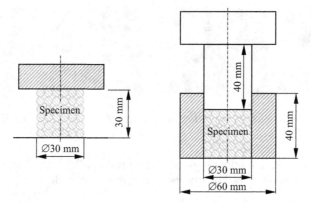

Figure 4.16 Experimental set-up for compressive tests with free boundaries (left) and radial constraints (right).

4.5.1 Compressive Loading of Confined APM Foam Elements without Bonding

During the first set of experimental tests the APM foam elements were compressed inside the radial constraints tube with inner diameter $d = 30$ mm without any adhesive bonding. The average engineering stress–strain diagram for the compressive loading of confined APM foam elements without bonding is shown in Figure 4.17. The smooth characteristic compressive behaviour of cellular materials can be observed from the figure. At first the non-bonded APM spheres are initially compacted at low external forces in the initialisation region before the

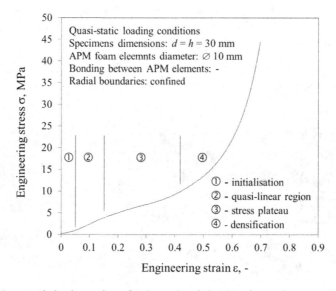

Figure 4.17 Average behaviour of confined non-bonded APM foam elements under quasi-static loading.

Figure 4.18   Compressive behaviour of partial APM composites with free boundaries.

stiffness increases and contributes to the quasi-linear elastic region, stress plateau and densi-fication at the end of deformation process. The average strain energy density of non-bonded Ø10 mm APM spheres with confined boundaries until 50% of deformation is 2.99 J/mm^3. The initialisation (compaction) region also influences (decreases) the measured energy absorption capability. Without consideration of the initialisation region the strain energy density would increase by 24%.

4.5.2   Partially Bonded APM Foam Elements

In partially bonded APM composites the adhesive is applied only at contact points between the APM foam elements. The specimens with partial morphology have been tested with confined and free boundaries. During the preparation of partial composites using APM spheres with smaller diameter (Ø5 mm) the viscosity of the fluid epoxy proved to be too high to produce high-quality specimens. Therefore, only the behaviour of partial composite APM foam specimens with Ø10 mm spheres was studied. The compressive behaviour of partial APM composites with free boundaries is represented in Figure 4.18.

The APM foam elements were gradually compressed during the loading. However, in some cases the adhesive forces of the partial bonded composite proved to be insufficient to retain the APM elements as part of the composite structure during the compressive loading. As shown in Figure 4.19 some APM spheres were released from the unconfined APM structure at specimen free surfaces during testing, resulting in a sudden decrease of compressive load carrying capacity. This effect was even more pronounced at dynamic loading.

During compressive loading of partially bonded APM composites using the radial constrains, all APM foam elements contributed to the energy absorption. The deformation mechanism during the loading procedure could not be observed during the enclosed specimen within the radial constraints. The final compressed and compacted structure is shown in Figure 4.20.

Figure 4.19 Low adhesive forces between APM foam elements in partial bonded composite.

Figure 4.21 illustrates the average behaviour of the partially bonded composites with confined and free boundary condition. As expected, the specimens with confined boundaries exhibit much stiffer behaviour and experience higher stress levels. They have much higher load carrying capacity and also possess much better energy absorption capabilities. The global stress in specimens with free boundaries increases up to around 10% and then decreases. The reason for such decrease in global stiffness is the adhesive – epoxy resin. After a certain deformation stage, the adhesive is not strong enough to keep the APM foam elements in the composite structure. Therefore some APM elements are released from the specimens, resulting in a significant global stiffness decrease of the composite structure. The average strain energy density until 50% of deformation of partial APM foam with free boundaries is 1.39 J/mm^3, while for partial APM foam with confined boundaries it is 4.49 J/mm^3. This

Figure 4.20 Compressed partial APM composites after loading using the confined boundaries.

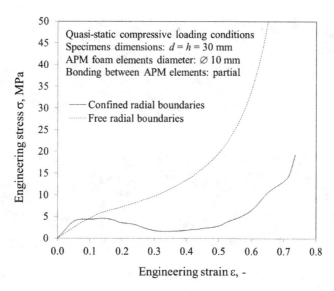

Figure 4.21 Average behaviour of partial APM foam composites with free and confined boundaries under quasi-static loading conditions.

shows that the APM foam material with confined boundaries is able to absorb significantly more strain energy.

4.5.3   Fully Bonded APM Foam Elements – Syntactic Structure

The syntactic APM composite specimens consisted of APM foam elements fully embedded in the epoxy matrix. The specimens were tested only with free radial boundaries because in the case of the constrained radial boundaries the loading force exceeded the maximal achievable force of the hydraulic testing machine (50 kN). The syntactic composite structure was subjected to compressive loading under quasi-static loading conditions and dynamic loading conditions with strain rate up to 10/s. Deformation behaviour of the syntactic APM foam composite with Ø5 and Ø10 mm APM spheres without radial constraints is shown in Figures 4.22 and 4.23, respectively. The compressive behaviour shows typical layer-wise collapse mechanics, common for cellular structures subjected to low velocity loading [27]. After collapse of all sphere layers the stiffness of the composite structure significantly increases due to compaction and densification.

Figure 4.24 represents the average response of the syntactic APM composites with free radial boundaries under compressive loading conditions. It can be observed that the composite specimens subjected to quasi-static compressive loading conditions experience a characteristic cellular material behaviour. A comparison between the structure with Ø5 and Ø10 mm spheres shows a very low deviation in the quasi-linear region up to yield stress. After yielding, the APM composite with Ø5 mm spheres experiences higher stresses in comparison to the APM composite with Ø10 mm spheres, which can be attributed to their outer porosity. This response

Figure 4.22 Compressive behaviour of syntactic APM foam composite with Ø5 mm APM spheres under quasi-static loading.

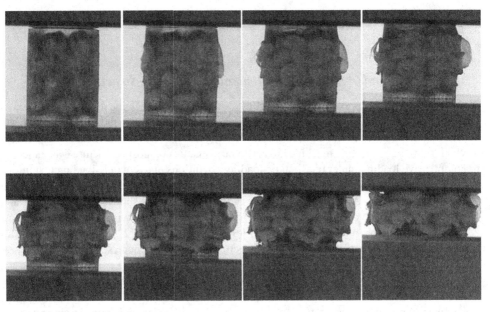

Figure 4.23 Compressive behaviour of syntactic APM foam composite with Ø10 mm APM spheres under quasi-static loading.

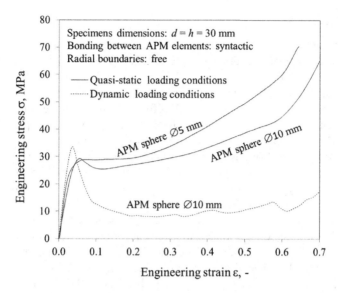

Figure 4.24   Response of syntactic APM foam composites with free radial boundaries under quasi-static and dynamic compressive loading conditions.

difference increases at higher strains. The outer porosity of Ø5 mm spheres is 22% and the outer porosity of Ø10 mm spheres is 38%. In the case of an APM composite with Ø5 mm spheres, the spheres are much more compacted than in the structure with Ø10 mm spheres. This results in a higher stiffness and consequently also a shifted densification strain.

The response of the syntactic APM composite subjected to dynamic loading shows distinctively different behaviour in comparison to the quasi-static loading response. After initial stress increase (the yield stress is as expected higher than in the case of quasi-static loading), the structure abruptly loses its load capacity. As observed in Figure 4.19 this might be a result of the adhesive which is not strong enough to keep the APM foam elements in the composite structure. The measured average strain energy density until 50% of deformation is shown in Table 4.1.

The reason for higher energy absorption capabilities of syntactic APM structure with Ø5 mm spheres is due to its higher degree of compaction. The strain energy density of syntactic composites with Ø10 mm APM spheres during dynamic testing is 59.1% lower in comparison to quasi-static loading. The dynamic energy absorption capabilities are significantly lower due to weak adhesion between APM spheres and their observed release from the structure at specimen free surfaces during testing.

Table 4.1   Average strain energy density of syntactic APM composites.

	Ø5 mm APM elements	Ø10 mm APM elements
Quasi-static loading	16.8 J/mm³	14.2 J/mm³

4.6 Conclusion

This chapter focuses on cellular materials and their composites and discusses their applicability as elements for mechanical energy absorption in the automotive industry as well as in general engineering. Different aspects influencing the behaviour and energy absorption of cellular composites materials are presented, with the main focus on the newly developed APM foam elements and their composites. The APM foam elements can be combined in different composite structures using epoxy resin, which results in composites with partial and syntactic morphology. Their behaviour has been determined using compressive experimental tests with quasi-static and dynamic loading regimes. Additionally, two different boundary conditions (free and constrained radial boundaries) are considered during experimental testing. Single APM elements show typical cellular material behaviour. It has been observed that larger foam elements experience lower densification strain and higher energy absorption capabilities. The deformation behaviour of syntactic APM foam composite with free radial boundaries shows characteristic layer-wise collapse mechanics, typical for cellular structures subjected to low velocity loading. A comparison between syntactic structures with Ø5 and Ø10 mm APM spheres shows a very low deviation in the quasi-linear region, which then increases towards larger strains due to different outer porosity. The advantageous behaviour of composite structures with partial morphology and confined boundaries has been observed. Such specimens prove to have better energy absorption capabilities which are of great advantage in the case of filling the construction part with APM foam elements. Further, the adhesive can be regarded as an additional design parameter to influence the behaviour of APM composites.

Acknowledgements

This work was supported by the Slovenian Research Agency. The support of Dr. Dirk Lehmhus and Dr. Karsten Stöbener is gratefully acknowledged.

References

[1] Banhart, J. (2001) Manufacture, characterisation and application of cellular metals and metal foams. *Progress in Materials Science*, **46**(6), 559–632.
[2] Gibson, L.J. and Ashby, M.F. (1997) *Cellular Solids: Structure and Properties*, Cambridge University Press, Cambridge, UK.
[3] Yu, J.L., Li, J.R. and Hu, S.S. (2006) Strain-rate effect and micro-structural optimization of cellular metals. *Mechanics of Materials*, **38**(1–2), 160–170.
[4] Körner, C. and Singer, R.F. (2000) Processing of metal foams - challenges and opportunities. *Advanced Engineering Materials*, **2**(4), 159–165.
[5] Avalle, M., Belingardi, G. and Montanini, R. (2001) Characterization of polymeric structural foams under compressive impact loading by means of energy-absorption diagram. *International Journal of Impact Engineering*, **25**(5), 455–472.
[6] Yu, H., Guo, Z., Li, B. *et al.* (2007) Research into the effect of cell diameter of aluminium foam on its compressive and energy absorption properties. *Material Science Engineering: A*, **454–455**, 542–546.
[7] Ochsner, A. and Lamprecht, K. (2003) On the uniaxial compression behavior of regular shaped cellular metals. *Mechanics Research Communications*, **30**(6), 573–579.
[8] Davies, G.J. and Zhen, S. (1983) Metallic foams: their production, properties and applications. *Journal of Materials Science*, **18**(7), 1899–1911.
[9] Vesenjak, M., Fiedler, T., Ren, Z. and Öchsner, A. (2008) Behaviour of syntactic and partial hollow sphere structures under dynamic loading. *Advanced Engineering Materials*, **10**(3), 185–191.

[10] Nakajima, H. (2007) Fabrication, properties and application of porous metals with directional pores. *Progress in Materials Science*, **52**(7), 1091–1173.

[11] Vesenjak, M., Gačnik, F., Krstulović-Opara, L. and Ren, Z. (2011) Behavior of composite advanced pore morphology foam. *Journal of Composite Materials*, **45**(26), 2823–2831.

[12] Lehmhus, D., Baumeister, J., Stutz, L. *et al.* (2010) Mechanical characterization of particulate aluminium foams—strain-rate, density and matrix alloy versus adhesive effects. *Advanced Engineering Materials*, **12**(7), 596–603.

[13] Fiedler, T., Öchsner, A. and Grácio, J. (2010) Numerical investigations on the mechanical properties of adhesively bonded hollow sphere structures. *Journal of Composite Materials*, **44**(10), 1165–1178.

[14] Vesenjak, M., Ren, Z., Fiedler, T. and Öchsner, A. (2009) Impact behavior of composite hollow sphere structures. *Journal of Composite Materials*, **43**(22), 2491–2505.

[15] Ashby, M.F., Evans, A.G., Fleck, N. *et al.* (2000) *Metal foams: a design guide*, Butterworth-Heinemann, Boston.

[16] Vesenjak, M., Öchsner, A. and Ren, Z. (2008) Characterization of open-cell cellular material structures with pore fillers. *Materials Letters*, **62**(17–18), 3250–3253.

[17] Vesenjak, M., Krstulović-Opara, L., Ren, Z. *et al.* (2009) Experimental study of open-cell cellular structures with elastic filler material. *Experimental Mechanics*, **49**(4), 501–509.

[18] Vesenjak, M., Öchsner, A., Hriberšek, M. and Ren, Z. (2007) Behaviour of cellular structures with fluid fillers under impact loading. *International Journal of Multiphysics*, **1**, 101–122.

[19] Ohrndorf, A., Schmidt, P., Krupp, U. and Christ, H.J. (2000) Mechanische Untersuchungen Eines Geschlossenporigen Aluminiumschaums. Werkstoffprüfung 2000. Deutscher Verband für Materialforschung und-prüfung e.V., Bad Nauheim.

[20] Lankford, J. and Dannemann, K.A. (1998) Strain rate effects in porous materials. *Materials Research Society Symposium Strasbourg*, **521**, 103–108.

[21] Shkolnikov, B.M. (2002) Honeycomb modeling for side impact moving deformable barrier (MDB). Seventh International LS-DYNA Users Conference, Dearborn, Michigan, USA, 19–21 May, Livermore, California: Livermore Software Technology Corporation.

[22] Zhang, W., Xu, Z., Wang, T.J. and Chen, X. (2009) Effect of inner gas pressure on the elastoplastic behavior of porous materials: a second-order moment micromechanics model. *International Journal of Plasticity*, **25**(7), 1231–1252.

[23] Elzey, D.M. and Wadley, H.N.G. (2001) The limits of solid state foaming. *Acta Materialia*, **49**(5), 849–859.

[24] Shim, V.P.W. and Yap, K.Y. (1997) Modelling impact deformation of foam-plate sandwich systems. *International Journal of Impact Engineering*, **19**, 615–636.

[25] Ren, Z., Vesenjak, M. and Öchsner, A. (2008) Behaviour of cellular structures under impact loading: a computational study. *Material Sciences Forum*, **566**, 53–60.

[26] Vesenjak, M. (2006) Computational Modelling of Cellular Structure under Impact Conditions. Ph.D. thesis. Faculty of Mechanical Engineering, Maribor.

[27] Vesenjak, M., Krstulovič-Opara, L., Öchsner, A. *et al.* (2006) Experimental and numerical modelling of open-cell cellular structures, in *Fifth International Congress of Croatian Society of Mechanics* (ed. F. Matejiček), Croatian Society of Mechanics, Trogir, Croatia.

[28] Montanini, R. (2005) Measurement of strain rate sensitivity of aluminium foams for energy dissipation. *International Journal of Mechanical Sciences*, **47**(1), 26–42.

[29] Wang, Z., Ma, H., Zhao, L. and Yang, G. (2006) Studies on the dynamic compressive properties of open-cell aluminium alloy foams. *Scripta Materialia*, **54**(1), 83–87.

[30] Deshpande, V.S. and Fleck, N.A. (2000) High strain rate compressive behaviour of aluminium alloy foams. *International Journal of Impact Engineering*, **24**(3), 277–298.

[31] Xue, Z., Vaziri, A. and Hutchinson, J.W. (2005) Materials with application to sandwich plate cores. *CMES Computer Modeling in Engineering*, **10**(1), 79–95.

[32] Lee, S., Barthelat, F., Moldovan, N. *et al.* (2006) Deformation rate effects on failure modes of open-cell al foams and textile cellular materials. *International Journal of Solids and Structures*, **43**(1), 53–73.

[33] Hosford, W.F. (2005). *Mechanical Behaviour of Materials*, Cambridge University Press, Cambridge, UK.

[34] Tan, P.J., Reid, S.R., Harrigan, J.J. *et al.* (2005) Dynamic compressive strength properties of aluminium foams. Part I: experimental data and observations. *Journal of Mechanics and Physics of Solids*, **53**(10), 2174–2205.

[35] Su, X.Y., Yu, T.X. and Reid, S.R. (1995) Inertia-sensitive impact energy-absorbing structures. Part I: effects of inertia and elasticity. *International Journal of Impact Engineering*, **16**(4), 651–672.

[36] Su, X.Y., Yu, T.X. and Reid, S.R. (1995) Inertia-sensitive impact energy-absorbing structures. Part II: effect of strain rate. *International Journal of Impact Engineering*, **16**(4), 673–689.

[37] Tan, P.J., Reid, S.R., Harrigan, J.J. *et al.* (2005) Dynamic compressive strength properties of aluminium foams. Part II: "Shock" theory and comparison with experimental data and numerical models. *Journal of Mechanics and Physics of Solids*, **53**(10), 2206–2230.

[38] Vesenjak, M., Veyhl, C. and Fiedler, T. (2012) Analysis of anisotropy and strain rate sensitivity of open-cell metal foam. *Material Science Engineering: A*, **541**, 105–109.

[39] Dannemann, K.A. and Lankford, J. (2000) High strain rate compression of closed-cell aluminium foams. *Material Science Engineering: A*, **293**(1–2), 157–164.

[40] Mukai, T., Kanahashi, H., Miyoshi, T. *et al.* (1999) Experimental study of energy absorption in a close-celled aluminium foam under dynamic loading. *Scripta Materialia*, **40**(8), 921–927.

[41] Mukai, T., Kanahashi, H., Yamada, Y. *et al.* (1999) Dynamic compressive behavior of an ultra-lightweight magnesium foam. *Scripta Materialia*, **41**(4), 365–371.

[42] Hallquist, J.O. (2006) *LS-DYNA Theoretical Manual*, Livermore Software Technology Corporation, Livermore, CA, USA.

[43] Hallquist, J. (2007) *LS-DYNA Keyword User's Manual*, Livermore Software Technology Corporation, Livermore, CA, USA.

[44] Altenhof, W., Harte, A.M. and Turchi, R. (2002) Experimental and numerical compressive testing of aluminium foam filled mild steel tubular hat sections. 7th International LS-DYNA Users Conference, Dearborn, Michigan, USA, 19. - 21. 5., Livermore Software Technology Corporation (LSTC), Livermore, California.

[45] Bodner, S.R. and Symonds, P.S. (1962) Experimental and theoretical investigation of the plastic deformation of cantilever beams subjected to impulsive loading. *Journal of Applied Mechanics*, **29**, 719–728.

[46] Jacob, P. and Goulding, L. (2002) *An Explicit Finite Element Primer*, NAFEMS, Glasgow, UK.

[47] Mamalis, A.G., Robinson, M., Manolakos, D.E. *et al.* (1997) Crashworthy capability of composite material structures. *Composite Structures*, **37**(2), 109–134.

[48] Andersen, O., Waag, U., Schneider, L. *et al.* (2000) Novel metallic hollow sphere structures. *Advanced Engineering Materials*, **2**(4), 192–195.

[49] Augustin, C. and Hungerbach, W. (2009) Production of Hollow Spheres (HS) and Hollow Sphere Structures (HSS). *Materials Letters*, **63**(13–14), 1109–1112.

[50] Vesenjak, M., Gačnik, F., Krstulović-Opara, L. and Ren, Z. (2012) Mechanical properties of advanced pore morphology foam elements. *Mechanics of Advanced Materials and Structures*, accepted for publication.

[51] Stöbener, K. (2007) Advanced Pore Morphology (APM) – Aluminiumschaum. Ph.D. Thesis, University of Bremen, Bremen.

[52] Busse, M., Rausch, G., Stöbener, K. and Baumeister, J. (2007) *Advanced Pore Morphology (APM) Metal Foams – brochure (in German)*. Fraunhofer Institut Fertigungstechnik Materialforschung (IFAM), Bremen, Germany.

[53] Stöbener, K., Lehmhus, D., Avalle, M. *et al.* (2008) Aluminium foam-polymer hybrid structures (APM aluminium foam) in compression testing. *International Journal of Solids and Structures*, **45**(21), 5627–5641.

[54] Stöbener, K. and Rausch, G. (2009) Aluminium foam–polymer composites: processing and characteristics. *Journal of Materials Science*, **44**(6), 1506–1511.

[55] Stöbener, K., Baumeister, J., Rausch, G. and Rausch, M. (2005) Forming metal foams by simpler methods for cheaper solutions. *Metal Powder Report*, **60**(1), 12–16.

[56] Farley, G.L. (1983) Energy absorption of composite materials. *Journal of Composite Materials*, **17**(3), 267–279.

[57] Hohe, J., Hardenacke, V., Fascio, V. *et al.* (2012) Numerical and experimental design of graded cellular sandwich cores for multi-functional aerospace applications. *Materials and Design*, **39**(0), 20–32.

[58] Avalle, M., Lehmhus, D., Peroni, L. *et al.* (2009) AlSi7 metallic foams – aspects of material modelling for crash analysis. *International Journal of Crashworthiness*, **14**(3), 269–285.

[59] Song, B., Chen, W. and Frew, D.J. (2004) Dynamic compressive response and failure behavior of an epoxy syntactic foam. *Journal of Composite Materials*, **38**(11), 915–936.

5

Automotive Composite Structures for Crashworthiness

Dirk H.-J.A. Lukaszewicz

Research and Innovation Centre, BMW AG, Knorrstrasse 147, 80788, Munich, Germany

5.1 Introduction

This chapter will investigate the potential benefits of composite materials in a crash considering the changing landscape of safety and sustainability requirements. The current state of the art for traffic safety, and more specifically passive safety, is studied and current safety requirements are highlighted. With the advent of novel drivetrains future vehicles have to be increasingly lightweight, for example to make battery electric vehicles viable, which may in turn result in novel structural designs. One such option is explored further, where automotive structures utilise composite materials, exhibiting better weight specific mechanical properties than their metallic counterparts, to achieve weight savings. Passenger safety for front and side impacts through structural design is then discussed. The concept of load paths is introduced to highlight the function of the vehicle structure in different crash events, such as front and side impacts. While composites are regularly employed in stiffness critical structures in industry sectors such as aerospace and renewable energy, a unique property of composites for automotive passive safety is their high specific energy absorption. Consequently, a review of the current research into energy absorbing composite structures is presented and some conclusions for future vehicle design using composites are made.

5.2 Traffic Safety

Traffic accidents can deeply harm and affect the involved individuals, both psychologically and physiologically. Throughout the world, traffic related deaths and injuries are a major factor of cost for society as well as the individual. In the western world the total number of vehicles (and their density) has increased continuously since the Second World War; and Asia, South

Advanced Composite Materials for Automotive Applications: Structural Integrity and Crashworthiness,
First Edition. Edited by Ahmed Elmarakbi.
© 2014 John Wiley & Sons, Ltd. Published 2014 by John Wiley & Sons, Ltd.

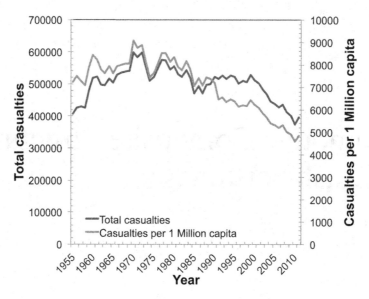

Figure 5.1 Evolution of road casualties over time in Germany since 1955.

America and Africa have followed suit. In Western Europe and America, this resulted in a peak of traffic related deaths and injuries in the 1970s (see Figure 5.1). As an example, in Germany the number of road accidents with serious or fatal injuries decreased from 414 362 in 1970 to 306 266 in 2011. The total number of fatal accidents decreased even more dramatically, from 14 204 in 1970 to 3724 in 2011. This reduction is even more impressive when it is compared to the total number of vehicles in Germany, which rose from 20 816 802 to 52 945 127 in the same timeframe and to the number of accidents per 1 000 000 capita, which decreased from 6266 to 3746 [1]. Similar trends can be observed in all western societies, such as the United Kingdom, United States and others. As an example, the average annual reduction of fatal road accidents was 7% annually for both the United Kingdom and Germany between 2000 and 2010 [2]. This reduction since the 1970s can be attributed to legislative measures that have been implemented, for example the requirement for seat belts, to reduce the severity of accidents, or improvements to road safety and signage. Additional customer expectations have resulted in a further increase in passive safety, for example the widespread adoption of airbags. More recently, in the last two years the number of accidents and fatalities is increasing again. This may be surprising considering the improving levels of vehicle safety, but an ageing population in western societies, general distractions in traffic, deteriorating road quality due to lack of investment and relatively mild winters leading to increased vehicle usage could be cited as possible explanations.

The majority of accidents in western societies are car to car accidents. In Germany accidents involving at least one car accounted for 71% of all accidents in 2011. Of these 67% involved other cars, vans or no other crash partner, while the remaining 33% of crashes involved vulnerable road users, such as pedestrians, cyclists and motorbike users (see Figure 5.2).

To conclude, vehicle safety may significantly reduce the impact of accidents and save the lives of both the occupants and other traffic partners. Crash requirements for occupant safety

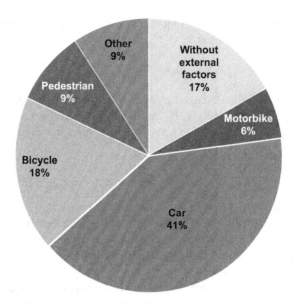

Figure 5.2   Distribution of crash partners for accidents where cars were the accident initiators.

have to account for the areas of front, side and rear impact with static objects, other vehicles and vulnerable road users. Within the discipline of vehicle safety, the area of passive safety aims to reduce the severity of impact through structural design and it is in this area where novel materials, such as composites, may yield improved vehicle performance.

5.3   Alternative Vehicles

The term "alternative vehicles" broadly encompasses the areas of alternative engines and alternative materials for the body in white (BIW). The two are also linked and the first, alternative engines, is also the driving force for alternative materials for BIW and this section will thus follow and explain this argument.

Fuel consumption is already an important driver for designing conventional internal combustion engine (ICE) vehicles, due to increasing sustainability requirements. For battery electric vehicles (BEVs), range extender (REX) vehicles and plug-in hybrids (PHEVs) some or all of the necessary energy of the vehicle is stored in the car, mostly using batteries. Additionally, batteries may require a significant amount of time to recharge, which currently detrimentally affects the usability of a vehicle. It is thus imperative that the day to day operability of modern electric vehicles is comparable to ICE vehicles to gain customer acceptance.

However, the reduced power density of current batteries implies that, in order to obtain the same operability as current ICE vehicles, the weight of electric vehicles has to increase disproportionally due to the increase in drivetrain weight. This also results in a further increase in weight of the auxiliary BIW, for example due to stiffness requirements, which again increases the total required energy and weight of the batteries. This parasitic weight cycle is a major challenge to alternative vehicles as it directly increases the cost of the vehicle and reduces their efficiency.

The fuel consumption or total power required by a vehicle can be calculated from the forces resisting the forward motion of a car

$$F_{tot} = F_{roll} \cdot F_{acc} \cdot F_{pot} \cdot F_{aero} \qquad (5.1)$$

with F_{tot} being the total force generated by the vehicle, F_{roll} the rolling resistance, F_{acc} the force required for further acceleration, F_{pot} the force required due to elevation and changes in height and F_{aero} the force required to compensate aerodynamic drag. Extending the above equation we have

$$F_{tot} = m \cdot k_T \cdot g \cdot \cos(\alpha_{ele}) + \mathrm{m} \cdot k_m \cdot a + m \cdot g \cdot \sin(\alpha_{ele}) + \frac{\rho}{2} \cdot c_w \cdot A \cdot v^2 \qquad (5.2)$$

with m being the mass of the vehicle, k_T the efficiency coefficient of the tyre, g the gravity, α_{ele} the angle of an elevation, k_m the efficiency coefficient of the vehicle during acceleration, ρ the current density of air, c_w the aerodynamic efficiency of a vehicle, A the frontal projection area and v the instantaneous speed.

Of these, the first three components are directly proportional to the mass of a vehicle. Other important factors for the total power requirement of a vehicle are the rolling resistance of the tyres, aerodynamic drag and internal losses in the aggregates, such as the engine or air-conditioning. Depending on the use scenario, the individual components of resistance contribute differently to the total forces acting on a vehicle (see Figure 5.3). During everyday use in a city environment almost 90% of the energy consumption is due to rolling resistance and acceleration, assuming no elevations. This is in contrast to typical motorway energy consumption at constant speed where the majority of the energy consumption is due to aerodynamic losses. For electric vehicles the required range, that is the distance that can be travelled with a single charge, is defined by customer requirements and varies between

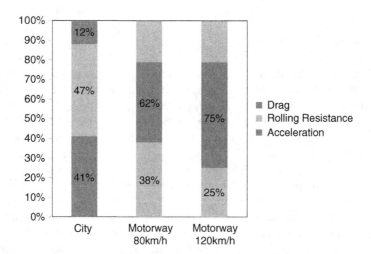

Figure 5.3 Typical normalised energy consumption for three different use scenarios [3]. For urban mobility approx. 90% of the energy consumption are proportional to the vehicle weight.

countries. Generally, a vehicle with a range of 150–200 km per charge will cater for the vast majority of everyday use using fully electric energy, which is emission free at the point of use. Consequently, BEVs need to carry the necessary energy for a given range stored in batteries, which are both heavy and expensive. By reducing the weight of the BIW the battery can be made smaller for the same range, again reducing the required weight for auxiliary structures such as undercarriage, and starting a positive weight spiral.

Often, the potential for weight savings is evaluated in terms of the extra cost that can be accepted for 1 kg of saved weight for a given component. The additional cost that can be accepted for a vehicle depends on the total energy savings over the life cycle of a vehicle and the resulting reduction of cost of ownership, for example due to fuel savings. For ICE vehicles this target value ranges from 3 to 7 €/kg depending on the component since petrol is relatively cheap and existing vehicle architectures are highly optimised. For BEVs the break-even point is shifted to approximately 16 €/kg for a midsize car [3], due to the aforementioned parasitic weight cycle and the battery cost and weight. This opens potential new applications and materials, such as CFRP, while REX vehicles and HEVs are in between this range at 3 to 16 €/kg. For bigger cars and sport vehicles the acceptable cost increase may again be higher due to a higher retail price and smaller volume, and conversely it could be lower for smaller, high volume cars.

While this provides a global estimate for acceptable cost increases for vehicle weight savings, these are not equally desirable for all components of a car, that is some applications and components pay off quicker and have an even higher point of break even, while others are lower. Structures higher above the centre of gravity of the vehicle (e.g. the roof) pay off quicker because of the lowering of the centre of gravity.

Similarly, structures in the vehicle front may be more desirable for weight savings than in the rear, due to desirable changes in the weight distribution for front and rear axles; however the inverse may be true for a rear drive architecture. New materials and changing drivetrain concepts are altering the landscape for the passive safety design of future vehicles. However, customer expectations with respect to safety are going to remain high and this implies that future vehicles have to achieve comparable performance in legal and customer protection tests, which are explored in the following.

5.4 Selective Overview of Worldwide Crash Tests

Generally, vehicle safety can be divided into two disciplines, active and passive safety. Active safety aims to prevent the occurrence and likelihood of a crash, while passive safety aims to reduce or mitigate the severity of a crash for the vehicle structure, the occupants and other involved crash partners.

During a crash incidence the BIW of a vehicle must protect the passengers and other crash partners, and reduce the crash severity by allowing controlled energy dissipation through structure deformation. When evaluating real-world crashes involving at least one car, about 50% of all crash events are frontal impacts, while side impacts account for 33.7% of all accidents and rear impacts make up 15.9% of accidents (Figure 5.4). Frontal impact with full coverage amounts to only 31.1% of all typical frontal crash accidents (Figure 5.5), while the majority of crashes cover only a small proportion of the vehicle front. This makes it necessary not only to design the frontal section of the vehicle to absorb the kinetic energy of the vehicle

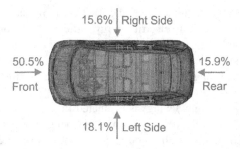

Figure 5.4 Relative distribution of crash positions for cars, data from [4].

in an accident, but also to design different zones into the vehicle front which are able to cope with various crash cases. In the last two decades passive safety requirements have continuously increased and today vehicles need to pass a certain set of legally defined tests. In addition new car assessment programs (NCAPs, e.g. Insurance Institute for Highway Safety, IIHS) have been implemented that aim to define tests that emulate real-world crash cases while allowing a comparison between different vehicles. Obviously, there are regional differences, due to the differences in typical vehicle use, for example United States tests focus on sport utility vehicles (SUVs), vans and limousines while the European Union tests focus on normal passenger cars with additional requirements for pedestrian and child safety.

The minimum level of protection is defined by governmental test procedures, while additional consumer protection tests have been implemented by NCAPs to define an enhanced level of protection and to differentiate vehicle safety further. Figure 5.6(a) and (b) provides an overview of current European and American legislative, NCAP and IIHS test procedures for

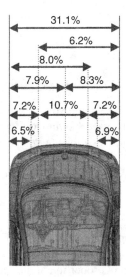

Figure 5.5 Distribution of frontal crash positions, data from [4].

(a)

Test Name	ECE R94	EuroNCAP	FMVSS 208	FMVSS 208	FMVSS 301; 305	USNCAP	IIHS	IIHS
Load case	Offset front	Offset front	Full width front	Offset front	30° front	Full width frontal	Offset front	Small Overlap
Test Speed	56 km/h	64 km/h	32 - 56 km/h	40 km/h	48 km/h	56 km/h	64,4 km/h	64,4 km/h
Barrier	progressive	progressive	rigid	starr	rigid barrier	rigid barrier	progressive	rigid barrier
Impact Angle	0°	0°	0°	0°	0°	0°	0°	0°
Overlap	40 - 50%	40 - 50%	100%	40%	100%	100%	40%	25%
Belted / Unbelted	B	B	B & U	B	B	B	B	B
Symbol								

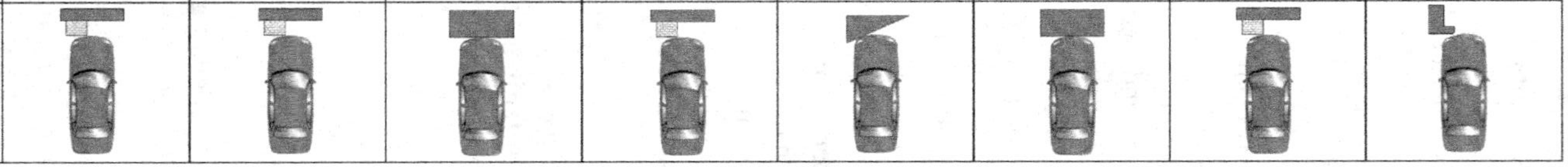

(b)

Test Name	ECE R 95	EuroNCAP	EuroNCAP	FMVSS 214 (new)	FMVSS 214 (new)	USNCAP	USNCAP	IIHS
Load case	Side barrier	Side pole	Side barrier	Crabbed barrier	Oblique pole	Crabbed barrier	Oblique pole	Side barrier
Test Speed	50 km/h	29 km/h	50 km/h	54 km/h	32 km/h	62 km/h	32 km/h	50 km/h
Barrier	deformable	rigid	deformable	deformable	rigid	deformable	rigid	deformable
Impact Angle	90° / 270°	90° / 270°	90° / 270°	90° / 270° (27°)	90° / 270° (75°)	90° / 270° (27°)	90° / 270° (75°)	90° / 270°
Barrier description	950 kg	Ø 254 mm	MDB EEVC 950 kg	1368 kg	Ø 254 mm	1368 kg	Ø 254 mm	MDB 1500 kg
Belted / Unbelted	Belted	Belted	Belted	Belted	Belted	Belted	Belted	Belted
Symbol								

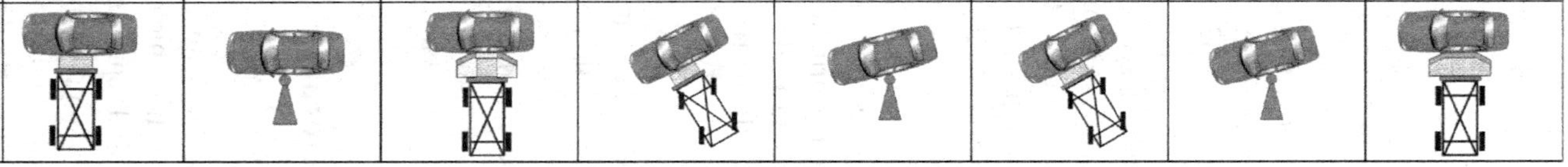

Figure 5.6 Overview of European and United States crash tests for (a) front events and (b) side events.

front and side impacts. The European and United States NCAPs as well as IIHS will serve as a reference here to illustrate the requirements for occupant safety.

The severity of the impact during a test is measured using one or several test devices akin to human bodies, which are commonly referred to as "dummies". The discussion of dummies and measuring technology will be omitted, though it should be noted that tests that may seem identical in Figure 5.6(a) and (b) may differ in the employed dummies and their positioning. The ECE R94 test procedure is a legal certification test where the vehicle is accelerated to 56 km/h and into a deformable barrier covering 40% of the frontal section. The corresponding EuroNCAP employs increased test speeds. Legal certification tests for the United States are defined in the FMVSS208 test protocol. A particularly noteworthy aspect of these tests is the fact that the vehicle has to pass a test where the dummy is unbelted. The FMVSS301, 305 are test procedures specifically designed to test hybrid and electric vehicles for their electric safety, and similar test procedures exist in Europe. The corresponding USNCAP employs a similar configuration to the legal test protocol but only tests at high speed, while the IIHS (a second consumer protection test) protocol employs a barrier configuration similar to the EuroNCAP but with different requirements and test conditions.

For side impact events European legislation ECE R95 requires an impact with a 950 kg barrier at a speed of 50 km/h, which impacts into the structure just above the doorsill. European NCAP procedures use the same barrier test and a side pole test in addition, where a rigid pole is impacted into the structure at 29 km/h. This is comparable to United States legislative testing, that requires both a side barrier and pole test, but with different test configurations. According to FMVSS214 the barrier hits the vehicle while traversing at an angle of 27° at 54 km/h, while the barrier weight is significantly higher than the European test procedure, with 1368 kg. This is due to the fact that SUVs, vans and pickups, which are heavier than limousines and small cars are more common in the United States.

The legislative side pole test is impacting the vehicle at 32 km/h, again at a slight angle. USNCAP side barrier tests then specify higher test speeds, while the pole test is comparable. Again, it should be noted that these tests might still differ in the type and number of dummies and the crash weight of the vehicle. Finally, IIHS, uses a different and heavier barrier which impacts the side structure.

5.5 Structural Crash Management

5.5.1 Front Crash

The structural elements of a conventional vehicle architecture are shown in Figure 5.7. In a frontal impact the vehicle structure has multiple purposes depending on the event and these will be discussed representatively in the following.

In low speed impact events, which can occur often during parking, the front end module has to allow a cost effective repair and protection of the main structure. Additionally, the front end has to enable good pedestrian protection through defined deformation and kinematics in a frontal impact. The front end module is thus often joined to the main structure using a threaded assembly of the defo-box to the longitudinal beams. The reparability of the front end is normally tested according to the RCAR test procedure. This test impacts the structure at relatively low speed, 15–17 km/h, and rather than occupant protection, the reparability of the damaged structure after such an impact is evaluated.

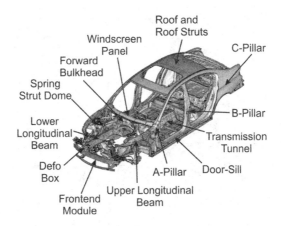

Figure 5.7   Structures of a conventional vehicle architecture.

In a high speed impact event, for example tested using the USNCAP or EuroNCAP pro-cedure, the crash event propagates deeper into the structure. Initially, the upper and lower longitudinal beams are deformed irreversibly. The incoming load is often taken up by the forward bulkhead and from there diverted into the undercarriage and into the A-pillar and door sill. After the barrier has reached the engine, this is pushed into the forward bulkhead, and the steering rack and front axle module have to deform in a controlled manner, to dis-sipated energy and mitigate the intrusions into the occupant space. Finally, the collapse of the forward bulkhead and the A-pillar has to be prevented and the incoming loads have to be transferred towards the rear of the vehicle.

The principle function of the vehicle architecture is easily demonstrated using the USNCAP frontal impact test, where the majority of the kinetic energy of the vehicle has to be dissipated in the structure to decelerate the vehicle. Lower deceleration, often referred to as crash pulse, reduces the likelihood and severity of secondary impacts of occupants in the interior – for example when connecting with the seat belt or the passenger airbag. In general the crash pulse has to be below a defined value for the occupant thresholds to protect the occupant from injury or death. The pulse is in principle defined by the filtered force–length curve of a vehicle. To achieve a good pulse a well defined load has to be maintained over a certain length of the vehicle. Achieving the necessary loads for good energy absorption and crash pulse mostly requires a simultaneous optimisation of cost, weight and necessary functionality.

Figure 5.8 shows for example the frontal loading paths of the BMW i3 [5]. The vehicle has three loading paths, upper, middle and lower, where the middle loading path absorbs most of the energy; but the i3 has no engine in the front, which is consequently shorter than typical limousines. In typical frontal impact cases the majority of the absorbed energy is dissipated through the longitudinal beams, seen as the middle loading path in Figure 5.8 and the loss of vehicle weight (about 15% of the total impact energy) once the engine contacts the barrier. By contrast, Figure 5.9 shows the proportions and energy absorption in relation to the area of the frontal section during a USNCAP frontal crash for an ICE vehicle. It can be seen that 50% of the total kinetic energy is absorbed in front of the engine; 80% of the total energy is absorbed in the front of the car including the bulkhead [6]. Whilst it is intuitive to assume that optimisation of these beams may yield improved crash functionality it is worth noting

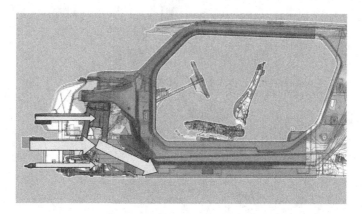

Figure 5.8   Schematic of load paths during a frontal impact [5]. Load paths shown with aple arrows.

that the best possible crash functionality is often achieved by optimising the whole system of structures to work in unison, which may have implications for the longitudinal beams, such as weakening of the structure at defined positions using crush initiators.

Overall, the front section of a vehicle in terms of crash performance is governed by requirements for stiffness, energy dissipation and controlled deformation to enable good occupant protection. All three goals can be simultaneously achieved using composite materials, however it is the energy dissipation of composites that has widely been researched due to the potential improvements in performance and weight and this will be discussed later in more detail.

5.5.2   Side Crash

In a side crash the sequence of events depends on the barrier, however, most side impacts are oblique impact test cases where the impactor hits the vehicle at an angle. In a pole impact

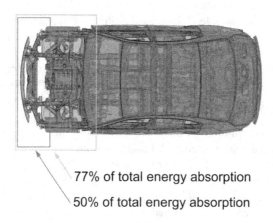

77% of total energy absorption

50% of total energy absorption

Figure 5.9   Top view of an ICE BMW vehicle. The main areas of energy absorption and their contribution are highlighted [6]. Reproduced from Ref. [6]. Copyright 2009 Springer.

the load is continuously introduced into the door sill, which has to be supported by the seat cross members and the undercarriage. These are often supported by the transmission tunnel, which tends to possess significant geometric stiffness. Simultaneously, the roof structure is loaded and the roof cross members have to take up the load from the impact and transfer it to the backside of the impact. With increasing intrusions the seat cross members may have to deform in a controlled manner to avoid excessive deformation of the door sill, which then continues to take up load and directs it into the A- and B-pillars. In a barrier test the B-pillar is impacted first, while the door sill sits just below the barrier. To protect the occupants the B-pillar has to allow a controlled deformation into the occupant space. While the B-pillar is deformed, increasing amounts of load are then taken up by the roof cross members and door sill, which are loaded in tension through the B-pillar. Simultaneously, the A-pillar is deformed, dissipating further energy.

In a side impact a certain survival space has to be maintained throughout the crash event. Typical examples of side impact crashes are car to pole (FMVSS214) and barrier to car (IIHS) where the barrier is often deformable. Typically, additional thorax airbags and curtain head airbags will be deployed to reduce the severity of the impact in these cases. As an example the i3 side crash concept is shown in Figure 5.10, with the load paths indicated by yellow arrows (note that the vehicle has a B-pillar integrated into the doors, which are not shown here).

Generally, the load from the impact is first distributed in a perpendicular plane to the impact axis using structures mostly loaded in tension and bending, such as the door reinforcements. The entire side frame such as A- and C-pillars acts to support the load transfer in the plane. After the initial load has been transferred away from the point of impact it is transferred away from the impact side to the opposite face, for example over the roof and forward bulk head. The goal of this concept is to maintain the survival space of the occupants until curtains and side airbags are deployed and to keep the maximum intrusions into the passenger cell below a maximum value throughout the entire crash.

In a typical barrier impact the car is initially stationary and the side barrier is accelerated into the car, while in a side pole test the car is thrown against a stationary pole using a flying floor. Consequently the design goals can be achieved using very stiff structures that deform very late

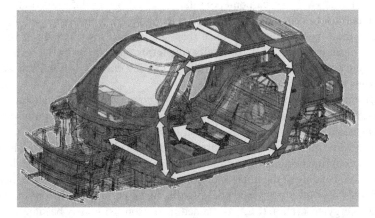

Figure 5.10   Schematic of typical load paths during a side crash for the BMW i3 [5].

into the crash when the car is already partially accelerated or by allowing defined deformation while transferring the load in the plane during the initial impact. Finally, rear impacts play an increasingly important role in modern vehicle crashes, because the battery of a hybrid vehicle or REX vehicle may be placed over the rear axle to achieve a neutral weight distribution. In a rear crash the intrusions into the passenger cell have to be minimised. Concepts for achieving this are similar to front crash and rear crashes will thus not be discussed in further detail.

Generally, effective crash structures have to fulfil two distinct requirements: (i) they need to be able to dissipate the energy of the impact into the structure (e.g. in a front crash through plastic folding of the longitudinal members), or (ii) they need to be able to resist a high crush load for an extended period of time (e.g. in a side impact minimise the impact on the occupants). A combination of the two events is also common; for example the forward bulkhead needs to be sufficiently stiff to resist the loading of the front structure, but may have to deform to dissipate further energy.

With the advent of more and more batteries in hybrids and electric vehicles, passive safety requirements are likely to change for both side and rear crashes. For example, in a side pole crash the maximum intrusion depth is currently constrained by the survival space that is necessary for the occupants. However, if batteries are placed in the under floor, the maximum intrusion depth has to ensure that the batteries are not penetrated, which could result in a lowering of the maximum intrusion depth. Similarly, for hybrid vehicles the batteries are often placed in the boot of the car and have to be equally protected, which may affect safety requirements. Further optimisation of existing vehicle architectures is going to yield diminishing improvements and it is thus necessary to find revolutionary means of improving passive safety. One way of meeting the conflicting goals of reducing weight and improving passive safety are composite materials, which possess better specific mechanical properties than the commonly employed metallic materials.

5.6   Composite Materials for Crash Applications

The possible benefits of composite materials were first realised for applications in helicopters and Formula 1 vehicles. In 1980 McLaren introduced a chassis as a safety cell made from an aluminium sandwich with carbon-epoxy face sheets [7]. In 1985 frontal crash tests were introduced and most Formula 1 teams adopted a composite monocoque and a nose cone to comply with these new regulations. Since then side impact tests have been introduced (in 1995), rear impact tests (in 1997) and finally side intrusion tests (in 2001). As a consequence the fatality rate decreased from 1 in 40 in 1980 to 1 in 250 crashes in 1992, and since 1994 no fatal accidents have occurred.

Deformation in composites differs from the more commonly used metallic materials, such as aluminium and steel. Figure 5.11 shows a comparison between an aluminium structure that deformed in plastic folding and a composite structure of similar shape that deformed in brittle fracture. In metallic materials, energy is dissipated through plastic folding, work hardening and adiabatic losses during heating. Composites, by contrast, dissipate energy through external and internal friction, bending of the fibres and kinematic dissipation through fragmentation. Composite materials may offer many potential benefits for improving crashworthiness. Primarily, the specific energy absorption (SEA, explained in Section 5.6.1) may be significantly higher, that is composites may achieve the same level of crashworthiness as current steel structures

Figure 5.11 Comparison between plastic folding in metals (left) and brittle fracture in FRP s (right).

at a fraction of the weight. A lighter vehicle may offer new crashworthiness benefits, for example in a rollover crash as it may be improved due to the lower overall weight. Finally, composites deform by brittle fracture, unlike metals that deform by plastic folding (see Figure 5.11). During brittle fracture the whole component can be disintegrated, while for structures exhibiting plastic folding the hinges limit the foldable length. This may have potential benefits for optimising the deceleration of the vehicle, which can occur over a longer distance, thus resulting in a lower overall deceleration, which reduces the risk of secondary impacts [7].

5.6.1 Performance Metrics for Energy Absorbing Structures

Most of the work in composite materials for crash applications has focused on several key metrics to access the relative performance of a given structure for energy absorption. Typical performance metrics are the specific energy absorption (SEA), the crush force efficiency (CFE) and the sustained crush stress (SCS).

The SEA can be calculated from Equation 5.3

$$SEA = \frac{W}{\rho A \delta} = \frac{\int_0^\delta F d\delta}{\rho A \delta} = SSCS \tag{5.3}$$

where W is the total absorbed energy, ρ the density of the material, A the cross-section, δ the total crush displacement and F the crush force. In most cases F is taken to be equal to the mean force (see Figure 5.12). In older publications the SEA is also referred to as the specific sustained crush stress (SSCS).

The CFE is the ratio of maximum to mean force, where the mean force can be inversely calculated from the SEA or fitted to the stable load after the initial peak (Figure 5.12). The peak force is often attributed to the elastic properties of the structure prior to fracture, while

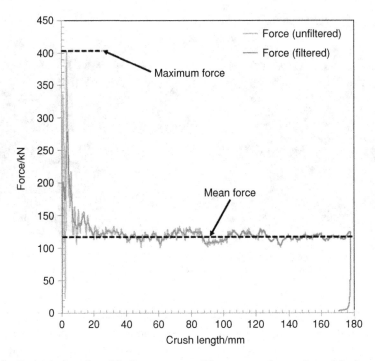

Figure 5.12 Example for a force/displacement record for a composite specimen. Maximum and mean forces are indicated.

the mean force is the result of stable permanent deformation events in the structure, leading to energy dissipation. The inverse of the CFE is often called the load uniformity index (LU) (Equation 5.4)

$$CFE = \frac{F_{Mean}}{F_{Peak}} = \frac{1}{LU} \tag{5.4}$$

where F_{Mean} is the mean force and F_{Peak} the peak force, as given in Figure 5.12. A CFE close to 1 is obviously most desirable, as it implies that, after achieving the elastic load limit of the structure, it fractures at a constant load, resulting in effective energy dissipation. For typical quasi-static tests in the literature values from 0.6 to 0.8 have been reported for optimised effective structures, while dynamic tests can yield values as low as 0.4. The CFE is thus an important metric for designing systems of crash structures, as the auxiliary structures may have to support the peak load, resulting in heavier structures. The sustained crush stress can be calculated from Equation 5.5

$$SCS = \frac{F_{Mean}}{A} \tag{5.5}$$

The SEA and CFE are most commonly reported and will be used to compare the relative performance of composite structures.

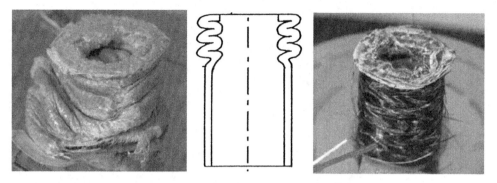

Figure 5.13 Examples of progressive folding deformation for aramid fibre composites (left) and aramid–carbon fibre hybrids (right) [9]. The principle mode of deformation is shown in the middle.

5.6.2 Energy Absorbing Deformation Mechanisms in Composite Profiles

Hull [8] has classified the principal deformation modes of energy absorbing composite structures into four categories:

- Global buckling of the structure. This is equivalent to the Euler buckling of a column.
- Progressive folding.
- Progressive crushing:
  - Progressive splaying.
  - Progressive fragmentation.

Generally, the first deformation mode of global buckling absorbs very little energy and is thus undesirable for energy absorption applications. As a result it will not be discussed further as it can be easily prevented.

In certain test cases structural integrity may be critical and in these cases composite structures with progressive folding may be an interesting alternative to existing metallic solutions. Progressive folding is a deformation mode that generally occurs in metallic structures, but may also be found in ductile FRPs and metal/FRP hybrid materials. As an example, Figure 5.13 shows the progressive folding deformation of aramid fibre reinforced structures and the principle mode of deformation. Progressive folding can be interpreted as a sequence of localised folding at the crush front, propagating away from the impact point through the structure. The SEA of such a structure can reach intermediate values, but does not reach the same levels as progressive crushing due to a lack of mechanisms for energy dissipation. The fact that this type of deformation has been observed for some aramid and glass fibre reinforced specimens, as well as some thermoplastic fibre reinforced composites, has led to the conclusion that a high strain to failure for the fibre and matrix is necessary for this deformation mode to occur.

Progressive crushing is the deformation mode exhibiting the highest SEA in composite structures. It can be divided into two main modes, progressive fragmentation and progressive splaying (see Figure 5.14).

Progressive splaying is initiated through a wedge that is pressed into the gap between the structure and the crush plate (Figure 5.15). Often, the initial wedge is the debris of the trigger,

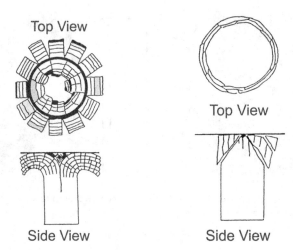

Figure 5.14 Illustration of different progressive crushing modes, progressive splaying (left) and progressive fragmentation (right) [8]. Reproduced from Ref. [8]. Copyright 1991 Elsevier.

such as a chamfer. Due to compressive forces imparted from the crush plate this debris is pushed down into the structure and continuously crushes the material in front of the wedge. Frictional forces act between the wedge and adjacent fibres while it propagates further into the structure. Longitudinal cracks are initiated, which also allow the fibres at the perimeter of the structure to be separated. These fibres are bent between the remaining structure and a crush plate and consequently splay out. Whilst the fibres are bent, a large amount of intralaminar cracks is generated between the plies. Additional frictional forces act between the fibres being bent by the crush plate and between the individual layers in the laminate as the layers are bent through different radii and thus need to slide relative to each other.

The individual plies in a structure permanently deformed in crushing are often referred to as fronds. During progressive fragmentation a stable crush zone is established that progresses down the structure, away from the point of initial contact. Again the first deformation may be

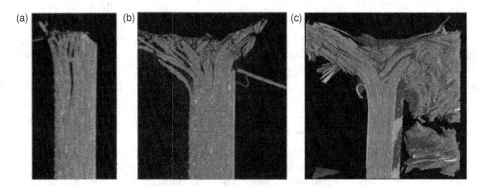

Figure 5.15 Sequence of deformation events during progressive crushing [10], showing (a) the beginning of the crushing process, (b) the beginning of frond bending and (c) the fully established crush front. Reproduced from Ref. [10]. Copyright 2010 Taylor and Francis.

due to a trigger breaking away by shear failure, leaving a sharp tip in front of the crush plate. This tip will be compressed and form new debris until the load reaches a critical limit, where further shear failure may occur in the structure. Consequently, new wedges break away from the structure, leaving a renewed tip in front of the crush plate and the process is repeated until the crash energy or the structure is consumed in this manner.

Principally, both deformation modes may occur in any structure and the exact sequence of events resulting in either splaying or fragmentation depends on the exact configuration of the specimen and the test conditions; thus a focus of early studies was to understand whether one of the crush mechanisms was more effective than the other and where and how the energy was dissipated in the structure. The importance of frictional effects was highlighted by Farley, Wolterman and Kennedy [11], who studied the crushing of identical structures using crush plates with different surface roughness and observed that the rough crush plate generally yielded higher SEA results. Mamalis *et al.* [12] studied the energy absorption of glass fibre reinforced frame rails made from vinylester resin in quasi-static and dynamic compression and later extended this approach to conical shells [13]. Based on their results they developed a simple analytical model that allowed the separation of the relative contribution of the aforementioned mechanisms for energy absorption. The total dissipated energy is

$$W_{diss} = W_{fric} + W_{bend} + W_{crack} + W_{split} \tag{5.6}$$

with W_{diss} being the total dissipated energy, W_{fric} the energy dissipated due to friction between the angular wedge and fronds and between fronds and platen, W_{bend} the energy dissipated due to frond bending, W_{crack} the energy dissipated due to crack propagation and W_{split} the energy dissipated due to axial splitting. From their data the authors observed that 45% of the total energy was dissipated due to friction, 40% dissipated due to frond bending, 12% dissipated due to crack propagation and 3% dissipated due to axial splitting. These results depend on the exact deformation mode, but the analysis showed that splaying deformation may exhibit higher SEA than progressive fragmentation deformation due to the contribution in energy dissipation of the bending of laminate layers.

It should be noted that all three deformation modes relevant for energy absorption may in principle occur (i) for all fibre types, such as carbon, glass or aramid, (ii) for all resin types, such as thermoset or thermoplastic and (iii) for all material architectures, for example different fibre volume fractions. The exact deformation mode and hence the corresponding energy absorption is a function of the micro-, meso- and macro-structural behaviour of a composite component and this is explored further in Section 5.7.

5.7 Energy Absorption of Composite Profiles

In similarity to the general design of composite structures, engineers or designers wishing to obtain maximum performance from a composite structure with respect to certain requirements need to be able to design the structure on multiple levels.

- A detailed understanding of the structural response of the geometry itself is necessary.
- The micro-structural features need to be well known as they affect the macroscopic properties.

- Knowledge of the manufacturing technology is necessary to understand the performance of the finished composite structure.
- A detailed understanding of the interaction between all three is necessary to fully understand the function of an energy absorbing structure.

The factors that may affect the crashworthiness of a composite component which are discussed in more detail are:

- Matrix and fibre material
- Material architecture and manufacturing technology
- Geometry
- Trigger mechanisms.

5.7.1   Fibre Material

Common fibre types used in crash structures include glass, carbon and aramid, but also steel wire, polyethylene fibre amongst others. Surprisingly, natural fibres such as cotton or hemp have seen only limited research, considering the strict recycling requirements in the automotive sector.

While glass and carbon fibre reinforced tubes tend to crush in brittle fracture, through splaying, fragmentation or a combination of the two, aramid, polyethylene and steel fibre reinforced structures tend to exhibit a more ductile deformation, leading to progressive folding. The more ductile deformation of these fibres has often been associated with their higher strain to failure. Table 5.1 provides an overview of some of the SEA values reported for CFRP specimens, while Table 5.2 contains SEA values for GFRP specimens. The data for the SEA have not been normalised with respect to the shape, matrix, fibre or architecture and the

Table 5.1 Overview of reported SEA values for CFRP specimens during axial loading.

Shape	Matrix	Fibre	Architecture	SEA	Source
Round	Epoxy/PEEK	Carbon	Unidirectional	57–127 kJ/kg	[15]
Round	Epoxy	Carbon	Knitted	26–60 kJ/kg	[16]
Round	PEEK	Carbon	Unidirectional	171 J/kg	[17]
Square	Epoxy	Carbon	Fabric	15 kJ/kg	[18]
Square	Vinyl ester/epoxy	Carbon	Braid	20 kJ/kg	[19]
Round/square	Epoxy	Carbon	Winding	35–43 kJ/kg	[20]
Conical	Epoxy	Carbon	Fabric	70 kJ/kg	[21]
Conical	Epoxy	Carbon	Weave	65 J/kg	[22]
Round	PA/PEEK	Carbon	Unknown	160 J/kg	[23]
Round	Epoxy	Carbon	Unidirectional/ Plain Weave	73–82 J/kg	[9]
Square	Epoxy	Carbon	Unidirectional	37 kJ/kg	[24]
Round	Epoxy	Carbon	Braid	51 J/kg	[14]
Square	Epoxy	Carbon	Braid	30 kJ/kg	[25]

Table 5.2 Overview of reported SEA values for GFRP specimens during axial loading.

Shape	Matrix	Fibre	Architecture	SEA (kJ/kg)	Source
Square/hexagonal	PA	Glass	Woven	40–49	[26]
Square/I-beam	Polyester	Glass	Unidirectional	36–44	[27]
Round	Polyester	Glass	Mats	32–51	[28]
Round	Polyester and vinyl ester	Glass	Random mat and NCF	35–77	[29]
Round	Vinyl ester	Glass	Chopped mat	62	[13]
Round	Epoxy	Glass	Unidirectional	58–76	[30]
Square	Polyester	Glass	Cloth	32	[31]
Conical	Epoxy	Glass	Random mat	68	[32]

tables are simply intended to provide a limited overview of the available data. From the data summarised in the tables it can be concluded that CFRP tends to yield better SEA than GFRP, but it should be noted that the best GFRP structures achieve a much higher SEA than poor or intermediate CFRP structures, due to deformation in different modes. Carbon fibre and kevlar fibre exhibit typical SEA values of 60–70 kJ/kg, while glass fibre may exhibit lower SEA values, around 40–50 kJ/kg. The difference between GFRP and CFRP is often explained by the higher density of glass fibre, which results in a lower specific energy. Hybrid materials, for example carbon–kevlar braiding, typically exhibit SEA values in between those of the individual constituents. With respect to aramid fibres, Chu, Tsai and Huang [14] studied the crush performance of round tubes made by triaxial braiding using carbon, aramid and carbon/aramid hybrids. They observed that the general deformation mode, such as splaying, was dependant on the braiding yarns only.

Aramid fibre specimens yielded very low energy absorption of 14 kJ/kg due to a folding deformation mode, comparable to metals, which incidentally also achieve similar SEA values, while carbon braidings yielded higher values of 51 kJ/kg. Hybrid architectures yielded values of 14–51 kJ/kg but produced very good post-crush integrity.

This has previously been mentioned to be desirable for some applications, such as roof crush. Another study by Kim, Yoon and Shin [9] focused on the correlation between the mechanical properties of CFRP and kevlar reinforced tubes and their respective energy absorption. Specimens were tested for their compressive strength, interlaminar shear strength and shear modulus – in all cases a good linear relationship was observed between the mechanical properties and the SEA and the maximum and mean loads. The compressive and ILS strength have often been attributed to the matrix properties and the manufacturing quality, that is the absence of defects such as voids, and the results can thus be interpreted as a demonstration that higher manufacturing quality and uniformity yields higher SEA values.

5.7.2 Matrix Material

Most of the research on matrix materials has studied either thermoset or thermoplastic matrices, as fibres with sizings for both thermoset and thermoplastic are readily available. Hamada, Coppola and Hull [33] studied the impact of sizing on the crush performance of glass fibre in an epoxy matrix, using two different types of sizing: one suitable for epoxy and the other not. They found a 25% increase in SEA between the unsuitable and suitable sizing and this was attributed to the splaying deformation mode of the amino-silane sized glass cloth, which produced more frictional losses at the crush interface.

The first studies on thermoplastic energy absorbing tubes were conducted by Hamada *et al.* [15] on PEEK /carbon and epoxy/carbon round tubes. For a similar layup, epoxy tubes yielded a SEA of 57 kJ/kg and PEEK tubes exhibited a SEA of 127 kJ/kg (Table 5.1). However, by changing the orientation of the carbon fibre in the PEEK specimen to predominantly 0°, values of up to 170 kJ/kg could be obtained [17]. This was later followed up with a comparison of different thermoplastic matrix systems [34].

It has been argued that the higher fracture toughness of thermoplastic matrices is the prime source for their improved energy absorption capability compared to thermoset and this was studied in more detail by Cauchi Savona and Hogg [35] using glass fibre reinforced epoxy flat plates. Mode I and mode II fracture toughness were correlated with SEA. It was observed that high mode I values yielded better SEA but could also decrease the CFE, while high mode II fracture toughness only increased the SEA. This led to the conclusion that shear failure events play an important role in composite energy absorption, for example during splaying when the plies slide relative to each other. Ghasemnejad, Hadavinia and Aboutorabi [36] studied the SEA and mode I and mode II fracture toughness of rounded rectangular box sections made from unidirectional and woven CFRP. They reported a linearly increasing SEA with increasing mode I and mode II fracture toughness. In a further study by Cauchi Savona and Hogg [37] NCF laminates were stitched to improve their toughness using various configurations, such as stitch density and yarn type. An improvement of SEA of up to 30% was reported for some configurations however other configurations also yielded lower performance than their unstitched equivalents, particularly for carbon fibre NCFs. This leads to the conclusion that toughness is indeed an important factor for improving energy absorption correct and tailoring of the constituents and micro-structure results in higher SEA values.

5.7.3   Fibre Volume Fraction

The effect of fibre volume fraction on the SEA is difficult to separate from other effects governing the relative performance of composite structures. This is illustrated in Figure 5.16 that shows an overview of some of the SEA values reported in the literature in comparison to the fibre volume fraction of the specimens for different reinforcing fibres and material configurations. While it may be concluded that certain fibres and fibre volume fractions yield better SEA, such as carbon fibre compared to glass fibre, the spread is quite large.

For comparison, typical values for the SEA of both steel and aluminium are shown and range from 15 to 30 kJ/kg with automotive structures performing at the lower end of this range. Generally, composite structures exhibit higher energy absorption, as the fibre volume fraction increases. Tao, Robertson and Thornton [38] reported that the SEA of glass fibre rods increased with increasing fibre volume fraction from 10 to 40%, where it reached saturation. While the above conclusions may in general be true for continuous or long fibre, for chopped short fibre reinforced specimens, Jacob *et al.* [39] observed that increasing the fibre volume fraction beyond a certain level resulted in a reduction of the SEA, so care should be taken when interpreting experimental results. In serial production components a certain level of

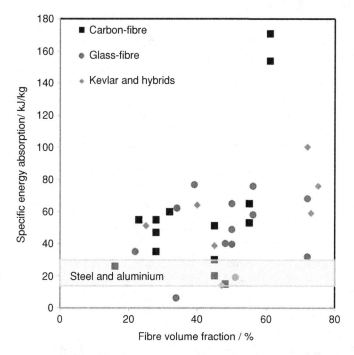

Figure 5.16 Reported SEA from the literature as a function of fibre volume fraction for different reinforcing fibre configurations. Typical performance of metallic structures given for comparison.

fibre volume fraction variability has to be expected and this may result in some undesirable variability of the function of a composite component. Engel *et al.* [40] recently reported that a scatter of 1.7% in fibre volume fraction for their CFRP specimens could partially explain a variation of 13% in their SEA data, which indicates that the fibre volume fraction can have a significant impact on the SEA. In conclusion a vast range of properties can be obtained for nominally identical samples and this depends on the deformation mode and it is thus the interaction between fibre volume fraction and SEA that governs the absolute value of the SEA.

5.7.4 Fibre Architecture

Fibre architecture refers to the impact of the fibre arrangement, such as unidirectional, woven or braided on the SEA of composite structures. Bisagni *et al.* [41] studied the SEA of conical CFRP specimens made from unidirectional and woven prepreg and observed that woven specimens exhibited a roughly 10% higher SEA. Kim, Yoon and Shin [9] conducted a study on unidirectional and woven carbon fibre, kevlar fibre and carbon–kevlar fibre prepreg and reported a similar increase of 12–35%.

This has generally been observed for both woven and braided material and has been attributed the through-thickness reinforcing effect of the crimp of the individual layers. If we consider the typical deformation in progressive folding, the plies are separated and then bent to the sides or

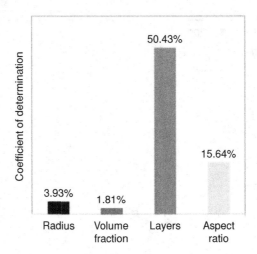

Figure 5.17 Squared Pearson's correlation coefficient for braided carbon /epoxy composites [40]. Higher values imply stronger effect on SEA, either positive or negative. Reproduced with permission from Ref. [40].

crushed in the middle. As a result, woven or braided specimens should yield a higher SEA for progressive crushing deformation modes than unidirectional specimens of comparable layup. Rüger [42] studied the influence of various braiding parameters on the SEA of braided CFRP tubes using quasi-static testing and reported SEA values of up to 86 kJ/kg for glass/carbon braiding and 104 kJ/kg for carbon/carbon braiding. Clearly, these are very high values when compared to typical values for UD CFRP, which are around 60–70 kJ/kg [8]. Hull studied different ratios of axial to transverse plies in round specimens and reported that the best ratio of axial to transverse fibres is around 40–50%. Similar results were found by Engel *et al.* [40] who analysed the factors affecting the SEA of a braided carbon/epoxy composite structure in order to identify the main factors to achieve a high SEA. Figure 5.17 shows the squared Pearson's correlation coefficient for three geometric and one fibre architecture variable. Note that the sum of the correlation coefficients does not add up to 100%, because interaction effects were also studied but not included in the plot. The main factors affecting the SEA are the fraction of fibres in the axial direction and the number of layers, that is the thickness of the specimen, in the range studied.

A variation of braiding is braid pultrusion where rods are inserted longitudinally into a structure that is then overbraided. Such structures have been produced and tested by Saito *et al.* [43] using both glass and carbon fibre for the braiding and the rods.

The structures investigated can be considered to be highly optimised for energy absorption, as an example the braiding angle was 17° with respect to the sample axis and consequently SEA values of 59.1 kJ/kg were reported for glass, 76.0 kJ/kg for carbon rods overbraided with glass and 99.9 kJ/kg for carbon fibre rods with carbon braiding. Another interesting aspect is the tow count of the fibre material. Generally, it is accepted that lower tow counts tend to yield higher mechanic properties, for example in tensile testing. Similar results were reported for crushing experiments by Schultz [44] who found that carbon fibre reinforced specimens with similar thickness manufactured from tows with lower tow count yielded better energy absorption than those with higher tow count. By contrast Karbhari and Haller [45] studied glass fibre reinforced specimens with two different tow counts and observed that specimens with higher tow count yielded better SEA.

5.7.5   Trigger

The purpose of a trigger is to concentrate the crush force of the impact to initiate permanent deformation at a desired location. As such, almost all triggers that are commonly used are some variation of changing the effective cross-section. Non-geometric means of triggering a specimen are ply drops and changes in layup fibre orientation. Such triggers are relevant for structures where deformation is not meant to occur at an intersection, but rather in the structure, for example to achieve a favourable crash kinematic. Sigalas, Kumosa and Hull [46] conducted an early study on the effectiveness of chamfer triggers having different chamfer angles. Glass cloth/epoxy specimens were tested using various different triggers, yielding radically different crush events and load/displacement curves. In most cases a shear failure was observed propagating from the initial tip of the crack. For intermediate chamfer angles this was accompanied by fibre bending, while for small chamfer angles a wedge was formed. The study however provides no indication whether one chamfer angle was more effective than another. Further, this area of research has received surprisingly limited attention, even though the interaction between trigger and geometry may be critical to the SEA of a structure. This was demonstrated by Palanivelu *et al.* [47] who conducted a study into the triggering of round and rectangular tubes comparing chamfer and tulip trigger. While the chamfer trigger yielded higher mean loads for round specimens, the tulip trigger yielded higher mean loads for rectangular specimens. A number of non-geometric triggers were studied by Thuis and Metz [48], such as staggered ply drops, replacing UD layer segments with transversely orientated UD layers and removal of rectangular segments of a layer. In almost all cases the triggered geometry yielded a better CFE than specimens with no trigger. The best SEA and CFE was reported for a staggered ply drop, which effectively simulated an external chamfer.

5.7.6   Geometry

An important parameter for the shape of a crushing geometry is the ratio of diameter D to thickness t, as it characterises the tendency to buckle both locally and globally. For a non-circular structure comparable metrics that can be imagined are the ratio of thickness to contact area or the moment of inertia of the structure. Early on, it was established that a favourable t/D ratio could improve SEA. For example, Hamada and Ramakrishna [17] studied PEEK/carbon tubes of various cross-sectional shape for their energy absorption. The tubes had a 45° chamfer trigger and were tested using quasi-static test conditions. The authors observed that increasing the diameter influenced SEA very little, while increasing the thickness resulted in an improved t/D ratio, which yielded increasing SEA from $t/D = 0.05$–0.10; below and above this range SEA decreased again. This can be explained assuming unstable collapse below 0.05 and changes in crush zone morphology above 0.10. Such t/D ratios will result in either thick or small geometries, which may affect overall stiffness detrimentally and could thus require a trade-off between energy absorption and stiffness requirements.

When comparing the SEA of similar specimens with either round or rectangular geometry it can be observed that round specimens tend to have better SEA than rectangular ones. This has often been explained with the more uniform loading of round specimens and the hoop constraints imparted by transverse layers on the perimeter that can act to support axial fibres in the centre of the layup.

5.7.7   Test Speed

Composite materials exhibit a strain stiffening behaviour; the modulus during dynamic loading is higher than for static test conditions. This implies that, for the same strain to failure, the failure stress increases. This should then result in an increase of the load to first failure, which is the peak load, irrespective of the triggering mechanism being used.

From this it can be expected that the peak load during dynamic load may be significantly higher for a given structure and triggering method than for quasi-static testing. However, it was also observed that the SEA might change between static and dynamic loading. Mamalis *et al.* [12] studied the CFE and SEA of a glass/vinyl ester frame rail and found that the SEA increased during dynamic loading. This was explained with a change in friction coefficient from static to dynamic friction, where the dynamic friction coefficient had higher values. As discussed previously, it has been estimated that up to 45% of crush energy are dissipated through friction and this may thus explain this observation. Further, the authors observed an increase in peak load from 50 to 90% depending on specimen configuration. Finally, it was reported that the load response was significantly more serrated when compared to quasi-static results.

Karbhari and Haller [45] tested braided carbon, glass and aramid fibre specimens at various strain rates and found that a 10-fold increase in strain rate from quasi-static test conditions resulted in an increase in peak load from 12.9 to 18.5%, however the strain rate used there was still comparatively low. In addition it was reported that the mean load remained relatively constant irrespective of test speed.

While there is strong evidence that dynamic loading may yield results for both the SEA and CFE that can be vastly different from quasi-static tests, a vast majority of crush experiments are conducted using quasi-static testing. This is due to the fact that these tests can be easily performed on a general purpose servo-hydraulic test machine, while dynamic tests require dedicated test setups. For applications in automotive and helicopter structures, the dynamic results are more critical as they replicate the real application and further studies into dynamically loaded crash structures are necessary.

5.7.8   Test Direction

Most tests for the energy absorption of composite materials are conducted loading a circular or rectangular profile axially. In real crashes and NCAP ratings, a significant amount of the loading will occur at an angle or off axis, resulting in large bending moments in the main structure and ineffective deformation. This is often called "toppling", where the specimen deforms away from the impact zone and takes up no further energy.

Song and Du [30] studied the off axis crushing of glass fibre reinforced tubes made from polyester and epoxy resin. At small angles of up to 10° the load curve produced a decreasing slope until reaching a stable load, however the average crush load remained unchanged,

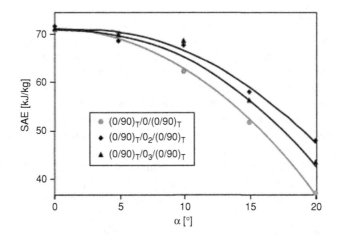

Figure 5.18   SEA as a function of the inclination angle of a cone structure [21]. Reproduced from Ref. [21]. Copyright 2009 Elsevier.

which actually resulted in a better SEA. In the regime from 10 to 15° the mean load became increasingly smaller, and at 20° and 25° the specimen was toppled with barely any energy absorption. The change in SEA was attributed to the change in deformation mode. Fleming and Vizzini [31] conducted similar experiments on carbon/epoxy tubes and observed an increase in SEA of 5.3% at 5° oblique impact angle and toppling deformation at angles above 5°. Similar results were reported by Greve, Pickett and Payen [49] who tested braided carbon tubes at an oblique angle of 15° and observed an unstable collapse of the tubes. The increase at smaller oblique angles is likely due to the increase in contact area and the resulting additional frictional energy dissipation. Ochelski and Gotowicki [21] studied the SEA of carbon and glass/epoxy tubes and cones at various inclination angles and observed a significant SEA reduction with increasing inclination angle (Figure 5.18). The tests were conducted using quasi-static testing and it has to be expected that, during dynamic testing, structures might fail catastrophically due to an increase in peak load.

The results show that, for variable loading, beams may not present a good solution for energy absorption, however small inclinations of a profile with respect to the general loading direction may improve SEA and improve the overall robustness of the design. Overall, it must be concluded that this area is understudied and presents a challenge to the application of composites in automotive structures. For a well defined loading, beams and profiles are ideally suited to dissipate energy in the case of a crash. Most load scenarios in real-world applications however are not well defined and are thus difficult to design for using beams and profiles only. As an example, the FMVSS301 test used a barrier of 1000 kg weight, which impacts the front structure at an angle of 30°.

In this case the normal and transverse loads on a structure are equal and, depending on the length of the structure, it may be necessary to design for static strength and not energy dissipation only, which could result in additional structural weight. In this case it is often beneficial to employ flat plates and sandwich structures that dissipate load consistently, independent of the loading direction, while also allow distributing the load over a larger contact area and transferring it away from the point of initial impact.

5.8 Conclusion

Over the past two decades passive safety has greatly improved the survivability of a crash. Existing approaches to occupant protection using metallic structures are well established and deliver an unprecedented level of safety. However, with the increasingly stringent sustainability requirements in the automotive sector, new drivetrains such as hybrid electric vehicles or battery electric vehicles are increasingly replacing conventional internal combustion engine drivetrains. This alone drastically changes passive safety requirements, for example in a rear impact test it may now be necessary to ensure that batteries, which may be placed in the boot of a vehicle, are not deformed during a crash event, which may result in a requirement for significantly lower overall deformation of the vehicle. Simultaneously, more and more crash tests have been introduced to validate the crash performance of vehicles in some extreme test scenarios, and existing vehicle architectures are increasingly difficult to adapt to meet this multitude of requirements.

For modern electric vehicles this change in passive safety requirements is coupled with an enhanced requirement to reducing the overall weight of the vehicle. This is due to the additional weight and cost of the batteries, customer expectations for range and convenience (e.g. when charging a vehicle) and changing use scenarios where the majority of the energy consumption is proportional to vehicle weight, for example during acceleration.

Composites, exhibiting higher specific properties than steel and aluminium, may allow further weight savings and can also be economically viable if the additional battery cost is taken into account. To this end, the BMW project i vehicles, which are either range extenders or battery electric vehicles, are made from a carbon fibre monocoque and an aluminium drive train. For conventional internal combustion engine vehicles composites are introduced in areas where either their high specific strength (e.g. in a B-pillar) or unique forming properties (e.g. in an aerodynamically shaped bonnet) are beneficial.

Consequently, the next step would be the exploitation of the high specific energy absorption properties of composites, which are discussed in more detail. It is noted that the highest energy absorption is obtained from composite structures exhibiting progressive crushing, where the composite structure is disintegrated during deformation. When comparing the SEA of glass, carbon and aramid fibre it can be concluded that carbon fibre exhibits the highest SEA, while glass fibre may prove to be a cost-effective solution for some applications. With respect to matrix materials, thermoset epoxies tend to be well researched and exhibit a good SEA around 60 kJ/kg, while thermoplastic matrices may exhibit an even higher SEA and are thus an area for further improvement. With respect to fibre volume fraction, results are inconsistent and, while it would be expected that a higher fibre volume fraction should yield a higher SEA, some researchers have reported decreasing SEA for high fibre volume factions. In general it can be concluded that, for fibre volume fractions above 50%, further increases in fibre volume result in diminishing improvements in SEA and may thus not be desirable from a cost perspective, since the fibre is generally much more expensive than the matrix. In terms of fibre architecture, arrangements with through-thickness reinforcement such as woven or braided tend to yield higher SEA than unidirectional materials. To ensure stable failure, global buckling of the structure has to be avoided, but it has also been observed that the shape of a crushing structure has an impact on SEA and the ratio of material thickness to diameter for round tubes can greatly influence SEA. To establish stable failure, triggers are often used. These aim to concentrate the impact load on a small area, for example chamfer triggers, but non-geometric triggers such as ply-drops have also been explored. With respect to test speed, most tests are conducted using quasi-static testing, while during an actual crash event the strain rates may be significantly higher. While this may also influence SEA more often, strain rate studies have focused on the CFE for composite structures and have reported significant reductions in CFE for dynamic testing, which are unfavourable. Finally, oblique impacts have been studied experimentally and it has generally been observed that oblique angles below 10° tend to have a limited impact on SEA, while for oblique angles above 10° the SEA tends to decrease significantly.

It can be concluded that there is a robust understanding of the mechanisms governing all elements of composite design for energy absorption. Currently, the biggest impediment to the more widespread adoption of composite crash structures is the lack of reliable predictive methods and significant efforts will be required make progress in this area.

Acknowledgements

The author would like to express his gratitude to his colleagues Dr. Moeller, Dr. Kerscher and Mr. Boegle (BMW Group) for invaluable discussions and suggestions. Particular gratitude is due to Dr. Carwyn Ward (University of Bristol) and Ms Sindy Engel (BMW Group) for proofreading the first draft and providing numerous suggestions for improving the manuscript.

References

[1] Anon (2012) *Verkehr: Verkehrsunfaelle 2011*, Statistisches Bundesamt, Wiesbaden.
[2] Brandstaetter, C., Yannis, G, Evgenikos, P. *et al.* (2011) *Annual Statistical Report 2011*, NTUA, Greece.
[3] Leohold, J. (2011) Chancen und Grenzen für einen nachhaltigen FVK-Einsatz im Automobil. CCeV Automotive Forum 2011, Ingolstadt, Germany, 30th June.
[4] Bakker, J. GIDAS http://www.gidas.org/en%3E (Accessed 5th June 2012).
[5] Anon (2012) *Nachhaltige Mobilität – Der integrale Sicherheitsansatz beim neuen BMW i3*, BMW, Munich, Germany.
[6] Kerstan, H. and Bartelheimer, W. (2009) Application of Innovative Processes and Methods for the Design of Structural Features for the Frontal Impact. 7th VDI-Tagung Fahrzeugsicherheit. Innovativer Kfz-Insassen- und Partnerschutz, Berlin, Germany, 22–23 October.
[7] Barnes, G., Coles, I., Roberts, R., Adams, D.O. and Garner, D.M.J, (2011) *Crash Safety Assurance Strategies for Future Plastic and Composite Intensive Vehicles (PCIVs)*, Volpe National Transportation Systems Center, Cambridge, MA, USA.
[8] Hull, D. (1991) A unified approach to progressive crushing of fibre-reinforced composite tubes. *Composite Science and Technology*, **40**, 377–421.
[9] Kim, J.S., Yoon, H.J. and Shin, K.B. (2011) A study on crushing behaviors of composite circular tubes with different reinforcing fibers. *International Journal of Impact Engineering*, **38**(4), 198–207.
[10] Johnson, A.F. and David, M. (2010) Failure mechanisms in energy-absorbing composite structures. *Philosophical Magazine*, **90**, 31–32.
[11] Farley, G.L., Wolterman, R.L. and Kennedy, J.M. (1992) The effects of crushing surface roughness on the crushing characteristics of composite tubes. *Journal of the American Helicopter Society*, **37**(3),53–60.
[12] Mamalis, A.G., Manolakos, D.E., Demosthenous, G.A. and Ioannidis, M.B. (1996) The static and dynamic axial collapse of fibreglass composite automtoive frame rails. *Composite Structures*, **34**, 77–90.
[13] Mamalis, A.G., Manolakos, D.E., Demosthenous, G.A. and Ioannidis, M.B. (1997) Analytical modelling of the static an dynamic axial collapse of thin-walled fibreglass composite conical shells. *International Journal of Impact Engineering*, **19**(5), 477–492.

[14] Chiu, C.H., Tsai, K.H. and Huang, W.J. (1999) Crush-failure modes of 2D triaxially braided hybrid composite tubes. *Composite Science and Technology*, **59**, 1713–1723.

[15] Hamada, H., Coppola, J.C., Hull, D., Maekawa, Z. and Sato, H. (1992) Comparison of energy absorption of carbon/epoxy and carbon/PEEK composite tubes. *Composites*, **23**(4), 245–252.

[16] Ramakrishna, S. and Hull, D. (1993) Energy absorption capability of epoxy composite tubes with knitted carbon fibre fabric reinforcement. *Composite Science and Technology*, **49**(4), 349–356.

[17] Hamada, H. and Ramakrishna, S. (1995) Scaling effects in the energy absoprtion of carbon-fiber/PEEK composite tubes. *Composite Science and Technology*, **55**, 211–221.

[18] Mamalis, A.G., Manolakos, D.E., Ioannidis, M.B. and Papastolou, D.P. (2005) On the response of thin-walled CFRP composite tubular components subjected to static and dynamic axial compressive loading: experimental. *Composite Structures*, **69**(4), 407–420.

[19] Xiao, X., McGregor, C.J., Vaziri, R. and Poursartip, A. (2009) Progress in braided composite tube crush simulation. *International Journal of Impact Engineering*, **36**(5), 711–719.

[20] Schultz, M.R. and Hyer, M.W. (2001) Static energy absorption capacity of graphite-epoxy tubes. *Journal of Composite Materials*, **35**(19), 1747–1761.

[21] Ochelski, S. and Gotowicki, P. (2009) Experimental assessment of energy absorption capability of carbon-epoxy and glass epoxy composites. *Composite Structures*, **87**, 215–224.

[22] Obradovic, J., Boria, S. and Belingardi, G. (2012) Lightweight design and crash analysis of composite frontal impact energy absorbing structures. *Composite Structures*, **94**(2), 423–430.

[23] Ramakrishna, S. (1997) Microstructural design of composite materials for crashworthy structural applications. *Materials and Design*, **18**(3), 167–173.

[24] Feraboli, P., Wade, B., Deleo, F. and Rassain, M. (2009) Crush energy absorption of composite channel section specimens. *Composites: Part A*, **40**(8), 1248–1256.

[25] McGregor, C., Vaziri, R. and Xiao, X. (2010) Finite element modelling of the progressive crushing of braided composite tubes under axial impact. *International Journal of Impact Engineering*, **37**(6), 662–672.

[26] Zarei, H., Kroeger, M. and Albertsen, H. (2008) An experimental and numerical crashworthiness investigation thermoplastic composite crash boxes. *Composite Structures*, **85**(3), 245–257.

[27] Jimenez, M.A., Miravete, A., Larrode, E. and Revuelta, D. (2000) Effect of trigger geometry on energy absorption in composite profiles. *Composite Structures*, **48**(1–3), 107–111.

[28] Solaimurugan, S. and Velmurugan, R. (2007) Progressive crushing of stitched glass/polyester composite cylindrical shells. *Composite Science and Technology*, **67**(3–4), 422–437.

[29] Warrior, N.A., Turner, T.A., Robitaille, F. and Rudd, C.D. (2004) The effect of interlaminar toughening strategies on the energy absorption of composite tubes. *Composites: Part A*, **35**(4), 431–437.

[30] Song, H.W. and Du, X.W. (2002) Off axis crushing of GFRP tubes. *Composite Science and Technology*, **62**, 2065–2073.

[31] Mamalis, A.G., Manolakos, D.E., Demosthenous, G.A. and Ioannidis, M.B. (1996) Energy absoprtion capability of fibreglass composite square frusta subjected to static and dynamic axial collapse. *Thin-Walled Structures*, **25**(4), 269–295.

[32] Mamalis, A.G., Manolakos, D.E., Demosthenous, G.A. and Ioannidis, M.B. (1997) Experimental determination of splitting in axially collapsed thick-walled fibre-reinforced composite frusta. *Thin-Walled Structures*, **28**(3–4), 279–296.

[33] Hamada, H., Coppola, J.C. and Hull, D. (1991) Effect of surface treatment on crushing behaviour of glass cloth/epoxy composite tubes. *Composites*, **23**(2), 93–99.

[34] Kompass, K., Domsch, C. and Kates, R.E. (eds) (2012) *Integral Safety*, Springer, Heidelberg, Germany.

[35] Cauchi Savona, S. and Hogg, P.J. (2006) Effect of fracture toughness properties on the crushing of flat composite plates. *Composite Science and Technology*, **66**, 2317–2328.

[36] Ghasemnejad, H., Hadavinia, H. and Aboutorabi, A. (2010) Effect of delamination failure in crashworthiness analysis of hybrid composite box structures. *Materials and Design*, **31**, 1105–1116.

[37] Cauchi Savona, S., Zhang, C. and Hogg, P.J. (2011) Optimisation of crush energy absorption on non-crimp fabric laminates by through-thickness stitching. *Composites: Part A*, **42**, 712–732.

[38] Tao, W.H., Robertson, R.E. and Thornton, P.H. (1993) Effects of material properties and crush conditions on the energy absorption of fiber composite rods. *Composite Science and Technology*, **47**, 405–418.

[39] Jacob, G.C., Starbuck, J.M., Fellers, J.F. and Simunovic, S. (2005) Effect of fiber volume fraction, fiber length and fiber tow size on the energy absorption of chopped carbon fiber-polymer composites. *Polymer Composites*, **26**(3), 293–305.

[40] Engel, S., Boegle, C., Majamaeki, J., Lukaszewicz, D. and Moeller, F. (2012) Experimental Investigation of Composite Structures During Dynamic Impact. ECCM 15, Venice, Italy, 24–28 June.

[41] Bisagni, C., Di Pietro, G., Fraschini, L. and Terletti, D. (2005) Progressive crushing of fiber-reinforced composite structural components of a formula one racing car. *Composite Structures*, **60**(4), 391–402.

[42] Rüger, O. (2010) Einfluss von Flechtparametern auf die Performance von Faserverbundcrashrohren. 1st Internationaler Fachkongress- Composites in Automotive & Aerospace, Munich, Germany, 20–21 October.

[43] Saito, H., Chriwa, E.C., Inai, R. and Hamada, H. (2002) Energy Absorption of Braiding Pultrusion Process Composite Rods. *Composite Structures*, **55**(4), 407–417.

[44] Schultz, M.R. (1998) *Energy Absorption Capacity of Graphite-Epoxy Composites Tubes*, Virginia Polytechnic Institute and State University, Virginia, VA, USA.

[45] Karbhari, V.M. and Haller, J.E. (1998) Rate and architecture effects on progressive crush of braided tubes. *Composite Structures*, **43**(2), 93–108.

[46] Sigalas, I., Kumosa, M. and Hull, D. (1991) Trigger mechanisms in energy-absorbing glass cloth/epoxy tubes. *Composites Science and Technology*, **40**, 265–287.

[47] Palanivelu, S., Van Paepegem, W., Degrieck, J. *et al.* (2010) Experimental study on the axial crushing behaviour of pultruded composite tubes. *Polymer Testing*, **29**(2), 224–234.

[48] Thuis, H. and Metz, V. (1994) The influence of trigger configurations and laminate lay-up on the failure mode of composite crush cylinders. *Composite Structures*, **28**, 131–137.

[49] Greve, L., Pickett, A.K. and Payen, F. (2008) Experimental testing and phenomenological modelling of the fragmentation process of braided carbon/epoxy composites tubes under axial and oblique impact. *Composites Part B*, **39**, 1221–1232.

6

Crashworthiness Analysis of Composite and Thermoplastic Foam Structure for Automotive Bumper Subsystem

Ermias Koricho[1], Giovanni Belingardi[1], Alem Tekalign[1], Davide Roncato[2] and Brunetto Martorana[2]

[1]*Dipartimento di Meccanica, Politecnico di Torino, Corso Duca degli Abruzzi, 24, 10129, Torino, Italy*
[2]*Centro Ricerche FIAT, Strada Torino, 50, 10043, Orbassano, Italy*

6.1 Introduction

According to the Global Road Safety Partnership [1], almost 1.2 million people are killed and 20–50 million injured or disabled in car accidents every year, and over 90% of the deaths occur in developing and transitional countries.

The Asia-Pacific region, which accounts for about 16% of the motor vehicles worldwide, is the site of 44% of all traffic deaths. China, with 71 495 deaths, and India, with 59 927 deaths, had the most traffic fatalities in the world in 1995. The rate of deaths to crashes is 2000 for every 10 000 in Kenya and 3000 for every 10 000 in Vietnam.

Beyond the enormous suffering they cause, rood traffic crashes can drive a family into poverty as crash survivors and their families struggle to cope with the long-term consequences of the event, including the cost of medical care and rehabilitation and all too often funeral expenses and the loss of the family breadwinner, for example. In developing countries, the costs of motor vehicle accidents represent 3–5% of the GDP and their estimated yearly cost exceeds US$100 billion.

Advanced Composite Materials for Automotive Applications: Structural Integrity and Crashworthiness,
First Edition. Edited by Ahmed Elmarakbi.
© 2014 John Wiley & Sons, Ltd. Published 2014 by John Wiley & Sons, Ltd.

Blincoe *et al.* [2] reported that there were 16.4 million motor vehicle crashes in the United States in 2000. Of these, 13.5 million (82%) were property damage only crashes; no occupant injury occurred. These collisions, which can be considered relatively minor, caused an average of US$1484 of vehicle damage per crash and an estimated society total cost of US$59.9 billion. In the United Kingdom, insurers report that the total annual cost of motor vehicle insurance is around US$10 billion. About 70% of this cost is related to damage repairs in low-speed crashes, with an average of US$2000 damage per incident [3]. Therefore, reducing vehicle damage in low-speed impacts could have a large global financial benefit, and this can be tackled by careful design of frontal and rear vehicle structure, since many of these relatively minor crashes take the form of low-speed front or rear impacts.

Clearly, vehicle bumper systems should be able to absorb the energy of these collisions and prevent damage of more expensive components located nearby. Unfortunately, most bumpers are designed to meet only the minimum standards required by the regulatory body for any given market [4]. As a result, new designs are regularly introduced which pass current bumper standards and evaluation tests, but which still do not prevent thousands of dollars in damage in minor collisions. In short, existing test procedures are not motivating enough manufacturers to improve these bumper designs.

Currently, there are four main types of test to evaluate vehicle bumper systems worldwide. In the United States, the National Highway Traffic Safety Administration (NHTSA) has established a bumper standard for passenger that includes a series of car against barrier tests and subsequent pendulum tests on each vehicle. Barrier and pendulum impacts are conducted at 2.5 mph (4 km/h) on the full width of the front and rear bumpers. Pendulum impacts are also conducted at 1.5 mph (2.4 km/h) on the corners of the bumpers. While no damage to other parts of the vehicle is allowed from these impacts, unlimited damage is permitted to the bumper itself. The test pendulum impacts the vehicle at a height of 16–20 inches (40.6–50.8 cm), effectively regulating the bumper heights of passenger cars. There is no NHTSA bumper standard for passenger vans, sport utility vehicles, or pickups [4].

In Europe, vehicle manufacturers must comply with the bumper standard specified by the United Nations Economic Commission for Europe (UNECE). In this test, a profiled pendulum strikes the vehicle at a height of 15.25–19.75 inches (38.7–50.2 cm) with a speed of 2.5 mph (4 km/h) on the front and rear bumpers. A similar corner test is performed at 1.5 mph (2.5 km/h). The standard does not dictate unacceptable amounts of bumper damage but only ensures the integrity of the headlamps, turn signals and steering controls [5].

The Insurance Institute for Highway Safety (IIHS) has a consumer information low-speed testing program designed to evaluate bumper performance based on the cost of repairing vehicle damage resulting from four test configurations. The four configurations are full width front and rear impacts into a flat barrier, a front impact into an angled barrier and a rear impact into a centred pole. All four tests are conducted at 5 mph (8 km/h). The first two configurations are patterned after the federal full-width impacts, while the latter two represent additional crash configurations that may be encountered in the real world [6].

The Research Council for Automobile Repairs (RCAR) uses a fourth test procedure for assessing bumper designs. It consists of two tests evaluating bumper performance as well as how easily and cheaply any damage can be repaired. The tests are a 40% overlap front impact into a flat rigid barrier and a 40% overlap rear impact by a 1000 kg mobile barrier. The test vehicle strikes the barrier at 15 km/h (9.3 mph) in the frontal test, and the moving barrier strikes the vehicle at the same speed in the rear test [7]. The RCAR test program is used by several test houses in Europe, Asia and South America.

To fulfil the existing above mentioned regulations, appropriate conceptual bumper subsystem should be implemented. Besides, it is worth incorporating the essential lightweight materials in order to reduce environmental pollution and fuel cost.

Generally, lightweighting of existing automobile models can be carried out in two ways: structural improvement and material change [8]. Research shows that structural optimisation can give up to 7% weight reduction, while material replacement such as full aluminium body can lead to weight reduction up to 50% [9]. If further weight reduction is needed, the ideal candidate materials are fibre reinforced materials. Beside weight reduction, fibre reinforced materials have good corrosion resistance, impact cushion, noise attenuation and allow for relevant part consolidation.

One possible application area that allows material replacement to achieve vehicle lightweighting is the bumper subsystem. Optimisation of the car bumper subsystem, particularly the bumper beam, can improve not only weight reduction but also structural energy absorption to meet pedestrian safety standards.

Over the last few years, some factors have made this bumper subsystem application more interesting for composite materials such as glass mat thermoplastic (GMT), which are as follows [10]:

- *Increasing demands for vehicle weight reduction*: Reduction in fuel consumption and vehicle handling, since the bumper is located far from the centre of gravity of the vehicle so its weight is also critical for the inertia effect.
- *Higher required energy absorption:* Achieving energy absorption at bumper mounting points to protect nearby components, in a low-speed crash.
- *Controllable fracture behaviour:* Part integrity and stabilisation function at very high-speed crashes. At these rates primarily the deformation behaviour is important.

Different researchers have implemented different types of composite materials, such as carbon fibre reinforced plastic (CFRP), glass fibre reinforced plastic (GFRP), sheet moulding compound (SMC) and GMT for the bumper beam, to improve its performance so that it can offer lightweight as well as improved vehicle safety [11, 12]. Currently, SMC and GMT are widely used because of easy formability and low material and manufacturing costs, even though CFRP and GFRP can offer better mechanical performances. GMT is more appreciated in automotive industries, because of its short shaping and curing cycles and, most importantly, it is recyclable material.

Currently, in the European Union (EU), about 75% vehicles at their end of life are recyclable, that is their metallic part. The rest (∼25%) of the vehicle is considered to be waste and generally goes to landfills [13]. EU legislation requires the reduction of this waste to a maximum of 5% by 2015.

To take into account this directive, this chapter intends to address material selection, material characterisations, design aspects and methods of analysis with particular reference to the application of composites (CFRP, polyamide 30% glass fibre), recyclable thermoplastic foam and adhesives materials to automotive front bumper design. Particular attention is also paid to reaction pick load, weight reduction, joining techniques and the manufacturability of the bumper transverse beam to evaluate the performance of the solutions proposed.

6.2 Materials for Automotive Applications

Currently, the vehicle body and under-body are mostly an assembly of steel based components. The processes used to manufacture the components and the methods used to join the components together have a significant impact on the corrosion resistance, durability and strength of the under-body. In general, flat rolled steels are versatile materials. They provide strength and stiffness with favourable mass to cost ratios, and they allow high-speed fabrication. In addition, they exhibit high energy absorption capacity, good fatigue properties, high work hardening rates, ageing capability and excellent paintability, which are required by automotive applications. These characteristics, plus the availability of high-strength low alloy (HSLA) and ultra-high-strength steel (UHS) in a wide variety of sizes, strength levels (UHS offer tensile strengths up to 1500 MPa), chemical compositions and surface finishes, with and without various organic and inorganic coatings, have made sheet steel the material of choice for the automotive industry. Even though steel will continue to be the dominant body material, interest in non-ferrous metals (e.g. aluminium) and reinforced polymer composites is increasing very rapidly. The main reasons behind this increasing demand are the drawbacks of steel-based components, which are associated with their rising costs, corrosion and high density.

Due to increasing customer needs, such us safety, comfort, drive performance, roominess, variability and quality, there has been a trend of vehicle weight increment over the past decades. But, it is extremely important to invert this trend, in order to reduce fuel consumption and consequently greenhouse gases.

There are a number of materials with higher stiffness to weight and strength to weight ratios with respect to the traditional and even the recently developed HS steel. However, their cost should be evaluated carefully, taking into account their manufacturability and component integration opportunities, besides reduction of fuel consumption during the vehicle life.

Aluminium offers a lightweight vehicle over steel, reducing the weight of an automobile body and potentially increasing the efficiency of the vehicle. Aluminium usage in automotive applications has grown substantially within past years. Examples of aluminium applications in vehicles cover power trains, chassis, body structure and air conditioning system. Aluminium castings have also been applied to various automobile parts for a long time. As a key trend, the material for engine blocks, which is one of the heavier parts, is being switched from cast iron to aluminium, resulting in a significant weight reduction. However the high cost of aluminium remains the biggest obstacle for its use in large-scale sheet applications. Polymer composite materials have been a part of the automotive industry for several decades. These materials have been used for vehicles with low production volumes, because of their shortened lead times and lower investment costs relative to conventional steel fabrication. Important stimuli for the growth of polymer composites have been the reduced weight and parts consolidation opportunities the material offers, as well as design flexibility, corrosion resistance and good mechanical properties. Although these advantages are well known by the industries, their use has been impeded so far, due to high material costs, slow production rates, concerns about crash energy absorption and recyclability. The cost of polymer composite materials is usually much higher (up to 10 times higher when using carbon fibres) than those of conventional metals. Therefore, the main targets for the future development of hybrid composites will require automated and rapid manufacturing processes. For example, the use of pre-impregnated composite fibres or prepregs, which are reinforced with carbon or glass in fibre and fabric forms coated with epoxy resins, are currently suitable for only limited automotive applications due to lower productivity. Some of the factors that make high-volume composite structure production difficult are long cycle times, high labour cost, and low productivity of the moulding tool investment.

In terms of strength to weight ratio and stiffness to weight ratio, composite and hybrid materials outperform most metallic alloys. Besides, fibres are flexible, so they can be shaped to any complex geometry, therefore this group of materials are suitable for any complex architecture, which is one of the most important subject for aircraft and vehicle designers, for improving the aesthetic value and, most importantly, reducing aerodynamic drag. These materials also have better corrosion and fatigue performances which allow a comparatively longer service life than conventional metallic materials. The anisotropic nature of the materials means that strength and stiffness can be tailored accurately to the expected loads, however this may result in a less damage tolerant structure.

Therefore multi-material mixes in the automotive body shop are increasingly gaining importance, and fibre reinforced composites like CFRP will become the next substrates to be combined with metals in the automotive body shop.

6.3   Composite and Thermoplastic Materials

It is worthwhile to focus specifically on the mechanical properties of composite materials to understand their best uses in engineering applications. The properties of composite materials (unidirectional or fabric type) depend on the mechanical properties of matrix and fibre. The strength and stiffness of a fabric depend on the type of fibre used, and they are also strongly affected by the type of weaving used (which changes the waviness and amount of friction between the fibres). Many fabrics are also composed of a strong fibre weave laminated with weaker polymer coatings. The fibres are critical, so it is useful to plot their specific strength and specific stiffness on a selection chart – this allows for a comparison of natural and synthetic fibres, and also a comparison of synthetic fibres with the same polymers in bulk form, as shown in Figure 6.1c. The chart shows that many fibres have excellent specific properties but, of course, these can only be exploited by using the fibres in a structured material like in a rope or a fabric. The material bubbles in red in Figure 6.1 show long fibre properties; the other materials and material classes show *bulk* properties. The strength for the bulk ceramics shown on the chart is compressive strength – the tensile strength is typically only 10% of this value; for the other materials the strength is similar in compression and tension. The strength for all fibres is for loading in tension.

In the composite class, carbon fibre reinforced polymer, CFRP, and glass fibre reinforced fibre polymer, GFRP, gained considerable attention in the aeronautic and automotive industries because of their high specific stiffness and specific strength. Figure 6.2 shows that how this group of materials, particularly CFRP, is becoming important in energy and transport sectors, that is Boeing 787, GEnx (General Electric next generation engine) and Toyota Prius [16].

However, long fibre reinforced thermoplastics (LFTs) have also gained an increasing market share in the automotive sector during the last few years because of their recyclability. The

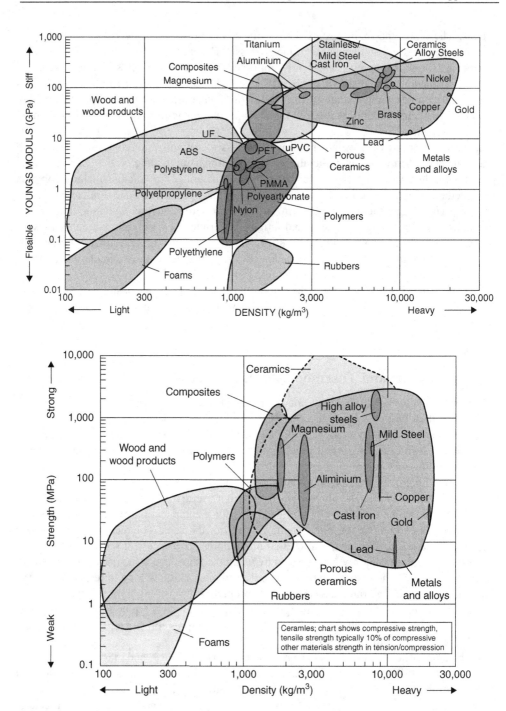

Figure 6.1   Material selection charts. (a) Young modulus versus density [14]. (b) Strength versus density [15]. (c) Specific stiffness versus specific strength [15].

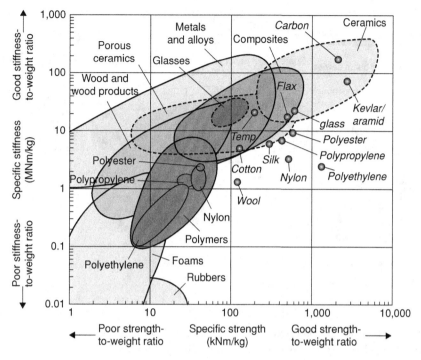

Figure 6.1   (*Continued*)

Weight => Fuel => Green

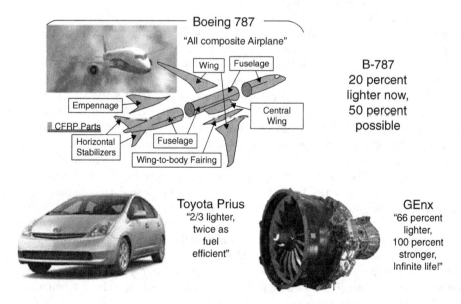

Figure 6.2   Application of CFRP in the aeronautic and automotive industries [16].

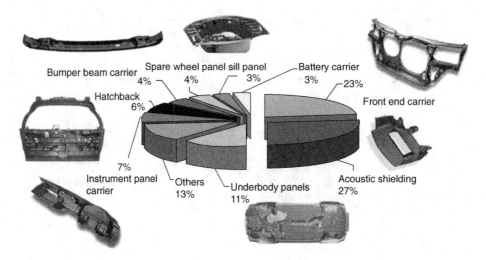

Figure 6.3 Automotive applications of long fibre reinforced PP [17]. Reproduced with permission from Ref. [17].

direct processes for LFT production (LFT-D) have also been established on the market. This processes avoid semi-finished materials by manufacturing products directly from the raw materials such as glass fibres, thermoplastic resins and, if necessary, the modifiers. They are well known to offer an excellent cost –performance ratio [17].

Currently, for most applications the LFT-D processing technology utilises polypropylene (PP) as the matrix material for the production of components for the automotive industries, as shown in Figure 6.3. Currently, the most commonly used thermoplastic matrices in the automotive industries is PP, whereas the potentiality of other engineering thermoplastics matrices, like polyamides, polyesters and so on, have not been exhaustively investigated [18].

However, the low melting temperature of PP against the engineering thermoplastics does not allow the PP/GF based LFT components to be implemented directly into the body in white (BIW) assembly prior to the e-coating process. The BIW assembly is sent through the e-coating process, painting process and the drying stages, where it is exposed to high temperature and chemical treatments. During such treatments the highest temperature can reach up to 215 °C and hence the implementation of the PP based composites into the BIW is not feasible prior to such an e-coating process. Some of the commercially available polyamides, such as polyamide (PA) 66 or PA6 and polyesters, such as polyethylene terephthalate (PET) or polybutylene terephthalate (PBT), could be considered as the potential thermoplastic materials for developing such applications. Particularly where temperatures can spike at above 215 °C but less than 250 °C, PA66 is preferred.

The current use of composite materials in automotive structures is limited to secondary exterior structures, such as body panels and wheel housings, and energy absorbing components, such as bumper and crash box. Body panels are commonly made from sheet moulding compound (SMC), which is based on a thermoset matrix reinforced by discontinuous glass fibre. Bumpers and front panels have been made from glass mat thermoplastic (GMT) which is commonly discontinuous glass fibre reinforced polypropylene.

Even though SMCs with random chopped glass fibre composites [15] are being used as a design solution in some vehicle structures, there is also a lot of evidence that GMT is applied mostly to the automobile body to achieve a lightweight vehicle. GMT gives a high strength to weight ratio, chemical/corrosion resistance and excellent impact properties at both low and high temperatures. Compared to metals, GMT offers greater design flexibility, lower tooling costs and opportunities for part consolidation. Compared with thermoset composites, GMT improves productivity with shorter moulding cycle time (e.g. compared with SMC), greater impact resistance, recyclability (melt reprocess ability) and elimination of controlled storage requirements [19]. There are three principal types of GMT, including continuous glass fibre, chopped glass fibre and unidirectional glass fibre [20].

As a new lightweight design approach, polymer reinforced composites with nano-clay are also one specific category of nanocomposites that attracted remarkable attention when Toyota central research and development laboratories introduced the first commercial nanocomposite made from nylon 6 in 1988 [21, 22]. Other types of nanoparticles have been introduced into polymeric matrixes to achieve materials with enhanced mechanical properties. Some commercially available nanoparticles include metals such as copper and aluminium, carbon based nanoparticles such as nanotubes and ceramics such as alumina and silica. The nanoparticles are incorporated into matrixes such as epoxies, nylon, polypropylene, polyamide and polystyrene, as well as other types of polymers. In both thermoplastic and thermosetting polymers, changes in mechanical, thermal and electrical properties are observed by using nano-modified composite materials [23–30].

For example, Fiat Research Centre (CRF) is currently working to implement extensively multifunctional composite and smart materials, as shown in Figure 6.4. Multifunctional composites could reduce vehicle weight by 20% owing to the reduction in heavy mechanical and electronic discrete components [31].

Smart materials are materials that have one or more properties that can be significantly changed in a controlled way by external stimuli, such as stress, temperature, moisture, pH, electric or magnetic fields.

6.4   Numerical Modelling of Fiat 500 Frontal Transverse Beam

The FIAT 500 transverse beam was studied to obtain a lightweight bumper subsystem. The reverse engineering process was applied on the actual frontal transverse bumper beam, which was taken from commercially available automotive spare parts, as show in Figures 6.5 and 6.6. Detailed dimensions of each section were taken carefully to avoid the influence of inappropriate geometry representation on the actual response of bumper subsystem during the impact phenomenon. A CAD model was developed using CATIA, based on the measured dimensions. Further, the geometries were edited in ABAQUS geometry editor to prepare the model for simulation. Besides joining techniques, the method of manufacturing process and type of materials in each profile were well studied, to incorporate appropriate design and material parameters inside the ABAQUS environment.

In this work, some of the challenging tasks to develop optimised numerical model were choice of element size and shape, contact definitions and assignment of appropriate constraints in order to save computational time. To deal with these problems, complex shapes were subjected to localised meshing to keep the desired shape of the bumper beam. Also

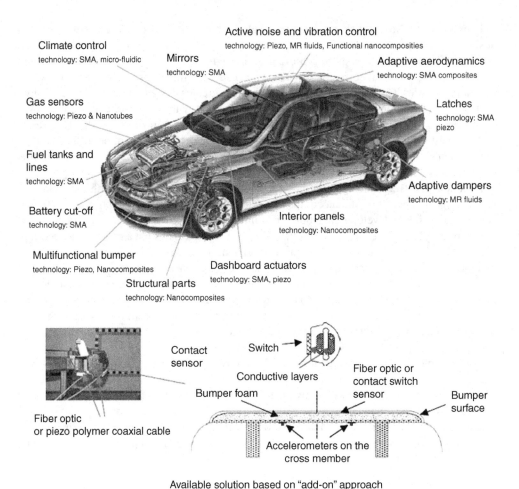

Figure 6.4 Application of smart and multifunctional materials: (a) on vehicle system, (b) on bumper subsystem [31]. Reproduced with permission from Ref. [31].

three-node triangular general-purpose shells and four-node doubly curved shells were considered, to avoid analytical and convergence problems due to unacceptable element aspect ratio. At the initial stage some preliminary analyses were performed to choice the element size and shape, the contribution of additional components on the global response and to understand the estimated computation time required. Based on the results found, new solutions were proposed: some parts such as crash box and joining techniques were represented by appropriate equivalent constraints without significant variation of dynamic behaviour, such as impact force, acceleration and kinetic energy. The final FEM model is shown in Figure 6.7.

To understand the performance of composite materials under low-velocity impact, two types of materials, CFRP twill fabric and polyamide 30% short glass fibres, were considered to substitute the existing steel bumper beam solution. In the case of polyamide material, two types of polyamide based thermoplastic materials, PA66 and PA66 with 30% short glass

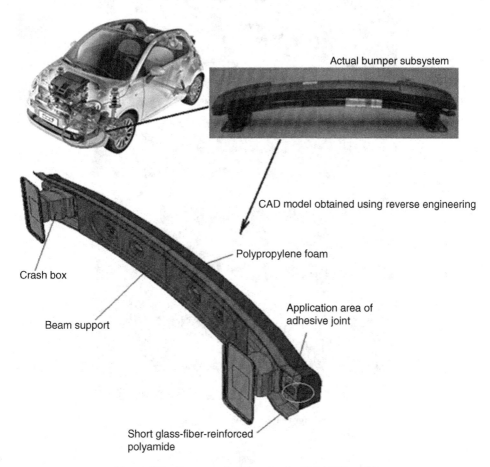

Actual bumper subsystem

CAD model obtained using reverse engineering

Polypropylene foam

Crash box

Application area of adhesive joint

Beam support

Short glass-fiber-reinforced polyamide

Figure 6.5 Bumper subsystem for the FIAT 500 model.

fibre, were manufactured using an injection moulding machine. At the initial stage, PA66 was chosen as a design solution for the bumper beam application. However, after material characterisations, the results revealed that the strength and stiffness of the polyamide plastic were much lower than the existing material solutions for bumper beam production. Hence, an alternative approach was implemented to improve the mechanical behaviour of PA66. In this regard, material and manufacturing costs were the main factor to choose a feasible solution for bumper beam application. Based on these factors, short glass fibre material was selected to be a filler inside the PPA66 matrix. Then, 30% of glass fibres were added inside the PA66 matrix and produced using an injection moulding machine. As a result, its tensile strength and modulus of elasticity were improved by 125.43 and 220.4%, respectively. Detailed material data are listed in Table 6.1.

CFRP and steel test data were taken from Refs. [32,33], respectively. Also the characteristic of polypropylene foam which was utilised as an energy absorber for pedestrian safety was taken from previous work performed in the department [34]. Regarding adhesive material, Prodas

Figure 6.6 FIAT 500 model.

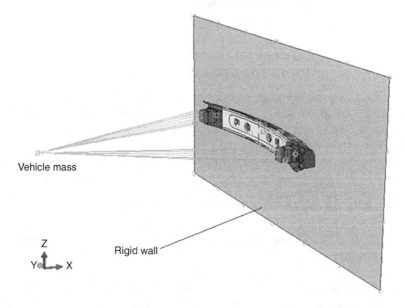

Figure 6.7 Numerical bumper subsystem model.

Table 6.1 Elastic properties for polyamide 30% glass fibre.

Composite	Fibre weight fraction	Maximum tensile stress (MPa)	Maximum compressive stress (MPa)	Young's modulus, E (GPa)	v_{12}
Polyamide 30% glass fibre	0.3	143.08	154.63	10.35	0.405

1400 hot melt adhesive was chosen and characterised for joining the composite transverse beam with the polypropylene foam [35].

6.5 Standards for Low-Speed Frontal Impact

To set up the appropriate boundary conditions and the desired general variables of the bumper subsystem, it is worth a survey of the existing standards related to the design of bumpers under impact load condition. Currently, there are three low-speed impact regulations to check the performance during crashing condition: the National Highway Traffic Safety Administration (NHTSA) Code 49 part 58 [36], the ECE Regulation No. 42 [37], and the Canadian Motor Vehicle Safety Regulation (CMVSR) [38]. The NHTSA safety regulation has the same limitation and safety damage requirement as the CMVSR, however the speed is reduced by half. In this chapter, the NHTSA standard was chosen to perform car into barrier impact tests. The impact test against the barrier was conducted at 4 km/h on the full width of the frontal bumper, as shown in Figure 6.7. This standard requires that the light system, bonnet and doors can be operated after the impact as in normal operation conditions, also all essential features should be still appropriately functional or serviceable.

6.6 Bumper Beam Thickness Determination

To study the crash behaviour of the above mentioned bumper beam materials, two approaches were adopted: a material comparison with a thickness equal to the reference solution (steel) and a material comparison with deflection (bending stiffness) equal to the reference solution. The first approach is the simpler one and allows material substitution without altering the geometry profile of the bumper beam. The second approach needs to calculate the thickness of each proposed material to obtain the same stiffness value as the reference solution [39, 40]. The calculated bumper beam thickness for each material type is depicted in Table 6.2.

Table 6.2 Bumper beam thickness for each material type.

	CFRP [0/90]$_{4s}$	30% glass fibre polyamide
Thickness, h_c (mm)	3.2	5.45

6.7 Results and Discussion

As shown in Figure 6.8, where the performance of the bumper is compared based on material substitution, the steel bumper beam performed appreciably since it absorbs the impact energy with a smaller peak force, which is a dangerous input for the occupant and needs to be controlled. As shown in the figure, at the initial stage of the impact scenario, the steel bumper beam was stiffer than the CFRP and polyamide 30% glass bumper beams. This means that the steel beam resisted the applied load better. However, after approximately 20 ms, it started to be weaker due to large plastic deformation. For this reason the remaining impact phenomenon requires the crash box to be involved. As the target of the study is the bumper beam, to reduce computational time and to single out the targeted component, the crash box was modelled as a rigid body. Hence, when the beam collapsed and the rigidly modelled crash box impacted with the rigid barrier, the reaction force increased instantly. In contrast, the CFRP and polyamide 30% glass fibre bumper beam responded with a lower rate at the beginning, which is important failure behaviour for energy absorbing components. During energy absorbing component design, not only the amount of energy absorbed, but also the rate of energy absorption has to be controlled, since a higher rate leads to higher deceleration.

To obtain an equivalent performance between the reference steel bumper beam and the remaining composite beams, the thickness of each composite beam was modified on the basis of an equal stiffness criterion. Figure 6.9 shows the impact force response of the modified composite beams. It can be clearly seen that some improvements were observed at the initial stage of the impact scenario, especially in the case of the CFRP bumper beam. Regarding the peak reaction load, polyamide with 30% glass shows a reduced peak reaction force. However, no significant improvement was observed for the CFRP beam.

To improve the composite bumper beam performance, the stress distribution and mode of deformation were analysed, particularly at the initial stage of impact. The results revealed that

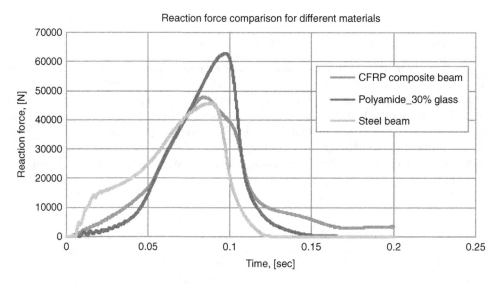

Figure 6.8 Reaction force versus time curve of bumper beam on the basis of equal thickness, 2 mm (without reinforcement).

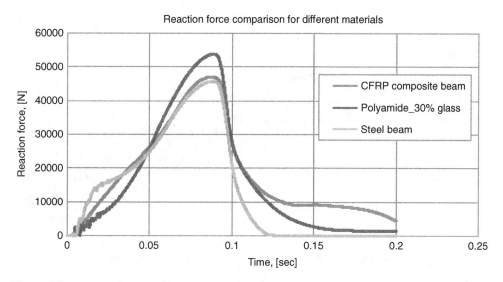

Figure 6.9   Reaction force versus time curve of bumper beam on the basis of equal stiffness (without reinforcement).

high deformation and localised stress distribution occurred near the end of the back support beam, which is adjacent to the crash box, as shown in Figure 6.10. In order to redistribute the concentrated stress, reinforcements were incorporated as a design solution on both ends of back beam support, as shown in Figure 6.11. The reinforcement developed considerable stress distribution and reduced the peak stress on the main bumper beam. By applying this new design solution, an improvement was made regarding peak reaction load reduction, as shown in Figure 6.12.

Therefore, through additional reinforcement, the CFRP beam solution results reduced the peak reaction load by 20.3% but no significant reduction was observed for 30% glass polyamide. Besides, the introduction of reinforcement yielded progressive failure mode, which

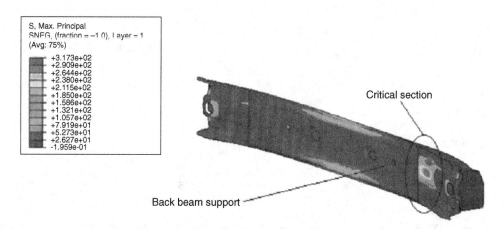

Figure 6.10   Local stress formation at the bumper critical section.

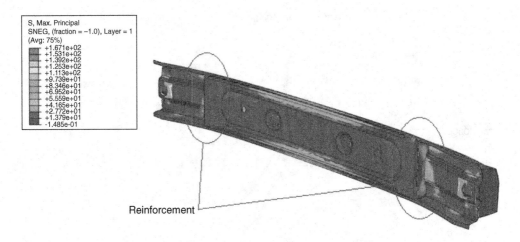

Figure 6.11 Introduction of reinforcements to reduce stress concentration.

is one of the most important factors that have to be considered when the energy absorbing component is designed. This is due to the fact that the energy absorbing characteristics of components made from composite materials mainly depend on their failure mode; and the most challenging task is to obtain this progressive failure mode, since it is essential for energy absorption to improve occupant safety.

Despite the structural modifications described above, the 30% glass polyamide material solution still needed the involvement of the crash box to absorb energy during this impact. This behaviour can be explained by its lower strength, stiffness and strain to failure properties. Therefore, with the same material volume, it will not be feasible to replace the existing metallic

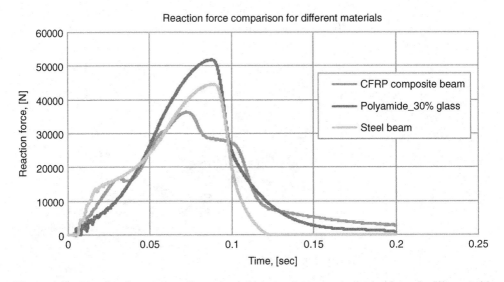

Figure 6.12 Reaction force versus time curve of bumper beam on the basis of equal stiffness (with reinforcement).

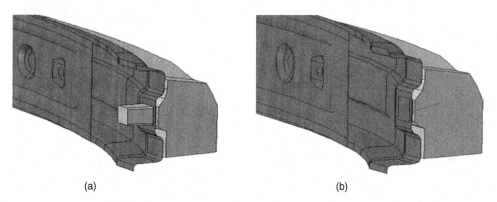

(a)   (b)

Figure 6.13 Joint techniques between bumper beam and energy adsorber: (a) mechanical fitting, (b) adhesive joint.

beam with this new material. Perhaps with the cost of weight, all the above mechanical behaviours can be improved by increasing the material volume (e.g. the section thickness) and by re-designing the section profiles, to obtain a progressive mode of failure and to use it as an alternative engineering material with the above-mentioned peculiar features.

As explained above, part of the study was conducted with a simply material substitution, meaning the existing metallic material was replaced by a composite material with the same dimensions. In this respect, the CFRP and 30% glass fibre polyamide bumper beams reduced the weight by 67.8 and 45.0%, respectively. Of course, for practical applications, the figure will be changed according to the functionality and the strength needed.

Regarding joining techniques, through an analysis of beam and foam connecting techniques, the currently used mechanical press fitting approach, as shown in Figure 6.13, is compared with an adhesive solution. As is well known, mechanical press fitting needs holes on the beam, which usually causes stress concentration and ultimately beam strength reduction. For the proposed adhesive joining, as is clearly shown in Figure 6.14, in addition to preventing the above-mentioned problems, it slightly reduces the reaction force peak, which is one of the main factors to be controlled during crashworthiness analysis.

6.8 Conclusion

As a key design factor, the material selection process takes the major part of the lightweight design process, since >50% of a vehicle production cost is material cost [39]. Hence, generally, the material selection process shall include the entire design problems and utilise novel materials to produce a cost effective, easily manufacturable and maintainable, recyclable and reliable lightweight vehicle. Composites, nanomodified composites and thermoplastic materials are some of the ideal candidate materials to achieve a lightweight vehicle and improve the performance of the vehicle under different loading conditions, such as static and crash loadings.

In the current study, the redesign of a front bumper subsystem has been developed. Alternative solutions have been considered by substituting steel with other suitable materials. FE

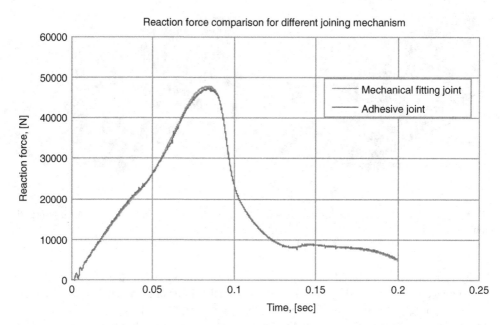

Figure 6.14 Comparison of reaction force versus time curves of bumper beam using mechanical fit and adhesive joint.

simulation results illustrate how the choice of material can significantly affect the performance of the bumper subsystem. The introduction of local reinforcements at the stress concentration point enhances the composite bumper beam performance by redistributing the stress and preventing local failures. However the PA66 solution, even if reinforced with short glass fibres, does not reach a comparable result with respect to the CFRP solution. Looking at the results from another point of view, the polyamide with 30% glass solution leads to better results in terms of possible material recycling for end of life vehicles, while CFRP still has a problematic perspective. Therefore, taking into account the EU recycling requirements, the polyamide with 30% glass solution can be considered an alternative solution. Some improvements can certainly be obtained with further geometry optimisation.

Finally, regarding the joining techniques, the proposed adhesive joint using a hot melt adhesive, in addition to preventing high stress concentrations and an inconvenient manufacturing process, reduces slightly the reaction force peak, which is one of the main factors to be controlled during crashworthiness analysis.

References

[1] UN (2009) http://www.un.org/ar/roadsafety/pdf/roadsafetyreport.pdf (Accessed 8th February 2012).
[2] Blincoe, L., Seay, A., Zaloshnja, E. *et al.* (2002) The Economic Impact of Motor Vehicle Crashes 2000 (DOT HS-809-446). National Highway Traffic Safety Administration, Washington, DC, USA.
[3] Association of British Insurers (2002) Annual Report 2002, Association of British Insurers, London, UK.
[4] Schuster, P. (2006) Current Trends in Bumper Design for Pedestrian Impact A Review of Design Concepts from Literature and Patents. SAE 2006 World Congress and Exhibition, SAE2006-01-0464.

[5] United Nations (1980) Agreement Concerning the Adoption of Uniform Conditions of Approval and Reciprocal Recognition of Approval for Motor Vehicle Equipment and Parts. Regulation 42, Uniform Provisions Concerning the Approval of Vehicles with Regard to their Front and Rear Protective Devices (Bumpers, etc.). United Nations Economic Commission for Europe, Geneva, Switzerland.

[6] Insurance Institute for Highway Safety. (2002) Low-Speed Crash Test Protocol: Version V, Arlington, VA. 2002, Available at http://www.iihs.org/vehicle_ratings/test_protocol_low.pdf (Accessed: November 3, 2003).

[7] Research Council for Automobile Repairs. (1999) The Procedure for Conducting a Low Speed 15 km/h Offset Insurance Crash Test to Determine the Damageability and Repairability Features of Motor Vehicles. Available at http://www.rcar.org/papers/craft_test.pdf (Accessed: November 18, 2003).

[8] Li, Y., Lin, Z., Jiang, A. and Chen, G. (2004) Experimental study of glass-fibre mat thermoplastic material impact properties and lightweight automobile body analysis. *Materials and Design*, **25**, 579–585.

[9] Jambor, A. and Beyer, M. (1997) New cars – new materials. *Materials and Design*, **18**(4/6), 203–209.

[10] Quadrant (2010) Structural Bumper Beams made in GMTexTM, Technical Information, Quadrant Plastic Composites site, http://www.quadrant.com/GMTexTM.pdf (Accessed: November 3, 2003).

[11] Davoodi, M.M., Sapuan, S.M., Ahmad, D. *et al.* (2011) Concept selection of car bumper beam with developed hybrid bio-composite. *Materials and Design*, **32**(10), 4857–4865.

[12] Marzbanrad, J., Alijanpour, M. and Kiasat, M.S. (2009) Design and analysis of an automotive bumper beam in low-speed frontal crashes. *Thin Walled Structures*, **47**, 902–911.

[13] Kanari, N., Pineau, J.-L. and Shallari, S. (2003) End-of-Life Vehicle Recycling in the European Union, The minerals, metals and material society, http://tms.org/pubs/journals/jom/0308/kanari-0308.html (Accessed 8th February 2011).

[14] Excellence Gateway (2010) http://tlp.excellencegateway.org.uk/tlp/cpd/puttingcpdintoa/futuretechepack/bioengineering/module_2/index.html (Accessed 18th February 2012).

[15] Materials Society (2011) http://www-materials.eng.cam.ac.uk/mpsite/interactive_charts/stiffness-density/IEChart.html (Accessed 6th February 2012).

[16] Tsai, S.W., Sung, H. and Miyano, Y. (2008) *Strength and Life of Composites*, JEC Composites, Hong Kong.

[17] Geiger, O., Henning, F. and Eyrer, P. (2006) LFT-D: materials tailored for new applications. *Reinforced Plastics*, **50**(1), 30–35.

[18] Heide, J. (2009) E-Coat Sustainable Long-Fibre Thermoplastic Composites for Structural Automotive Applications. Conference on Innovative Developments for Lightweight Vehicle Structures, Wolfsburg, Germany, May 2009.

[19] Steve Ickes, M. (2000) Development of Low Density Glass Mat Thermoplastic Composites for Headliner Applications. SAE 2000 World Congress, SAE 2000-01-1129.

[20] *Glass Mat Reinforced Thermoplastics, Processing Guidelines,* Quadrant Plastic Composite AG, GMT Parts, Edition 2.1.

[21] Okada, A., Fukushima, Y., Kawasumi, M. *et al.* (1988) Composite material and process for manufacturing dame, US Patent 4, 739,007, USA.

[22] Usuki, A., Kojima, Y., Kawasumi, M. *et al.* (1992) Synthesis of Nylon-6 clay hybrid. *Journal of Material Research*, **8**, 1179–84.

[23] Nikhil, G., Tien, C.L. and Mechael, S. (2007) Clay/epoxy nanocomosites: processing and mechanical properties. *Journal of Materials*, **59**(3), 61–65.

[24] Ray, S.S. and Okamoto, M. (2003) Polymer/layered silicate nanocomposites: a review from preparation to processing. *Progress in Polymer Science*, **28**, 1539–1641.

[25] Park, J.H. and Jana, S.C. (2003) The relationship between nano- and micro-structures and mechanical properties in PMMA–epoxy–nanoclay composites. *Polymer*, **44**(7), 2091–2100.

[26] Zhou, Y., Pervin, F., Biswas, M.A. *et al.* (2006) Fabrication and characterization of montmorillonite clay-filled SC-15 epoxy. *Materials Letters*, **60**(7), 869–873.

[27] Lau, K.T., Lu, M., Li, H.L. *et al.* (2004) Heat absorbability of single-walled, coiled and bamboo nanotube/epoxy nano-composites. *Journal of Material Science*, **39**, 5861–5863.

[28] Lau, K. T., Lu, M. and Hui, D. (2006) Coiled Carbon nanotubes: synthesis and their potential application in advanced composite structures. *Composites: Part B*, **37**(6), 437–448.

[29] Gojny, F.H., Wichmann, M.H.G., Köpke, U. *et al.* (2004) Carbon nanotube-reinforced epoxy-composites: enhanced stiffness and fracture toughness at low nanotube content. *Composites Science and Technology*, **64**(15), 2363–2371.

[30] Miyagawa, H. and Drzal, L. T. (2004) Thermo-physical and impact properties of epoxy nanocomposites reinforced by single-wall carbon nanotubes. *Polymer*, **45**(15), 5163–5170.

[31] Butera, R. A synergic approach for tomorrow's vehicles, FIAT Research Center (CRF), Turin, Italy.

[32] Belingardi, G. and Koricho, E.G. (2010) Implementation of Composite Material Car Body Structural Joint and Investigation of Its Characteristics with Geometry Modifications. Proceedings of the 14th European Conference on Composite Materials, Budapest, Hungary, 7–10 June.

[33] Belingardi, G. and Koricho, E.G. (2010) Implementation of Hybrid Solution in Car Body Structural Joints. Proceedings of XXXIX AIAS National Conference, Maratea, Italy 7–10 September.

[34] Avalle, M., Belingardi, G. and Montanini, R. (2001) Characterization of polymeric structural foams under compressive impact loading by means of energy-absorption diagram. *International Journal of Impact Engineering*, **25**, 455–472.

[35] Koricho, E.G. (2012) PhD thesis, Politecnico di Torino Department of Mechanical and Aerospace Engineering, Torino, Italy.

[36] NHTSA National Highway Traffic Safety Administration (1990) Laboratory Test Procedure for Regulation Part 581 Bumper Standard Safety Assurance.

[37] ECE (1980) Uniform provisions concerning the approval of vehicle with regard to frontal and rear protective device.

[38] CRC (2009) Canadian Motor Vehicle Safety Regulations.

[39] Belingardi, G., Koricho, E.G. and Martorana, B. (2011) Design Optimization and Implementation of Composite and Recyclable Thermoplastic Materials for Automotive Bumper. Proceedings of Fifth International Conference on Advanced Computational Methods in Engineering (ACOMEN 2011), Liège, Belgium, 14–17 November.

[40] Goede, M. (2007) Contribution of Lightweight Car Body Design to CO2 Reduction. Aachener Kolloquium Fahrzeug und Motorentechnik, Aachen, Germany, 2007.

7

Hybrid Structures Consisting of Sheet Metal and Fibre Reinforced Plastics for Structural Automotive Applications

Christian Lauter, Thomas Tröster and Corin Reuter

Automotive Lightweight Construction, University of Paderborn, Paderborn, Germany

7.1 Introduction and Motivation

Economic and ecological constraints have meant that the development of lightweight concepts for high-volume automotive applications has become extremely important [1,2]. The mass of a vehicle is one of the most important factors influencing its driving resistance (Figure 7.1). Reducing the mass also reduces the acceleration resistance, the roll resistance and the climbing resistance. The sum of these factors constitutes the energy requirement of a car. Hence, the mass is directly linked to the fuel consumption and the CO_2 emissions of an automobile [3]. Reducing the vehicle weight by 100 kg, for example, will give a reduction of some 8.5 g/km in CO_2 emissions and a reduction of about 0.3 l/100 km in fuel consumption.

Despite this correlation, the average weight of automobiles has increased over past years (Figure 7.2) [3]. There has been an increase of about 100 kg per decade. The reason for this is mainly the rising safety requirements and higher comfort standards [4]. In addition, driving performance has also been improved, for example by increasing the torsional stiffness of the automotive bodies.

The gross vehicle weight can be divided into a number of major drivers: the powertrain, the aggregates, the interior and the body. Against this background, the automotive body, weighing about 250–350 kg for a mid-class automobile, is one field that is eligible for lightweight construction.

Advanced Composite Materials for Automotive Applications: Structural Integrity and Crashworthiness,
First Edition. Edited by Ahmed Elmarakbi.
© 2014 John Wiley & Sons, Ltd. Published 2014 by John Wiley & Sons, Ltd.

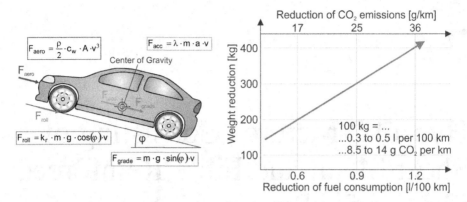

Figure 7.1 Effect of mass on the driving resistances (left) and correlation between vehicle weight, fuel consumption and CO_2 emissions (right).

In view of the excellent mechanical properties of continuous fibre reinforced plastics, this chapter will concentrate on these materials and their use in automotive structural parts. Reducing the weight of the body structure also plays an important role in reducing the weight of the vehicle as a whole [5].

7.2 Conventional Method for the Development of Composite Structures

Conventional methods for the product engineering of composite structures are based on a step by step procedure with work phases and milestones. One example of such a method can

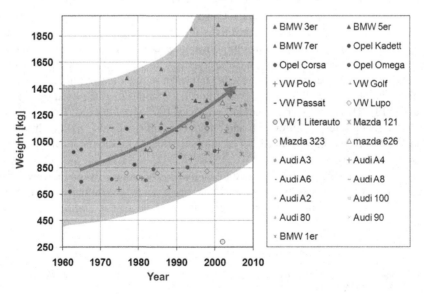

Figure 7.2 Increase in vehicle weight over the past years.

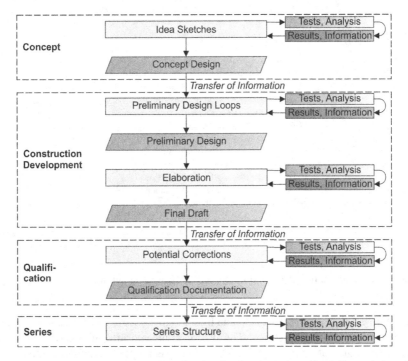

Figure 7.3 Current, incremental procedure for the development process of fibre reinforced plastic structures taking the example of VDI Standard No. 2014, according to [6–8].

be found in VDI Standard No. 2014 "Development of fibre reinforced plastics components" [6–8]. This method recommends dividing up the development process into four main steps: the concept phase, the construction development phase, the qualification phase and the series (Figure 7.3).

VDI Standard No. 2014 and other methods map the relevant steps for product development. They constitute the current state of the art in terms of development procedures. Nevertheless, new approaches are required for the holistic design of lightweight structures. Modern concepts for automotive lightweight construction take in not only the materials, the geometry and the testing, but also new manufacturing processes and new process strategies [3].

7.3   Approaches to Automotive Lightweight Construction

Lightweight design is a holistic approach towards minimising the weight of a structure while its overall performance remains the same or is even improved. As already mentioned, employing this construction principle makes it possible for technical systems to function with less energy, reduced emissions and the optimised use of material. When it comes to automotive applications, lightweight activities can be subdivided into a number of different aspects: material, design and function. For structural applications, material-based lightweight construction is the most important approach. This means improving the material properties, replacing a material by one that is a better fit, or combining different materials to achieve an optimum (Figure 7.4).

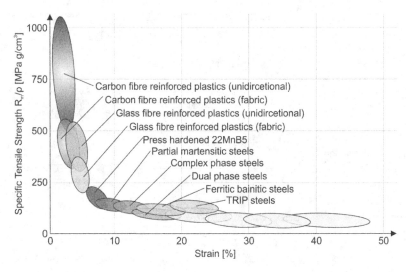

Figure 7.4 Mechanical properties of different materials used for automotive structural components.

Three main trends are currently evident in lightweight construction for automotive structural applications based on this approach. The use of high- and ultra-high-strength metal alloys with tensile strengths of up to about 1600 MPa is one example of an improvement in material properties [9–11]. By using high-strength metal alloys, it is possible to reduce the wall thickness of structures [12]. However, once a critical minimum thickness has been reached, stability problems can occur. The potential of high-strength metals for lightweight construction is thus limited [1]. Other developments in this field are tailor-welded blanks [2]. Nevertheless, automotive original equipment manufacturers (OEMs) mainly use steel for their body in white (BIW) designs because of its attractive properties [13].

The second approach involves replacing conventional construction materials with fibre reinforced plastics (FRPs), which permits considerable weight savings of up to some 50% [14,15] Materials like carbon fibre reinforced plastics (CFRPs) or glass fibre reinforced plastics (GFRPs) offer excellent specific mechanical properties and permit innovative solutions such as functional integration [16–18]. These parameters are required for load bearing or energy absorbing structures in the automotive, aerospace and energy sectors. In these and other applications, FRP is widely used to enhance the component performance of premium products [15, 19–21].

Research has shown that the use of FRP results in significant functional and economic benefits. These benefits range from increased strength and durability characteristics to lower fuel consumption and weight reduction. Research results show, for example the improvement in structural vehicle crashworthiness that is achieved by using FRP in specific automobile structures to serve as collapsible crash-energy absorbers [22]. The use of these structures is restricted to high-priced vehicles on account of the long cycle times required in the complex manufacturing process, the high material costs and other challenges [23,24]. The competitiveness of these structures thus depends on bringing down the costs by optimising the technologies involved, so as to permit simple and robust manufacturing processes or shorter cycle and lead times [25].

The third approach is the combination of different materials [26]. Here, the materials are used in a way in which their specific advantages can be exploited and their specific disadvantages concealed. Two approaches can be distinguished in this context. First, multi-material systems describe the combination of different materials at component level [27]. Examples of these include the so-called "Erlanger beam" (in this structure a thin-walled aluminium structure is reinforced with a FRP component) and circular CFRP rib construction [26]. Second, hybrid structures are created through the permanent connection of at least two different materials, such as sheet metal and CFRP, thereby achieving a transition between different classes of material. Hybrid materials are characterised by the combination of two classes of material to form a single structural material. The connection is usually achieved through extensive adhesive bonding. Hybrid materials consist of a sheet metal basic layer, a locally applied FRP reinforcement layer and an optional sheet metal covering layer. Advanced hybrid structures are characterised by a graded product design. Here, for example a sheet metal basic layer is only reinforced with a CFRP patch in the highly loaded areas [1, 28, 29]. The layered structure makes it possible to tailor components to their expected loading. Press hardened steels can be used as sheet metals, amongst other materials. The FRP reinforcements make it possible to reduce the wall thickness of the steel parts. Hybrid components can easily be integrated into existing vehicle production processes, because their metallic surface permits the use of conventional joining technologies like spot welding or clinching. Integration into existing body structures is also possible on account of the basic metal structure [30]. Corresponding combinations of CFRP and metals are increasingly being used in the automotive and aerospace industries and also in general mechanical engineering [28].

Structural components implemented in a multi-material or hybrid design are thus an attractive alternative [31]. As mentioned above, this construction technique makes it possible to exploit the advantages of different materials. For example, FRP attains the highest specific tensile strength of all the current automotive construction materials, while the elongation of metal-based materials is much higher, depending on the alloy. In this context, hybrid structures comprising a combination of high strength steel and a local CFRP reinforcement allow manufacturers to produce safety-relevant vehicle components, such as B-pillars, at a lower cost than components made of pure CFRP [31]. Employing this construction technique, an aluminium section reinforced with adhesively bonded CFRP was shown to be 33% lighter than a straightforward aluminium optimised structure [32].

For automotive applications, it is not just the mechanical properties that have to be taken into account. For high-volume manufacturing, in particular, other aspects such as integration into existing processes, production tolerances, quality and costs must also be evaluated. The higher the production volume, the more important the costs will be. The total costs of a construction can be seen as a function of the material, the manufacturing process and research and development costs (Figure 7.5). An optimal lightweight construction is attained at the minimum of this curve. Current conventional sheet metal structures are located on the right side of this field. The weight is adequate and the total costs are often minimal. FRP structures are characterised by a lower weight, but the total costs increase. Last but not least, hybrid structures can be classified between these two options. They permit a lower weight than conventional constructions, for just a slight increase in the total costs. Summarising, tailored hybrid structures not only hold a high potential for lightweight design, they also permit the optimised use of expensive FRP materials, which ultimately leads to cost-optimised lightweight constructions.

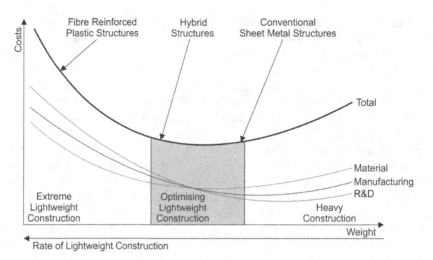

Figure 7.5   Cost related to different designs, weight and rate of lightweight construction, according to [33]. Reproduced from Ref. [33]. Copyright 2007 Springer.

7.4   Requirements for Automotive Structures

Automobiles have to meet various requirements deriving from such different sources as customer needs (e.g. comfort or driving performance), statutory provisions (e.g. with regard to environmental protection), safety aspects or requirements due to the manufacturing process itself (Figure 7.6) [34]. Global requirements such as driving performance or safety aspects, in particular, lead to specific requirements for a given structure or component within the automobile – ranging from the claimed mechanical properties for ensuring a high stiffness and crashworthiness, to quality or resistance against ageing and corrosion for ensuring a long

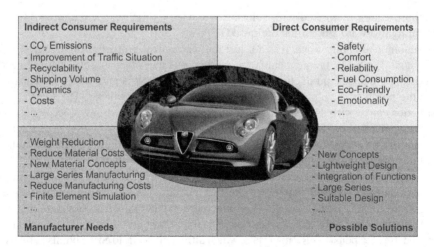

Figure 7.6   Examples of consumer requirements for an automobile.

vehicle lifetime [10]. The following section focuses on the mechanical properties and especially on the crashworthiness of automotive structural components since, in many cases, it is the mechanical properties that dictate the specific design of the components.

7.4.1 Mechanical Requirements

The key mechanical properties of a given material include the Young's modulus, the yield and tensile strengths and also the elongation at break. In order to ensure a high crashworthiness with a minimum weight, the strength of the material should be as high as possible. This is because one and the same deformation force can be achieved with a low yield strength and a high material thickness, or with a high yield strength and a low material thickness. A limiting factor might be the elongation at break (or better the forming limits), which usually decreases with higher strengths, thus resulting in the potential untimely failure of the component.

However, a high formability is required in order to manufacture complex component structures. Such structures are often necessary to ensure maximum stiffness and to fully utilise the available design space. This forming capability, however, often decreases as the strength of the material rises. Depending on the complexity of the component, therefore, there is actually a limit on the feasible strength level. Another important reason limiting the strength of the materials used results from the demand for optimal crash management. For example, a varying strength allows the deformation behaviour of the vehicle structures and the maximum passenger acceleration to be effectively controlled. It thus becomes possible to permit energy absorption solely in those areas where sufficient space is available for deformation.

In line with the considerations above, the components of an automotive body structure can be categorised according to their mechanical strengths (Figure 7.7). For example, soft materials are used for outer skin components, where complex geometries have to be achieved. Higher-strength materials can be found in areas intended for energy absorption during a crash. Finally, ultra-high strength materials are utilised for components subject to the highest requirements on structural integrity inside the passenger cabin.

A variety of different steel grades, aluminium alloys and FRPs are used within a vehicle, corresponding to the different strength levels. In the case of steel, the tensile strengths cover a range from approximately 300 MPa (deep drawing qualities) to 1600 MPa (press hardened steel). Currently, even steel grades with a tensile strength of around 1900 MPa are being developed. Aluminium alloys range from about 200 MPa (6000 series) to 700 MPa (7000 series) and, finally, FRP can have strengths ranging from approximately 80 MPa (sheet moulding compounds) to around 2000 MPa (unidirectional carbon fibre reinforced plastics).

7.4.2 Load Adapted Design

Structural automotive components are typically characterised by various loads generating a particular stress state within the components. In most cases, the resulting stress distributions are not constant over the structure. On the one hand, there are highly stressed areas (e.g. impact areas during a crash) and, on the other hand, areas with low loading – as a result of manufacturability, for instance (Figure 7.8). It can be stated that a load-adapted design is beneficial for these components, because the geometry, the material and the material utilisation will be effectively adjusted to the varying loads. The best-fitting material is positioned

Use of steels for body in white in a current middle class automobile*

Range of yield strength $R_{p0.2}$	Share	Tendency	Usage
☐ < 150 MPa	15 %	↓ -	Outer skin components, complex geometries
▣ 150...250 MPa	25 %	↓ -	Stress loaded components with complex geometries
▦ 250...300 MPa	25 %	↓ -	Structural components with higher strength requirements
▮ 300...450 MPa	15 %	↓ -	Energy absorption on a high stress level
▢ 450...1600 MPa	20 %	↑ +	Supporting components (highest deformation resistance)

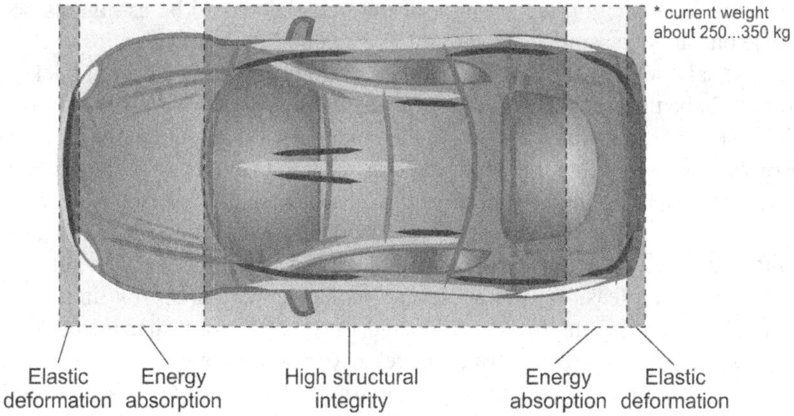

* current weight about 250...350 kg

Elastic deformation   Energy absorption   High structural integrity   Energy absorption   Elastic deformation

Figure 7.7   Use of different materials in a current automobile.

where it displays its specific advantages. For example, an expensive but cost effective CFRP reinforcement is inserted locally in highly loaded areas. The design principle of load adaption is conducive to holistic lightweight design because of the good material utilisation it permits.

Currently, there are four main trends for achieving load-adapted structures. First, many efforts have been undertaken to develop different types of tailored sheet metal products, including tailored welded blanks, tailored strips, patchworks and tailored tubes. Here, different types of sheet metal are typically combined in a single semi-finished product through welding operations [35]. Another possibility is the tailored tempering process. Press hardenable steel are

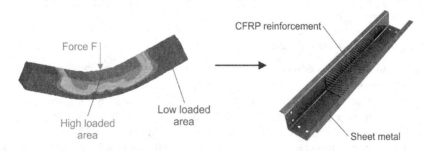

Force F

High loaded area

Low loaded area

CFRP reinforcement

Sheet metal

Figure 7.8   Load-adapted hybrid design for an automotive structural component.

only heated locally prior to the forming operation. Afterwards they are formed and quenched in the mould. This results in areas with dissimilar mechanical properties depending on the temperature. As a third alternative, mention must be made of the partial hardening process. In this process, the mechanical properties of a sheet metal structure are influenced by inductive heating and quenching, for example [4].

Fourth, sheet metal structures can be produced with a lower wall thickness and subsequently reinforced with an FRP structure. This combination leads to significant weight reductions in structural automotive components and is characterised by a variety of advantages compared with pure FRP solutions (see Section 7.3) [36, 37]. Hybrid structures can be seen as one possibility for future automotive lightweight concepts. During the development process for automotive structures, at least three basic steps have to be conducted: the designing and modelling phase, the manufacturing processes phase and the testing phase for the structures.

On the product side, the loads and the geometry, or the product design, are the main points. Here, for instance, it is the load levels and also the locations, the specifications and the characteristics that are of interest. The geometry and the product design influence the simulation, the design and construction and also the load situation and component behaviour. The combination of different materials in a single (hybrid) component results in a higher complexity for designing and modelling: the materials have to be adapted to each other, for instance, and the manufacturing processes have to be fine-tuned, or new concepts have to be found for material characterisation and simulation. A suitable methodology for the product engineering of hybrid lightweight structures can provide support in the development process.

Hybrid structures can be manufactured by several different technologies. The manufacturing process for FRP structures, in particular, is an important issue, because the manufacturing route strongly influences the properties of the component. For example, in the field of automotive construction, aspects like availability, flexibility, cycle times, quality, process stability or reproducibility and recycling are of interest, and these can frequently only be fulfilled by a high level of process automation [38]. These factors are, once again, influenced by process parameters, geometry aspects and the material behaviour. Autoclave processed CFRP parts have the best mechanical properties, but neither the cycle times nor the costs are suitable for series automotive production. Hence, new approaches for manufacturing these components need to be developed and investigated.

The final step deals with the testing of hybrid structures. Here, standard tests have to be passed, which show potential for improvement by comparison to conventional constructions.

7.4.3 Derivation of Reference Structures

It does not usually make sense to use series automotive structures for basic investigations. These structures have already been adapted to the specific load situation and design requirements of the specific vehicle. For a given class of components, however, such as B-pillars or rocker panels on middle-class vehicles, it is possible to derive a virtual common design and a typical set of loads, which then can be used as a reference structure for basic investigations.

For example, a typical but simple geometry of this type can be derived from B-pillars, as shown in Figure 7.9. The cross-section of the pillar corresponds to a hat profile with a sheet closing the open side. To produce the hat profile requires a tool and a press and hence, from the manufacturability viewpoint, it is preferable to further simplify the profile. This simplification

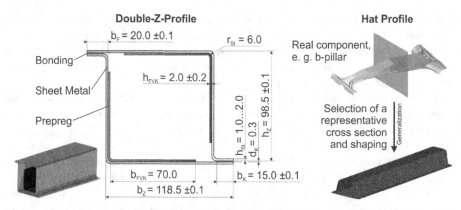

Length of the profiles e. g. 250...320 mm (Compression load) and 1000 mm (Bending load)

Figure 7.9 Typical reference structures for automotive structural components, according to Ref. [29].

finally results in the double-Z profiles shown on the right side of Figure 7.9. This geometry can be manufactured by a simple bending operation and is adequate for basic studies on the manufacturability of hybrid structures comprising a metallic component reinforced with FRP prepregs. This structure, for instance, permits investigations into the pre-curing of the resin system or the bonding properties of the different materials. Advanced studies of the influence of the profile shape (e.g. radii, angles) and the position and orientation of fibres after the forming operation can then be performed using the more complex hat profile (Figure 7.9).

In the studies presented here, both double-Z profiles and hat sections were used.

7.5 Simulation

In line with the evolutionary development of hybrid technology, the opportunities for CAE simulation have increased significantly. Further developments are making it possible to increasingly reconcile the demands in respect of simulation accuracy and calculation time.

When it comes to achieving an accurate simulation result, the choice of suitable software is of major importance. The decision should be based primarily on the existing load case. Although there is a whole range of commercially available software, there are few all-purpose programs. As a result, the engineer makes use of those areas of the individual programs that provide accurate results, such as crash loading, quasi-static load cases or optimisation.

To set up a FRP model, it is necessary to take the mechanical properties of the material into account. The key properties are the anisotropy and the inhomogeneity. Both are the result of fibre/matrix compounding. Calculations are mostly based on a rule of mixture [39] which homogenises the properties of a unidirectional (UD) material in each layer. The further calculation of a composite lamina is based first and foremost on the assumptions of the classical laminate theory [8]. In cases which include failure, the linear-elastic model has to be enhanced by a failure model. This mathematical model correlates the acting element strain or stress with the material properties. The common isotropic failure models cannot be applied to FRP material [40]. The complexity of failure models differs significantly, extending right through to the consideration of different failure modes in some cases. Failure in a FRP lamina can occur as fibre fracture, inter-fibre fracture (possibly sub-divided into different modes) or as delamination [41]. Delamination cannot be calculated straightforwardly if the FRP is considered on a layer by layer basis. However, the use of plate or shell elements is standard practice for in-plane stressed components. Modelling each layer with three-dimensional element formulations can potentially improve the accuracy of the simulation significantly if delamination is the predominant failure mode.

In practice, element orientation plays an important role. Element orientation is typically based on the node numbering. For an FRP laminate, an additional material coordinate system is introduced. In this way, the origin element orientation can be rotated. Where there are several FRP plies, the layered shell method is the most common method employed. A number of simplifying assumptions are made here that the engineer has to be aware of [42].

One of the most important factors in the simulation process is the material data. This data should be based on in-depth experimental work [42]. Figure 7.10 shows the testing that is required in order to determine the four constants for the description of the linear elasticity of an orthotropic material. If necessary, this program must be enhanced by individual tests to determine the required data for the failure model.

When simulating hybrid components, the connection between the two different materials has to be considered. In the case of a quasi-static simulation with a low deformation, in particular, this connection can be simplified through the ideal connection of the finite-element net while, in a crash simulation, it may be necessary to make allowance for its failure. In this case, the engineer is able to implement an adhesive model or to specify particular contact definitions.

Working with UD-FRP involves a whole range of possible material orientations and stacks. Optimisation processes thus reduce development time and costs. The optimisation process of a hybrid structure is shown in Figure 7.11.

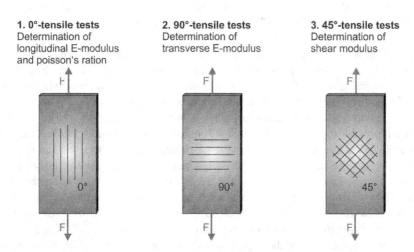

Figure 7.10   Required testing for the description of the linear elasticity of an orthotropic material.

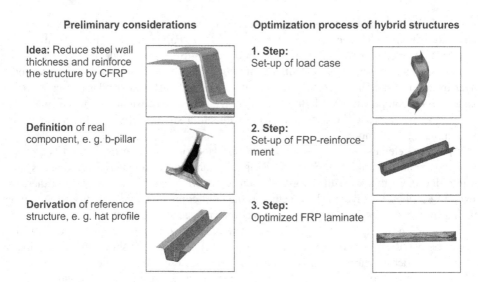

Figure 7.11   Optimisation process of a hybrid structure.

7.6   Manufacturing

7.6.1   Overview

In the field of FRP, many different production technologies are available, each of which has its own specific advantages and disadvantages. The manufacturing technology employed has a decisive influence on the characteristics of the FRP component [15]. In all the processes, only complete process control will give rise to good quality FRP components that are of interest for the automotive industry [43].

In the series production of automotive structures, use is currently made of several variants of the resin transfer moulding (RTM) process [44]. Cycle times of about 10–15 min can be achieved with this manufacturing technology, depending on the dimensions of the structure. For the RTM process, dry textile semi-finished products are employed. These are placed in a heated mould. As the next step, the matrix resin is injected and then cured in the closed mould [36, 44]. With the current RTM processes, it is not possible to manufacture hybrid structures in a single production step. A separate bonding stage is required.

Another upcoming manufacturing technique is prepreg press technology [1, 12]. With this process, it is possible to produce both pure FRP components and steel/FRP hybrid ones. Prepreg press technology is one approach to producing structural automotive parts in high volumes [45]. The process can be divided into several parts (Figure 7.12). First of all, prepregs (pre-impregnated, semi-finished fibre products) are produced continuously on special machines and shipped on coils [46]. Because epoxy resins are used in this case, special storage-related requirements have to be considered, such as cooling. Contrary to the case for thermoset resins, thermoplastic matrix systems offer advantages in this respect [47]. The layer structure is configured as a function of the expected loads in the component. The laminate is cut to match the subsequent geometry of the structure. Second, a robot handles the prepregs. After a

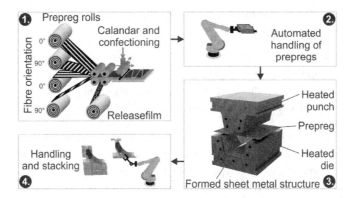

Figure 7.12   Prepreg press process, according to Benteler-SGL.

pre-formed steel structure has been placed in a heated steel tool, the prepreg is applied to the steel structure in an automated handling operation. The tailored prepreg is then pressed onto the sheet metal by a heated punch. Since the epoxy resin functions as an adhesive, the sheet metal and the CFRP are also joined together in this third step. After a pre-curing phase of about 90–120 s, depending on the thickness of the prepreg, the hybrid component is removed by a robot and stacked. The post-curing of the components is performed during a downstream cataphoretic painting process [46].

There are many different aspects of the prepreg press process to be studied (Figure 7.13). These aspects range from process parameters (temperature, time, force, velocity) and geometry issues (flange angle, symmetry, radius), via material behaviour (friction, resin flow, prepreg distortion, contour accuracy) and extend right through to mechanical properties (strength, stiffness, connection). Selected results are described in the sections that follow.

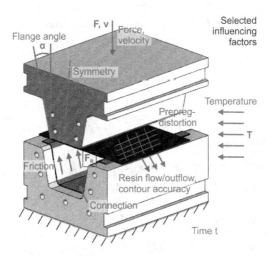

Figure 7.13   Aspects for the study of prepreg press technology.

7.6.2   Prepreg Press Technology: Basic Investigations
and Process Parameters

It is the curing time in the closed mould that is the chief factor that dictates the overall cycle time in prepreg press technology. A long cycle time reduces the suitability of prepreg pressing for high volume production. The right process parameters should permit low cycle times and good mechanical properties at one and the same time. The influence of different parameters is discussed below.

To investigate the process parameters, use was made of epoxy resin prepregs with nine layers. The layer structure of the bidirectional carbon fibre scrim was $(90/0/0/90/0/90/0/0/90)°$. The matrix resin employed was an SGL type E201. For the steel component, a DD11 and a 22MnB5 alloy (thickness $t = 2.0$ mm), respectively, were used. As the first step, test plates (sheet metal material: DD11) and hat sections (sheet metal material: 22MnB5) were manufactured by prepreg pressing. After a press process in which different parameters were used for the pre-curing time t_{prc}, the consolidation pressure p and the temperature T, the plates were post-cured in a furnace at 180 °C with $t_{poc} = 30$ min. As the next step, three-point bending samples were cut from the test plates and hat sections respectively and analysed on a testing machine. By using the matrix resin as an adhesive for bonding of the sheet metal and the CFRP an additional joining process can be avoided.

Figure 7.14 shows the results of three-point bending tests on hybrid samples cut out of test plates. In the force–displacement diagram, four curves are plotted for different process parameters. The consolidation temperature was not varied in this case. The parameters varied were the consolidation time ($t_{prc} = 60$ and 90 s) and the consolidation pressure ($p = 0.1$ and 0.6 MPa). A characteristic behaviour can be identified in the curves. After first rising, the curve flattens slightly. The main breakage occurs at the maximum force. Afterwards, single layers break step by step. In the end, the fibre component fails completely, and only the steel component carries the forces.

In addition to these characteristics, the influence of different process parameters is evident. In this case, only four different combinations of consolidation pressure and pre-curing time were depicted. A high consolidation pressure results in a large resin and fibre outflow. This determines a varying thickness of the cured prepreg and has an influence on the allocation of

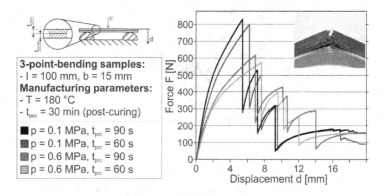

Figure 7.14   Force–displacement diagram for hybrid three-point bending samples.

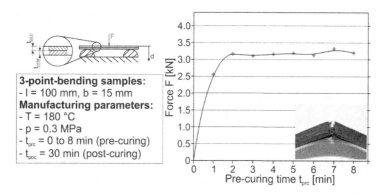

Figure 7.15 Force versus pre-curing time diagram for hybrid three-point bending samples.

fibres in the composite. This is especially important for bidirectional scrim, since aspects such as the fibre flow during the press process could prove to be problematical. The curing time has an influence on the mechanical properties, too. A pre-curing time of 60 s leads to slightly lower mechanical performances.

It is the pre-curing time that is the most important process parameter for achieving short cycle times. Several hat sections were manufactured by prepreg pressing with a consolidation pressure of $p = 0.3$ MPa and a temperature of $T = 180$ °C. The pre-curing time t_{prc} ranged from 0 to 8 min. The hat sections were post-cured for $t_{poc} = 30$ min at $T = 180$ °C. Afterwards, three-point bending samples were cut out of these hat sections. The results of the tests are illustrated in a force versus pre-curing time diagram (Figure 7.15). The maximum forces for the bending samples attain a steady level for about $t_{prc} = 60$ s at least. A higher pre-curing time does not lead to higher maximum forces. The maximum forces vary between 3000 and 3300 N. The failure of the CFRP is analogous to the samples described above. Thus, a minimum pre-curing time of 90–120 s is realistic for a prepreg thickness of 2 mm and the use of standard epoxy resins.

7.6.3 Prepreg Press Technology: Bonding of Composite Material and Sheet Metal

The joining technology for CFRP and sheet metal is a decisive factor for the strength of hybrid materials. In order to characterise the joining properties, single lap-shear specimens were investigated to DIN EN 1465:2009 [48]. For the investigation, four different joining technologies were used: (a) the epoxy resin matrix as the adhesive, (b) the adhesive Dow Betamate 1620, (c) blind rivets and (d) thread cutting screws (Figure 7.16).

The samples consisted of sheet metal and CFRP. For the metal component, a S235JR with a thickness of $t = 2$ mm was used. The prepregs were standard semi-finished products from SGL. The epoxy resin from SGL (type E201) has carbon fibres embedded in a symmetric nine-layer bidirectional scrim $(0/90/0/90/0/90/0/90/0)°$. The prepregs were manufactured by prepreg press technology. To optimally reconcile the economic aims and the joint strength, the consolidation time and temperature in the prepreg press process were varied in line with the

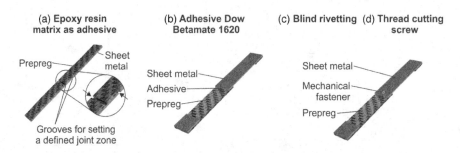

Figure 7.16   Sample geometry for different joining technologies.

reaction–velocity temperature rule (Arrhenius equation). The temperature was varied between 120 and 200 °C, while the highest strength was reached at a temperature of 180 °C with a curing time of 210 s. For bonding with epoxy resin as the adhesive, the prepreg was pressed directly onto the metal surface. Here, the influence of different types of surface treatment was investigated. The CFRP for the samples joined by Dow Betamate 1620, blind riveting (Gesipa PolyGrip 6.4 × 15 mm) and thread cutting screws (DIN 7513 M4 × 10 mm) was first pressed and cured. The CFRP was subsequently affixed to the sheet metal.

A comparison of the maximum force and the tensile energy absorption of the different joining technologies is shown in Figure 7.17. The epoxy resin matrix attains slightly lower values than the Dow Betamate for both parameters. The advantage of using the epoxy resin as an adhesive is that the complex and cost intensive joining process can be omitted. The values for the mechanical fasteners are significantly worse, because of the high stress concentration at the joining element. This leads to failure of the CFRP [49].

The analysis of the failure behaviour of the different bonding technologies revealed certain characteristics (Figure 7.18). While, for the pressed samples, failure occurred between the boundary layer and the second fibre layer of the CFRP, the adhesive-bonded samples failed due to delamination of the first fibre layer. The joint area itself remained undamaged. The mechanical fasteners were predominantly pulled out of the CFRP component.

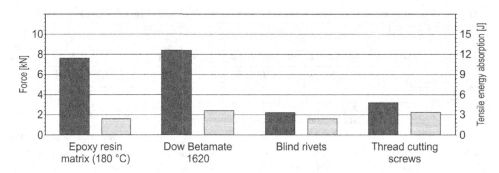

Figure 7.17   Comparison of maximum force and tensile energy absorption of different bonding technologies for hybrid structures.

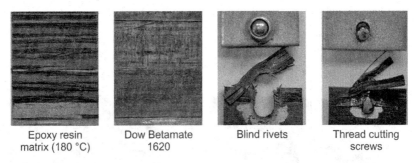

| Epoxy resin matrix (180 °C) | Dow Betamate 1620 | Blind rivets | Thread cutting screws |

Figure 7.18   Characteristic failure behaviour of different bonding technologies.

The fibre orientation of the boundary layer had the greatest influence on the properties of the joint (Figure 7.19). Samples with fibres vertical to the direction of loading (90°) had the highest strength. Samples with fibres parallel to the loading direction in the contacted layer failed at only 75% of the maximum load of the samples with fibres vertical to the direction of loading.

The inhomogeneous structure of CFRP is reflected in the properties of the joint. The special properties of these materials need to be considered in the design of hybrid structures in order to achieve the intended properties in the compound.

7.7   Testing

In the automotive industry, finite element analysis has become increasingly important over the past few decades. However, the testing of reference and original structures still constitutes an important step in the development of new concepts and products. This is the case for new

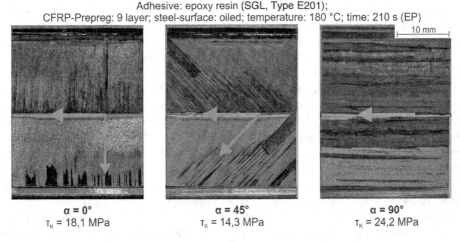

Adhesive: epoxy resin (SGL, Type E201);
CFRP-Prepreg: 9 layer; steel-surface: oiled; temperature: 180 °C; time: 210 s (EP)

| $\alpha = 0°$ | $\alpha = 45°$ | $\alpha = 90°$ |
| $T_K = 18{,}1$ MPa | $T_K = 14{,}3$ MPa | $T_K = 24{,}2$ MPa |

Figure 7.19   Influence of fibre orientation in the contacted layer on the bond strength.

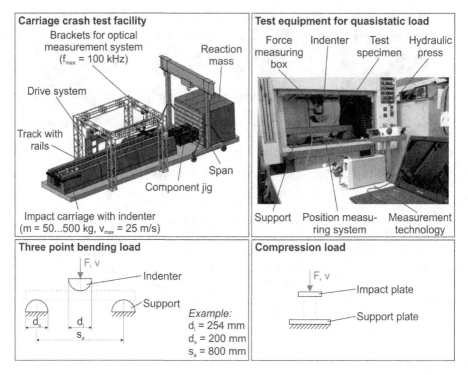

Figure 7.20   Examples of test equipment.

materials, in particular, where the mechanical properties are not clear or where no material models are available for simulation. Furthermore, the results of the tests are used to verify the simulation of component behaviour.

As mentioned above, automotive structures can have different requirement specifications. Applications for hybrid structures manufactured using prepreg press technology include crash-relevant automotive structural components. Here, the quasi-static mechanical properties have to be taken into account too with regard to the torsional stiffness of the automotive body.

There are several test facilities that can be used to analyse the behaviour of reference structures under crash and quasi-static loads (Figure 7.20). One example is a carriage crash test facility. The impact carriage is accelerated by a hydraulic drive system and guided on special rails. Different component jigs can be attached to a span for tests with pressure or bending loads, for example. The test facility used for the tests mentioned below is characterised by an impact carriage mass of between 50 and 500 kg. A maximum speed of 25 m/s can be achieved. An optical measurement system with a frequency of up to 100 kHz gauges the deformation and strain of the component. The force, the displacement and the acceleration can also be measured by default. Crash tests can also be performed on drop weight machines [50].

Quasi-static tests can be performed on a hydraulic press, for example. The test configuration is inserted and fixed on the spans. On the upper side, the indenter and a force measuring box are applied. On the lower side, the support and the position measuring system are installed. The measurement technology is based on standard components.

For both load characteristics, different test arrangements allow a range of different load cases. Typical load cases for automotive structural components are bending and compression loads. A simple setup for three-point bending consists of just two supports and one indenter. More complex buildups also map the stiffness of the remaining body. For this task, spring damper systems can be used, for example, while the specimen is fixed on the supports. Compression loads can be achieved with a support and an impact plate. Depending on the length, the stiffness and the geometry of the specimens, it may be necessary to use a guide, such as a bolt.

The results of the tests are usually force displacements or force time diagrams. To analyse, evaluate and compare the results of a number of crash tests, use can be made of different characteristic key figures [22, 51]. These parameters are the average specific force (Equation 7.1), the maximum energy absorption (Equation 7.2), the total specific energy absorption (Equation 7.3) and the specific energy absorption (Equation 7.4) for the tested profiles. The specific energy absorption can only be used for compression loads.

$$Fa = \frac{1}{\Delta s} \int_0^{\Delta s} F(s)ds \tag{7.1}$$

$$W_{\max} = \int_0^{s_{\max}} F(s)ds = F_m \cdot \Delta s_{\max} \tag{7.2}$$

$$E_{tot} = \frac{W_{\max}}{m} \tag{7.3}$$

$$E_{spec} = \frac{W(s)}{\Delta m(s)} \text{ and } \Delta m(s) = \frac{\Delta s}{l_0} \tag{7.4}$$

7.7.1 Quasi-Static Tests

Quasi-static simulations and tests on structural components are often easier to perform than crash loads. With knowledge of the quasi-static behaviour of components, it will be possible to investigate aspects such as stiffness, which is an important factor for certain components.

For the quasi-static tests, use was made of hat sections in press hardened 22MnB5 sheet metal, inter alia. The wall thickness was varied between 1.4 and 2.0 mm. The length of the hat sections was 1000 mm. Adapted and optimised CFRP patches were inserted into the sheet metal structure as reinforcement. The samples were manufactured with a pre-curing time of $t_{prc} = 300$ s, and the post-curing time was $t_{poc} = 30$ min. The temperature employed for both curing processes was $T = 180\,°C$. The consolidation pressure exceeded $p = 0.3$ MPa. In a final step, the covering plate was bonded in the flange area of the hat section using a Betamate 1620 adhesive.

Because of the optimisation target, which was to achieve a similar stiffness in the different profiles, the test specimens displayed nearly identical elastic behaviour (Figure 7.21). The highest forces of up to 75 kN were achieved with a combination of sheet metal and a 4.0 mm thick layer structure. The force displacement curves of sheet metal and hybrid components show significantly dissimilar characteristics. While the sheet metal structure displays high energy absorption over the full displacement, due to the isotropic and ductile material

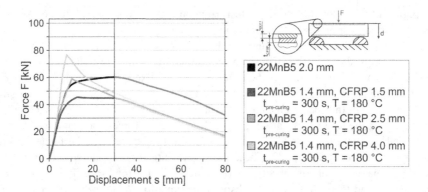

Figure 7.21 Results for quasi-static tests on hat profiles under bending load.

behaviour, the CFRP reinforcement breaks at a critical combination of force and displacement. The failure cut through the CFRP structure, so there is virtually no reinforcement effect left. The rest of the curves are similar to a purely steel solution with the same wall thickness as the sheet metal basic layer of the hybrid. Nevertheless, the weight saving potential is about 15–20% in this case, when it is taken into account that a defined deformation is specified for automotive structures.

7.7.2   Crash Tests

A key requirement for structural automotive components relates to crash behaviour. Passenger safety is conditional upon suitably developed components. Having an extensive knowledge about material behaviour is necessary for the development process, for example for crash simulation. To reduce the costs of experimental work and to improve on the significance, use was made of reference structures. The following section sets out the results of crash tests on different double-Z profiles.

The crash tests with a pressure load were performed using a 280 mm long double-Z profile. The cross section had a width of $w = 100$ mm and a height of $h = 100$ mm without the bonding flanges. The sheet metal was a DD11 with a wall thickness of $t = 1.0$, 1.5 and 2.0 mm. To manufacture the samples, sheet metal in a length of 320 mm was bent as the first step. Second, a 300 mm long and 140 mm wide CFRP reinforcement was pressed, centred, into the sheet metal. The epoxy resin from SGL (type E201) has carbon fibres embedded in a nine-layer bidirectional scrim (90/0/0/90/0/90/0/0/90°). The samples were manufactured with a pre-curing time of $t_{prc} = 120$ s. The post-curing time was $t_{poc} = 30$ min. $T = 180$ °C was employed as the temperature for both curing processes. The consolidation pressure exceeded $p = 0.3$ MPa. Third, the half-shells were cut to a length of 280 mm. This step was meant to achieve a complete hybrid material over the whole length of the profile. In the final step, the steel components were bonded in the flange area using a Betamate 1620 adhesive. The thickness of the bonding was 0.3 mm. The tests were performed with several structures: (1) steel in 1.0 mm, (2) steel in 1.5 mm, (3) steel in 2.0 mm, (4) a hybrid consisting of 1.0 mm steel and 2.0 mm CFRP, (5) a hybrid consisting of 1.5 mm steel and 2.0 mm CFRP and (6) a hybrid consisting of 2.0 mm steel and 2.0 mm CFRP.

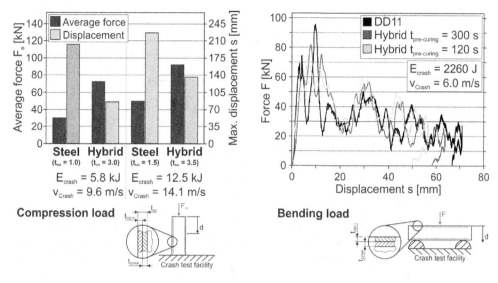

Figure 7.22 Results for different crash tests on double-Z profiles.

The three-point bending crash tests were performed with 1000 mm long double-Z profiles. A 300 mm long CFRP reinforcement was inserted, centred, into the sheet metal structure. The layer structure, the bonding technique and the process parameters were similar to those for the pressure load tests.

The curves in a force–displacement diagram are characterised by a peak force at the beginning (Figure 7.22). This point is called the stability peak. After this peak, the curves fall off and adopt an oscillating behaviour due to the elastic material properties [18, 52]. The average force decreases slightly up to the end of the deformation. The average forces were calculated on the basis of this information.

The influence of the reinforcement with lower steel wall thicknesses is more pronounced on account of the increasing amount of CFRP [37]. In this context, the same reinforcement patch was used for each of the samples. Characteristic key figures were calculated in order to compare the results of the different crash tests (Table 7.1).

Table 7.1 Average specific forces, specific energy absorption and total specific energy absorption for the tested double-Z profiles under compression load ($E_{Crash\ 1.0} = 5800$ J and $E_{Crash\ 1.5/2.0} = 12\,500$ J).

Number	t_{steel} (mm)	t_{CFRP} (mm)	F_a (kN)	E_{spec} (J/g)	E_{tot} (J/g)	Increase of E_{tot} compared to same t_{steel}
1	1.0	—	29.30	6.09	7.15	—
2	1.5	—	50.27	8.15	8.16	—
3	2.0	—	93.03	4.86	11.35	—
4	1.0	2.0	67.07	4.82	12.85	×1.80
5	1.5	2.0	91.08	7.77	12.51	×1.50
6	2.0	2.0	131.41	4.96	14.10	×1.25

Table 7.2 Average specific forces for the double-Z profiles tested under bending load ($E_{Crash\ 1.0} = 1270$ J and $E_{Crash\ 1.5/2.0} = 2260$ J).

Number	t_{steel} (mm)	t_{CFRP} (mm)	$F_{a,\ specific}$ (N/g)	Increase compared to same t_{steel}
1	1.0	—	0.45	—
2	1.5	—	1.29	—
3	2.0	—	3.16	—
4	1.0	2.0	1.69	×3.75
5	1.5	2.0	2.91	×2.25
6	2.0	2.0	3.91	×1.25

The results for the average specific force (7.1), the total specific energy absorption (7.3) and the specific energy absorption (7.4) are illustrated here. Pure steel structure 3 ($t_{steel} = 2.0$ mm, m = 2050 g) displays the same average force of about 90 kN as hybrid structure 5 ($t_{tot} = 3.5$ mm, m = 1820 g), for example. Although the amount of CFRP in the hybrid sample is only 280 g, an increase of 45 % in the average specific force in comparison with pure steel structure 2 ($t_{steel} = 1.5$ mm, m = 1540 g) can be observed. In this case, the hybrid structure offers a weight saving potential of about 12.5%. Another example is the specific energy absorption of the samples. Here, pure steel solution 3 ($t_{steel} = 2.0$ mm, m = 2050 g) can be compared with hybrid sample 4 ($t_{tot} = 3.0$ mm, m = 1305 g), displaying a lightweight potential of about 36%.

In addition, the average specific forces were calculated for the double-Z profiles tested under bending load (Table 7.2). Here, it was possible to determine an increase in the average specific force of the hybrid structures compared to the pure steel solutions by a factor of up to 3.75.

It can be stated that hybrid structures offer a high lightweight potential of up to 25–35% compared with pure steel structures. The weight saving potential is determined by several parameters, such as the load case or the parameters taken to optimise the structure (including the stiffness or the strength, respectively F_a or E_{spec}). The selection of the sheet metal alloy also plays an important role.

7.8 New Methodology for the Product Engineering of Hybrid Lightweight Structures

From the previous sections, the high complexity of developing, testing and manufacturing FRP and hybrid structures has become clear. The wide range of influencing factors and their interdependencies, which results in an as yet unknown complexity, calls for a new methodology for the product engineering of hybrid lightweight structures. This methodology constitutes a current object of research at the University of Paderborn. The methodology divides up into four parts: a procedure model, methods and IT tools, a specification technique and a knowledge base (Figure 7.23).

The procedure model defines the main steps that have to be checked during the product engineering process. For example, determining the mechanical characteristics of materials, the finite element simulation and the optimisation or experimental work are different aspects of the procedure model. This model coordinates the work of the developers involved from the different domains over the entire development process for hybrid lightweight structures.

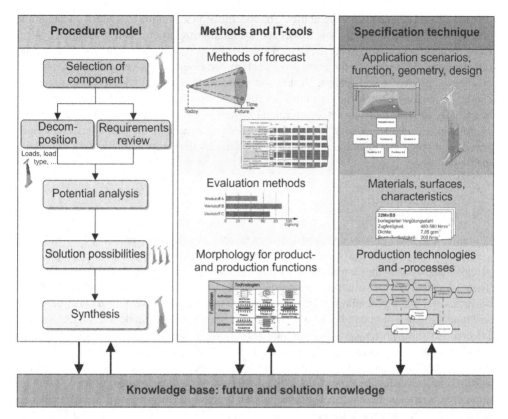

Figure 7.23   Approach for a new methodology for the product engineering of hybrid lightweight structures.

Further, it governs the use of the methods and IT tools. The procedure model describes the activities to be carried out in a readily comprehensible form. It is divided up into several main steps. In the first step, the component under consideration is analysed with regard to automotive requirements such as corrosion or surface quality, loads or mechanical characteristics. This includes the breakdown into the function owner and the specification of the requirements. For hybrid systems, in particular, it is necessary to specify the nature of loads and their direction (isotropic or anisotropic) in order to achieve an optimal lightweight structure. In the next step, partial solutions for the functional owners are identified. Hence, the material, the geometry and also the manufacturing and assembly technologies need to be taken into account. The partial solutions are then combined to give an overall solution. The large number of possible combinations ensures that this step results in a number of alternative solutions. Finally, the alternative solutions have to be evaluated to establish the best fitting solution.

The methods and IT tools support the product engineering process. Forecasting methods, such as the scenario technique, trend analysis or the Delphi method, make it possible to anticipate the product life cycle and identify the chances and risks. Evaluation methods permit the early assessment of manufacturing and life cycle costs, the robustness of the product and production system and also recyclability. The specification technique is used for an integrative description of the product and the associated manufacturing process and production system. Consideration is paid here to the interdependencies of the product and production systems, and information is provided for the evaluation, such as correlations between process parameters, mechanical properties and quality. The knowledge base can be used for the synthesis of process chains. It contains knowledge on future developments, such as a material price forecast, the availability of materials, market, environment scenarios and regulatory requirements, as well as domain-specific knowledge on lightweight design including, for example design rules, possible material combinations, manufacturing technologies and finite element simulation methods.

7.9 Conclusion

Automotive lightweight design is one approach that can be adopted to reducing CO_2 emissions and fuel consumption. It can be achieved on the basis of several different design principles. One promising solution is the combination of different materials in a single material or component. These so-called hybrid structures, consisting for example of sheet metal and fibre-reinforced plastics, offer major potential for lightweight design in the automotive industry. The designing and modelling is more complex than for conventional components made of a single, isotropic material. By using prepreg press technology, FRP prepregs can be formed into steel structures, which permit short process chains and cycle times. Bonding is achieved by using the epoxy resin as an adhesive. It has also been shown that the structures offer good crash performance in different load cases. In view of the high complexity of the development process for hybrid structures, a new methodological approach to the product engineering of these structures has been presented. This methodology comprises a procedure model, methods and IT tools, a specification technique and a knowledge base. It supports the entire development process for hybrid structures.

References

[1] Lauter, C., Dau, J., Tröster, T. and Homberg, W. (2011) Manufacturing processes for automotive structures in multi-material design consisting of sheet metal and CFRP prepregs. 16th International Conference on Composite Structures, Porto, Portugal.

[2] Shi, Y., Zhu, P., Shen, L. and Lin, Z. (2007) Lightweight design of automotive front side rails with TWB concept. *Thin-Walled Structures*, **45**, 8–14.

[3] Merklein, M. and Geiger, M. (2002) New materials and production technologies for innovative lightweight constructions. *Journal of Materials Processing Technology*, **125–126**, 532–536.

[4] Thomas, D., Block, H. and Tröster, T. (2011) Production of load-adapted lightweight designs by partial hardening. Third International Conference on Steel in Cars and Trucks, Salzburg, Austria.

[5] Zhang, Y., Zhu, P. and Chen, G. (2007) Lightweight design of automotive front side rail based on robust optimisation. *Thin-Walled Structures*, **45**, 670–676.

[6] The Association of German Engineers (VDI) (1989) VDI 2014: Entwicklung von Bauteilen aus Faser-Kunststoff-Verbund. Grundlagen. *Part 1*, Berlin, Beuth.

[7] The Association of German Engineers (VDI) (1993) VDI 2014: Entwicklung von Bauteilen aus Faser-Kunststoff-Verbund. Konzeption und Gestaltung. *Part 2*, Berlin, Beuth.

[8] The Association of German Engineers (VDI) (1993) VDI 2014: Entwicklung von Bauteilen aus Faser-Kunststoff-Verbund. Berechnungen. *Part 3*, Berlin, Beuth.

[9] Kolleck, R. and Veit, R. (2011) Current and future trends in the field of hot stamping of car body parts. 3rd International Conference on Steel in Cars and Trucks, Salzburg, Austria.

[10] Zhang, Y., Lai, X., Zhu, P. and Wang, W. (2006) Lightweight design of automobile component using high strength steel based on dent resistance. *Materials and Design*, **27**, 64–68.

[11] So, H., Faßmann, D., Hoffmann, H. *et al.* (2012) An investigation of the blanking process of the quenchable boron alloyed steel 22MnB5 before and after hot stamping process. *Journal of Materials Processing Technology*, **212**, 437–449.

[12] Asnafi, N., Langstedt, G., Andersson, C.H. *et al.* (2000) A new lightweight metal-composite-metal panel for applications in the automotive and other industries. *Thin-Walled Structures*, **36**, 289–310.

[13] Mayyas, A.T., Qattawi, A., Mayyas, A.R. and Omar, M. (2012) Quantifiable measures of sustainability: a case study of materials selection for eco-lightweight auto-bodies. *Journal of Cleaner Production*, **45**, 670–676.

[14] Jacob, A. (2010) BMW counts on carbon fibre for its megacity vehicle. *Reinforced Plastics*, **September/ October**, 29–30.

[15] Adam, H. (1997) Carbon fibre in automotive applications. *Materials and Design*, **18**, 349–355.

[16] Luo, R.K., Green, E.R. and Morrison, C.J. (1999) Impact damage analysis of composite plates. *International Journal of Impact Engineering*, **22**, 435–447.

[17] Sultan, M.T., Worden, K., Pierce, S.G. *et al.* (2011) On impact damage detection and quantification for CFRP laminates using structural response data only. *Mechanical Systems and Signal Processing*, **25**, 3135–3152.

[18] Abrate, S. (ed.) (2011) *Impact Engineering of Composite Structures*, Springer, ISBN 978-3-7091-0522-1.

[19] Iqbal, K., Khan, S.U., Munir, A. and Kim, J.K. (2009) Impact damage resistance of CFRP with nanoclay-filled epoxy matrix. *Composites Science and Technology*, **69**, 1949–1957.

[20] Phonthammachai, N., Li, X., Wong, S. *et al.* (2011) Fabrication of CFRP from high performance clay/epox nanocomposite: preparation conditions, thermal-mechanical properties and interlaminar fracture characteristics. *Composites: Part A*, **42**, 881–887.

[21] Vassilopoulos, A.P. and Keller, T. (2011) *Fatigue of Fiber-reinforced Composites*, Springer, ISBN 978-1-84996-180-6.

[22] Mamalis, A.G., Manolakos, D.E., Ioannidis, M.B. and Papapostolou, D.P. (2004) Crashworthy characteristics of axially statically compressed thin-walled square CFRP composite tubes: experimental. *Composite Structures*, **63**, 347–360.

[23] Yanagimoto, J. and Ikeuchi, K. (2012) Sheet forming process of carbon fiber reinforced plastics for lightweight parts. *CIRP Annals – Manufacturing Technology*, **61**, 247–250.

[24] Zhang, J., Chaisombat, K., He, S. and Wang, C.H. (2012) Hybrid composite laminates reinforced with glass/carbon woven fabrics for lightweight load bearing structures. *Materials and Design*, **36**, 75–80.

[25] Fink, A., Camanho, P.P., Andrés, J.M. *et al.* (2010) Hybrid CFRP/titanium bolted joints: performance assessment and application to a spacecraft payload adaptor. *Composites Science and Technology*, **70**, 305–317.

[26] Kopp, G., Beeh, E., Schöll, R. *et al.* (2012) New lightweight structures for advanced automotive vehicles - safe and modular. *Procedia – Social and Behavioral Sciences*, **48**, 350–362.

[27] Cui, X., Wang, S. and Hu, S.J. (2008) A method for optimal design of automotive body assembly using multi-material construction. *Materials and Design*, **29**, 381–387.

[28] Möller, F., Thomy, C., Vollertsen, F. *et al.* (2010) Novel method for joining CFRP to aluminium. *Physics Procedia*, **5**, 37–45.

[29] Grasser, S. (2009) Composite-Metall-Hybridstrukturen unter Berücksichtigung großserientauglicher Fertigungsprozesse. Symposium Material Innovativ, Ansbach, Germany.

[30] Lauter, C., Frantz, M. and Tröster, T. (2011) Großserientaugliche herstellung von hybridwerkstoffen durch prepregpressen. *Lightweight Design*, **4**, 48–54.

[31] Homberg, W., Dau, J. and Damerow, U. (2011) Combined Forming of Steel Blanks with Local CFRP Reinforcement. 10th International Conference on Technology of Plasticity, Aachen, Germany.

[32] Broughton, J.G., Beevers, A. and Hutchinson, A.R. (1997) Carbon-fibre-reinforced plastic (CFRP) strengthening of aluminium extrusions. *International Journal of Adhesion and Adhesives*, **17**, 269–278.

[33] Klein, B. (2007) *Leichtbau-Konstruktion. Berechnungsgrundlagen und Gestaltung*, Vieweg, ISBN 978-3-8348-1604-7.

[34] Weber, J. (2009) *Automotive Development Processes. Processes for Successful Customer Oriented Vehicle Development*. Springer, ISBN 978-3-642-01252-5.

[35] Kleiner, M., Chatti, S. and Klaus, A. (2006) Metal forming techniques for lightweight construction. *Journal of Materials Processing Technology*, **177**, 2–7.

[36] Maciej, M. (2011) Faserverbundkunststoffe: Von der Kleinserienfertigung von Sichtbauteilen zur Großserienproduktion von Strukturteilen. InnoMateria – Interdisziplinäre Kongressmesse für innovative Werkstoffe, Cologne, Germany.

[37] Lauter, C., Niewel, J., Siewers, B. *et al.* (2012) Crash worthiness of hybrid pillar structures consisting of sheet metal and local CFRP reinforcements. 15th International Conference on Experimental Mechanics, Porto, Portugal.

[38] Feldmann, K., Müller, B. and Haselmann, T. (1999) Automated assembly of lightweight automotive components. *Annals of the CIRP*, **48**, 9–12.

[39] Soboyejo, W. (2003) *Mechanical Properties of Engineered Materials*, Marcel Dekker, ISBN 978-0-8247-8900-8.

[40] Mallick, P. (2007) *Fiber-Reinforced Composites, Materials, Manufacturing and Design*, Taylor and Francis Group, ISBN 978-0-8493-4205-9.

[41] Knops, M. (2008) *Analysis of failure in Fiber Polymer Laminates. The Theory of Alfred Puck*, Springer, ISBN 978-3-540-75764-1.

[42] Schürmann, H. (2007) *Konstruieren mit Faser-Kunststoff-Verbunden*, Springer, ISBN 978-3-540-72189-5.

[43] Brouwer, W.D., van Herpt, E.C. and Labordus, M. (2003) Vacuum injection moulding for large structural applications. *Composites: Part A*, **34**, 551–558.

[44] Simacek, P., Advani, S.G. and Iobst, S.A. (2008) Modeling flow in compression resin transfer molding for manufacturing of complex lightweight high-performance automotive parts. *Journal of Composite Materials*, **42**(23), 299–310.

[45] Reuter, C., Frantz, M., Lauter, C. *et al.* (2012) Simulation and testing of hybrid structures consisting of press-hardened steel and CFRP. 1st International Conference on Mechanics of Nano, Micro and Macro Composite Structures, Turin, Italy.

[46] Dau, J., Lauter, C., Damerow, U. *et al.* (2011) Multi-material systems for tailored automotive structural components. 18th International Conference on Composite Materials, Jeju Island, South Korea.

[47] Kanellopoulos, V.N., Yates, B., Wostenholm, G.H. *et al.* (1989) Fabrication characteristics of a carbon fibre-reinforced thermoplastic resin. *Journal of Materials Science Letters*, **24**, 4000–4003.

[48] Deutsches Institut für Normung e.V. (2008) DIN EN 1465: Klebstoffe. Bestimmung der Zugscherfestigkeit von Überlappungsklebungen. Berlin, Beuth.

[49] Lauter, C., Sarrazin, M. and Tröster, T. (2012) Joining technologies for hybrid materials consisting of sheet metal and carbon fibre reinforced plastics. First International Conference of the International Journal of Structural Integrity, Porto, Portugal.

[50] Obradovic, J., Boria, S. and Belingardi, G. (2012) Lightweight design and crash analysis of composite frontal impact energy absorbing structures. *Composite Structures*, **94**, 423–430.

[51] Kröger, M. (2002) *Methodische Auslegung und Erprobung von Fahrzeug-Crashstrukturen*. Dissertation. University of Hanover.

[52] Morello, L., Rosti Rossin, L., Pia, G. and Tonoli, A. (2011) *The Automotive Body. Volume II: System Design*, Springer, ISBN 978-94-007-0515-9.

8

Nonlinear Strain Rate Dependent Micro-Mechanical Composite Material Model for Crashworthiness Simulation

Ala Tabiei

School of Advance Structures, College of Engineering and Applied Science, University of Cincinnati, Cincinnati, OH 45221-0071, USA

8.1 Introduction

In this chapter a micro-mechanical model is developed for laminated composite materials and implemented using the explicit finite element method. The micro-mechanical model implemented in the explicit finite element code can be used for simulating the behaviour of composite structures under various loads such as impact and crash. The stress–strain relation for the micro-model is derived for a shell element. A micro-failure criterion (MFC) is presented for each material constituent and failure mode. The model is implemented using the finite element code LS-DYNA. Examples of the crushing of an E-glass/epoxy composite square tube are presented to validate the model.

8.2 Micro-Mechanical Formulation

8.2.1 Equations for Micro-Mechanical Model

8.2.1.1 Constitutive Equations for Composite Materials

Tabiei and Chen [1] developed a micro-mechanical model for laminated composite materials. The constitutive equations used to model the behaviour of composite materials may be classified as micro-mechanical or macro-mechanical in nature. The primary advantage of

Advanced Composite Materials for Automotive Applications: Structural Integrity and Crashworthiness,
First Edition. Edited by Ahmed Elmarakbi.

macro-mechanical equations is that they require less computational work. These equations are adequate to describe a response due to mechanical loading. But it still needs to be developed more for complex load histories, strain rate sensitivity, creep and so on. Micro-mechanical constitutive equations require more computational work, and each constituent property has to be determined by experimental tests. However, their advantage is that cases, like complex load history, strain rate sensitivity and creep, can be studied with less difficulty. In macro-mechanical analyses of laminate, a lamina is modelled as an anisotropic homogeneous material and therefore it cannot give information about the state of stress and strain in the constituents. Therefore, in this study, the micro-model is presented in more detail.

In implementation of the finite element method for analysis of the mechanical behaviour of thick-wall laminated composite structures, shell elements are usually used. The required constitutive laws are generally written as

$$\begin{Bmatrix} \varepsilon_{11} \\ \varepsilon_{22} \\ \gamma_{12} \end{Bmatrix} = \begin{bmatrix} S_{11} & S_{12} & 0 \\ S_{12} & S_{22} & 0 \\ 0 & 0 & S_{66} \end{bmatrix} \begin{Bmatrix} \sigma_{11} \\ \sigma_{22} \\ \sigma_{12} \end{Bmatrix} \tag{8.1}$$

and

$$\begin{Bmatrix} \gamma_{23} \\ \gamma_{13} \end{Bmatrix} = \begin{bmatrix} S_{44} & 0 \\ 0 & S_{55} \end{bmatrix} \begin{Bmatrix} \sigma_{23} \\ \sigma_{13} \end{Bmatrix} \tag{8.2}$$

In these equations the global normal stress in thickness direction is ignored. During the incremental-iterative solution scheme, as used in finite element analysis of nonlinear problems, a change of the nodal displacements takes place. The displacement increment causes an increment of strain $\Delta\{\bar{\varepsilon}_i\}$ at a material point. The material model is required to calculate the incremental stress $\Delta\{\bar{\sigma}_i\}$. Stiffnesses, strains and stresses are tracked at the material points within each element. This information is provided by the micro-mechanics composite material model, which interfaces with the nonlinear explicit finite element code. The heterogeneous nature of the composite material is hidden from the main analysis code. The actual laminated composite is replaced by an equivalent homogenous material whose properties are determined by requiring that the actual material and the equivalent material behave in the same way when subjected to certain stresses or strains. The interface consists of a transfer of stresses and strains between the material model and the analysis code. The main analysis code only sees this equivalent homogenous anisotropic material.

8.2.1.2 Micro-Mechanics Constitutive Model

In this study, the model proposed by Pecknold et al. [2] is employed. The micro-model considers the response of a unidirectional lamina, starting from the fibre and matrix constitutive descriptions. We can consider a lamina transversely isotropic with the 2–3 planes (normal to the fibres) as the planes of isotropy. In the present study, several assumptions were made in the micro-mechanics model. First, the fibre material is homogeneous and linearly elastic. Second, the matrix material is homogeneous and linearly elastic. Third, the fibre positioning in the matrix material is such that the resulting lamina material is a macro-mechanically

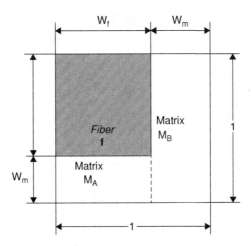

Figure 8.1 Representative unit cell of laminated composite.

homogeneous material with linearly elastic behaviour. Finally, there is a complete and strong bond at the interface of the constituent materials.

The proposed micro-model is based on the assumption that the internal micro-structure of the lamina consists of square fibres. Figure 8.1 shows a representative unit cell, which is an assumed geometry of the idealised composite. The unit cell is divided into three sub-cells. One sub-cell is fibre, denoted as f, and there are two matrix sub-cells, denoted as M_A and M_B respectively. The three sub-cells are grouped into two parts: *material part A* consists of the fibre sub-cell f and the series or parallel connected matrix sub-cell M_A; *material part B* consists of the remaining matrix M_B. The dimensions of the unit cell are 1×1 unit square. The dimensions of the fibre and matrix sub-cells are denoted by W_f and W_m respectively, as shown in Figure 8.1 and defined as

$$W_f = \sqrt{V_f}$$
$$W_m = 1 - W_f$$

(8.3)

where V_f is the fibre volume fraction. The effective stresses and strains in the lamina are determined from the sub-cell values in two phases: first, fibre f and matrix M_A construct part A; then, part A and part B construct the unidirectional lamina. The homogenised stresses and strains in part A are given by

$$
\begin{Bmatrix} \varepsilon_{11} \\ \sigma_{22} \\ \sigma_{12} \end{Bmatrix}_A = \begin{Bmatrix} \varepsilon_{11} \\ \sigma_{22} \\ \sigma_{12} \end{Bmatrix}_f = \begin{Bmatrix} \varepsilon_{11} \\ \sigma_{22} \\ \sigma_{12} \end{Bmatrix}_{M_A}
$$

$$
\begin{Bmatrix} \sigma_{11} \\ \varepsilon_{22} \\ \varepsilon_{12} \end{Bmatrix}_A = W_f \begin{Bmatrix} \sigma_{11} \\ \varepsilon_{22} \\ \varepsilon_{12} \end{Bmatrix}_f + W_m \begin{Bmatrix} \sigma_{11} \\ \varepsilon_{22} \\ \varepsilon_{12} \end{Bmatrix}_{M_A}
$$

(8.4)

Therefore the homogenised stresses and strains in the unit cell are given by

$$
\left\{\begin{matrix} \varepsilon_{11} \\ \varepsilon_{22} \\ \varepsilon_{12} \end{matrix}\right\}_C = \left\{\begin{matrix} \varepsilon_{11} \\ \varepsilon_{22} \\ \varepsilon_{12} \end{matrix}\right\}_B = \left\{\begin{matrix} \varepsilon_{11} \\ \varepsilon_{22} \\ \varepsilon_{12} \end{matrix}\right\}_A
$$

$$
\left\{\begin{matrix} \sigma_{11} \\ \sigma_{22} \\ \sigma_{12} \end{matrix}\right\}_C = W_f \left\{\begin{matrix} \sigma_{11} \\ \sigma_{22} \\ \sigma_{12} \end{matrix}\right\}_A + W_m \left\{\begin{matrix} \sigma_{11} \\ \sigma_{22} \\ \sigma_{12} \end{matrix}\right\}_B
$$

(8.5)

8.2.1.3 Constitutive Matrices and Stress Update for the Micro-Model

Part B is a homogeneous isotropic matrix (resin) material. The compliance matrix is given by

$$
[S]_B = \left\{\begin{matrix} \dfrac{1}{E_m} & \dfrac{-\upsilon_m}{E_m} & 0 \\[2mm] \dfrac{-\upsilon_m}{E_m} & \dfrac{1}{E_m} & 0 \\[2mm] 0 & 0 & \dfrac{1}{G_m} \end{matrix}\right\}
$$

(8.6)

and therefore the stiffness matrix is

$$
[Q]_B = [S]_B^{-1} = \left\{\begin{matrix} \dfrac{E_m}{1-\upsilon_m^2} & \dfrac{\upsilon_m E_m}{1-\upsilon_m^2} & 0 \\[2mm] \dfrac{\upsilon_m E_m}{1-\upsilon_m^2} & \dfrac{E_m}{1-\upsilon_m^2} & 0 \\[2mm] 0 & 0 & G_m \end{matrix}\right\}
$$

(8.7)

The total stress is obtained from

$$
\{\sigma\}_B = [Q]_B \cdot \{\varepsilon\}_B
$$

(8.8)

Part A consists of an isotropic matrix sub-cell M_A and an orthotropic fibre sub-cell f. The compliance matrix for M_A is given by

$$
[S]_{M_A} = \left\{\begin{matrix} \dfrac{1}{E_m} & \dfrac{-\upsilon_m}{E_m} & 0 \\[2mm] \dfrac{-\upsilon_m}{E_m} & \dfrac{1}{E_m} & 0 \\[2mm] 0 & 0 & \dfrac{1}{G_m} \end{matrix}\right\}
$$

(8.9)

$$[Q]_{M_A} = [S]_{M_A}^{-1} = \left\{ \begin{array}{ccc} \dfrac{E_m}{1 - \upsilon_m^2} & \dfrac{\upsilon_m E_m}{1 - \upsilon_m^2} & 0 \\[12pt] \dfrac{\upsilon_m E_m}{1 - \upsilon_m^2} & \dfrac{E_m}{1 - \upsilon_m^2} & 0 \\[12pt] 0 & 0 & G_m \end{array} \right\} \qquad (8.10)$$

The total stress is obtained from

$$\{\sigma\}_{M_A} = [Q]_{M_A} \cdot \{\varepsilon\}_{M_A} \qquad (8.11)$$

Sub-cell f is the fibre portion of the unit cell and the compliance matrix is given by

$$[S]_f = \left\{ \begin{array}{ccc} \dfrac{1}{E_1} & \dfrac{-\upsilon_f}{E_1} & 0 \\[12pt] \dfrac{-\upsilon_f}{E_1} & \dfrac{1}{E_2} & 0 \\[12pt] 0 & 0 & \dfrac{1}{G_f} \end{array} \right\} \qquad (8.12)$$

$$[Q]_f = [S]_f^{-1} = \left\{ \begin{array}{ccc} \dfrac{1}{E_2 F_f} & \dfrac{\upsilon_f}{E_1 F_f} & 0 \\[12pt] \dfrac{\upsilon_f}{E_1 F_f} & \dfrac{\upsilon_f}{E_1 F_f} & 0 \\[12pt] 0 & 0 & G_f \end{array} \right\} \qquad (8.13)$$

where $F_f = \frac{1}{E_1 E_2} - \frac{\upsilon_f^2}{E_1^2}$

The total stress is obtained from

$$\{\sigma\}_f = [Q]_f \cdot \{\varepsilon\}_f \qquad (8.14)$$

So, finally for part A, we have

$$\begin{Bmatrix} \sigma_{22} \\ \sigma_{12} \end{Bmatrix}_A = \begin{Bmatrix} \sigma_{22} \\ \sigma_{12} \end{Bmatrix}_f \qquad (8.15)$$
$$\{\sigma_{11}\}_A = W_f \{\sigma_{11}\}_f + W_m \{\sigma_{11}\}_{M_A}$$

The total stresses for the unit cell are finally obtained from

$$\begin{Bmatrix} \sigma_{11} \\ \sigma_{22} \\ \sigma_{12} \end{Bmatrix}_C = W_f \begin{Bmatrix} \sigma_{11} \\ \sigma_{22} \\ \sigma_{12} \end{Bmatrix}_A + W_m \begin{Bmatrix} \sigma_{11} \\ \sigma_{22} \\ \sigma_{12} \end{Bmatrix}_B \qquad (8.16)$$

The transverse shear stresses are given by

$$\begin{Bmatrix} \sigma_{13} \\ \sigma_{23} \end{Bmatrix}_B = \begin{bmatrix} G_m & 0 \\ 0 & G_m \end{bmatrix} \begin{Bmatrix} \varepsilon_{13} \\ \varepsilon_{23} \end{Bmatrix}_B \tag{8.17}$$

$$\begin{Bmatrix} \sigma_{13} \\ \sigma_{23} \end{Bmatrix}_A = \begin{bmatrix} \dfrac{1}{\dfrac{W_f}{G_{12}^f} + \dfrac{W_m}{G_m}} & 0 \\ 0 & \dfrac{1}{\dfrac{W_f}{G_{12}^f} + \dfrac{W_m}{G_m}} \end{bmatrix} \begin{Bmatrix} \varepsilon_{13} \\ \varepsilon_{23} \end{Bmatrix}_A \tag{8.18}$$

and

$$\begin{Bmatrix} \sigma_{13} \\ \sigma_{23} \end{Bmatrix}_c = W_f \begin{Bmatrix} \sigma_{13} \\ \sigma_{23} \end{Bmatrix}_A + W_m \begin{Bmatrix} \sigma_{13} \\ \sigma_{23} \end{Bmatrix} \tag{8.19}$$

8.2.2   Failure Analysis

The failure analysis was carried out by comparing the stresses or strain with those permissible values of stresses or strain. Simple failure models for composite materials can be used to reliably predict the onset of failure, but not to predict the post-failure deformation, which are both important in the impact analysis. Complex composite damage models must be developed which account for a combination of several typical failure mechanisms: transverse matrix cracking, transverse matrix crashing, fibre breakage, fibre buckling and matrix crashing in the fibre direction. These failure modes can be accounted for by employing micro-mechanical failure criteria (MFC) to model the progressive damage in the laminae.

The advantage of the micro-mechanical model over a macro-mechanical model is that the stresses can be associated and related to each constituent (fibre and matrix). Therefore, failure can be identified in each of these constituents and the proper degradation in strength can be modelled. In this investigation the failure issue is addressed using a criterion for each failure mode so that the basic important phenomenon can be captured and correlated with test observations.

When failure occurs, materials will lose their load carrying capability in certain modes of deformation. To adequately model this behaviour, the compliance matrix and stresses are modified according to the failure modes. To simulate failure in the explicit finite element method, failure must be modelled by a gradual loss of stiffness in order to provide a stable solution instead of an instantaneous loss. A transition to the failed condition is assumed to occur during a finite time. The failure criteria used in the model are:

1. Fibre fracture in tension occurs when the axial stress in fibre σ_{11}^f is greater than the fibre tensile strength.
2. Matrix shearing occurs when $|\tau_{12}^{m_A}|$ or $|\tau_{12}^{m_B}|$ is greater than the matrix shear strength.
3. Matrix cracking in transverse tension occurs when $|\sigma_{22}^{m_A}|$ or $|\sigma_{22}^{m_B}|$ is greater than the tensile strength of the matrix material.

4. Matrix cracking in transverse compression occurs when $\left|\sigma_{22}^{m_A}\right|$ or $\left|\sigma_{22}^{m_B}\right|$ is greater than the compression strength of the matrix material.
5. Matrix cracking in axial tension occurs when $\left|\sigma_{11}^{m_A}\right|$ or $\left|\sigma_{11}^{m_B}\right|$ is greater than the tensile strength of the matrix material.
6. Matrix cracking in the transverse direction occurs when the max of $\left(\left|\sigma_{13}^{m_A}\right|, \left|\sigma_{13}^{m_B}\right|\right)$ or $\left(\left|\sigma_{23}^{m_A}\right|, \left|\sigma_{23}^{m_B}\right|\right)$ is greater the shear strength of the matrix material.
7. Fibre micro-buckling resulting in kink-banding occurs when $\left|\sigma_{11}\right|_{La\ min\ ate} \geq V_f G_{12T}$, where V_f is the fibre volume fraction and is the tangent shear modulus.

8.2.3   Finite Element Implementation

8.2.3.1   Equations in Incremental Form

In order to simulate the impact response of a composite structure, finite element methods are required. With that goal in mind, the matrix constitutive equations and the composite micro-mechanics model described above have been implemented into LS-DYNA, a commercially available transient dynamic finite element code. LS-DYNA uses explicit central difference integration methods to integrate the equations.

LS-DYNA currently has several material options for the analysis of composite materials. Several composite material models are provided in LS-DYNA, such as model 54, model 58 and model 59. They can be used to perform the damage simulation of composite materials. However, these models cannot be used to simulate all kinds of composite materials because of the varieties of composite materials. And their accuracy is not so good in some simulations.

Thus, LS-DYNA provides a user defined material option that allows users to implement material models that are not included with the finite element package. To use this option, a FORTRAN subroutine is created which can be linked to the finite element code. In the subroutine, strain increments for the current time step are passed in from the main program. Stresses and other history variables (such as state variables) from previous time steps are also available to the subroutine. From this information, the user defined material subroutine must compute the stress increments, total stresses and values of the history variables at the end of the current time increment.

To implement the micro-mechanical model discussed earlier into an LS-DYNA user defined material subroutine, the equations must be converted into an incremental format. For the shell element, the transverse shear stiffness should be considered, or the values of σ_{13} and σ_{23} should be given. The incremental stresses in subcell B could be

$$d\sigma_{11b} = \frac{E_m}{1 - v_m^2}(d\varepsilon_{11} + d\varepsilon_{22} \cdot v_m) \tag{8.20}$$

$$d\sigma_{22b} = \frac{E_m}{1 - v_m^2}(d\varepsilon_{22} + d\varepsilon_{11} \cdot v_m) \tag{8.21}$$

$$d\sigma_{12b} = d\varepsilon_{12} \cdot g_m \tag{8.22}$$

$$d\sigma_{13b} = d\varepsilon_{13} \cdot g_m \tag{8.23}$$

$$d\sigma_{12b} = d\varepsilon_{23} \cdot g_m \tag{8.24}$$

and the stresses in subcell A could be

$$d\sigma_{22a} = E_{12m} \cdot [d\varepsilon_{22} + (W_f \cdot v_f + W_m \cdot v_m) \cdot d\varepsilon_{11}] \tag{8.25}$$

where

$$E_{12m} = \frac{E_1 \cdot E_2 \cdot E_m}{[E_m \cdot W_f \cdot (E_1 - E_2 \cdot v_f^2) + E_1 \cdot E_2 \cdot W_m \cdot (1 - v_m^2)]} \tag{8.26}$$

$$d\sigma_{11a} = d\varepsilon_{11} \cdot E_m + d\sigma_{22a} \cdot v_m \tag{8.27}$$

$$d\sigma_{11f} = d\varepsilon_{11} \cdot E_1 + d\sigma_{22a} \cdot v_f \tag{8.28}$$

$$d\sigma_{12a} = \frac{d\varepsilon_{12}}{\dfrac{W_f}{g_f} + \dfrac{W_m}{g_m}} \tag{8.29}$$

$$d\sigma_{13a} = \frac{d\varepsilon_{13}}{\dfrac{W_f}{g_f} + \dfrac{W_m}{g_m}} \tag{8.30}$$

$$d\sigma_{23a} = \frac{d\varepsilon_{23}}{\dfrac{W_f}{g_f} + \dfrac{W_m}{g_m}} \tag{8.31}$$

Then the stresses in the whole cell could be

$$\sigma_{11} = W_f(W_m \cdot \sigma_{11ma} + W_f \cdot \sigma_{11f}) + W_m \cdot \sigma_{11b} \tag{8.32}$$

$$\sigma_{22} = W_f \cdot \sigma_{22a} + W_m \cdot \sigma_{22b} \tag{8.33}$$

$$\sigma_{12} = W_f \cdot \sigma_{12a} + W_m \cdot \sigma_{12b} \tag{8.34}$$

$$\sigma_{13} = W_f \cdot \sigma_{13a} + W_m \cdot \sigma_{13b} \tag{8.35}$$

$$\sigma_{23} = W_f \cdot \sigma_{23a} + W_m \cdot \sigma_{23b} \tag{8.36}$$

For the shell element formulation, the through the thickness strain increment is

$$d\varepsilon_{33} = d\varepsilon_{33} - \frac{(d\sigma_{11b} + d\sigma_{22b} + d\sigma_{11ma} \cdot W_m) \cdot v_m}{E_m} - \frac{d\sigma_{11f} \cdot v_f}{E_1}$$
$$- d\sigma_{22a} \left(\frac{v_m}{E_m} + \frac{v_{f23}}{E_2} \right) \tag{8.37}$$

The failure analysis was carried out by comparing the stresses or strain with those permissible values of stresses or strain. At every increment of the applied stress, the stresses are monitored for failure. Property degradation models are being utilised. For structural level modelling, the ability to degrade only certain material properties based on the local ply failure mechanisms will be desirable to provide an improved simulation of the stress transfer mechanisms. Further, in implementing the model into a finite element code, a gradual degradation of material properties might improve the stability of the finite element analysis. In this study, stresses will be decreased to a certain value in 100 steps if failure happens.

8.2.3.2 Localisation and Modification

During the finite element simulation, after the elements reach their maximal stress values, strain softening starts and due to minor numerical differences some elements are further deformed, whereas other elements are unloaded. Such effects known as localisation have to be expected with the strain softening damage model and introduce considerable mesh size dependencies into any finite element computation. These mesh dependencies have to be considered in each analysis and make any simulation of unknown problems questionable unless some adjustment with known results can be made.

To avoid the observed localisation effects in many cases a simple modification is introduced. The idea is to modify the damage evolution law such that the stress does not fall below a threshold value, the limit stress. A new parameter α with $0 \leq \alpha \leq 1$ is introduced to characterise this yield stress to have some relation to the strength value. For example, if the element stress passes the limit value X_t^f, the stress remains constant at $\alpha * X_t^f$, instead of decreasing to 0.

However, we have to note that, with a factor $\alpha < 1$, localisation is still present; damage growth is only in the first affected elements. Without softening ($\alpha = 1.0$) however, the strains grow in all elements in a similar fashion.

8.2.4 Verification Examples

Examples of the crushing of a composite square tube are presented to validate the micro-mechanics composite model described above. The materials used here are E-glass/epoxy laminated composites, one with [45°/−45°] fibre orientation and the other with [30°/−30°] fibre orientation. The composite tube is assumed to be fixed at the end and impacted by a rigid wall.

The geometry parameters of the experiment are:

Length of the tube = 254 mm

Cross-section = 50.8 × 50.8 mm

Thickness = 1.829 mm

Radius of the corner = 0.0 mm

Density = 1.8 gm/cm^3

Velocity of the rigid wall = 10 mm/ms

In this case, material model 58 of LS-DYNA is also used to simulate the crushing of the tube. Model 58 is one of the composite material models provided by the LS-DYNA library. Comparison is performed for the prediction results of the presented methodology,

Table 8.1 Material properties for unidirectional E-glass fibre.

	Longitudinal modulus (Gpa)	Trans modulus (Gpa)	In-plane shear modulus (Gpa)	Out of plane shear modulus (Gpa)	In-plane Poisson's ratio	Out of plane Poisson's ratio	Failure allowable (Gpa)
Symbol	E_L	E_T	G_{LT}	G_{TT}	v_{LT}	v_{TT}	X_{tf}
Value	41.40	3.381	5.244	5.244	0.0244	0.3	0.7866

Table 8.2 Material properties of epoxy resin.

	Young's modulus (Gpa)	Shear modulus (Gpa)	Poisson's ratio	Tensile strength (Gpa)	Compressive strength (Gpa)	Shear strength (Gpa)
Symbol	E_m	G_m	v_m	Y_{tm}	Y_{cm}	S_m
Value	3.45	1.3	0.35	0.1911	0.1911	0.05382

the experimental data and the simulation results using material model 58 in LS-DYNA. The material properties of unidirectional E-glass fibre are listed in Table 8.1; the material properties of epoxy resin are listed in Table 8.2; and Table 8.3 presents the other properties used in model 58.

Figure 8.2 shows the initial condition of the square composite tube, meshed with shell elements. In LS-DYNA, 16 types of shell elements are provided. In this case, the full integration shell element, which is type 16, is used to simulate the crushing of the tube. Figure 8.3 shows the shape of the tube 5 ms after impact, or the displacement of the rigid wall reaches 50 mm, in the case of [45°/−45°] fibre orientation.

Figure 8.4 shows the shape of the tube 10 ms after impact, or the displacement of the rigid wall reaches 100 mm. Figures 8.5 and 8.6 present the rigid wall force versus time during the simulation. Figure 8.7 shows the comparison of rigid wall force, for the results of the present

Table 8.3 Other properties used in LS-DYNA material model 58.

Symbol	SLIMT1	SLIMC1	SLIMT2	SLIMC2	SLIMS
Value	0.5	0.9	0.5	0.9	0.9
Symbol	E11C	E11T	E22C	E22T	GMS
Value	0.0190	0.0190	0.056	0.0565	0.05382

SLIMT1 is a factor to determine the minimum stress limit after stress maximum (fibre tension), SLIMC1 is a factor to determine the minimum stress limit after stress maximum (fibre compression), SLIMT2 is a factor to determine the minimum stress limit after stress maximum (matrix tension), SLIMC2 is a factor to determine the minimum stress limit after stress maximum (matrix compression), SLIMS is a factor to determine the minimum shear stress limit after stress maximum, E11C is strain at longitudinal compressive strength, E11T is strain at longitudinal tensile strength, E22C is strain at transverse compressive strength, E22T is strain at transverse tensile strength and GMS is strain at shear strength.

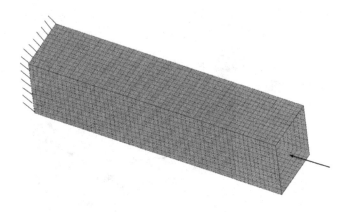

Figure 8.2   Composite tube [45/−45] in initial condition.

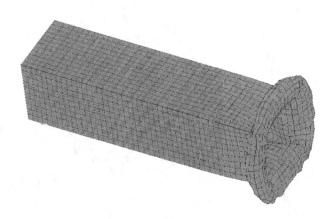

Figure 8.3   Composite tube [45/−45] under impact (after 5 ms).

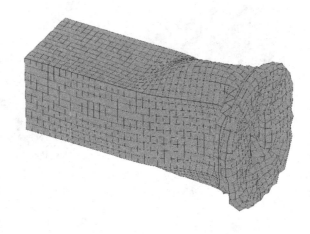

Figure 8.4   Composite tube [45/−45] under impact (after 10 ms).

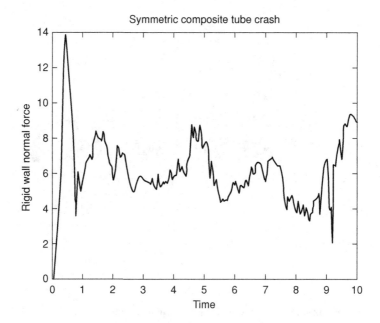

Figure 8.5   Rigid wall force versus time for E-glass/epoxy [30/−30] composite.

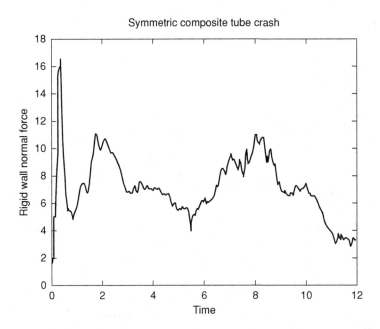

Figure 8.6   Rigid wall force versus time for E-glass/epoxy [45/−45] composite.

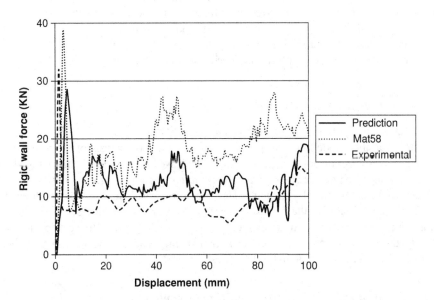

Figure 8.7   Comparison of prediction, experimental data and result of Mat58 for E-glass/epoxy [30/−30] composite.

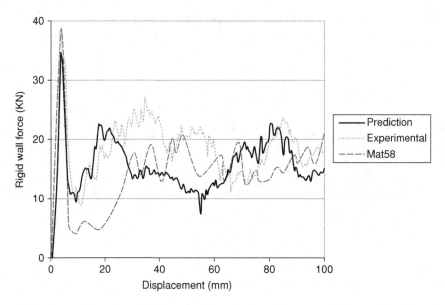

Figure 8.8 Comparison of prediction, experimental data and result of Mat58 for E-glass/epoxy [45/−45] composite.

model, material model 58 and experimental data, in the case of [30°/−30°] fibre orientation. Figure 8.8 shows the comparison of rigid wall force in the case of [45°/−45°] fibre orientation. It can be observed that the result of the present model fits the experimental data well. It is better than the result of material model 58 in this case. The results indicate that the failure criteria are able to predict ply failure for a variety of fibre orientations.

8.3 Strain Rate Dependent Effects

In this section, the polymer resin's strain rate effects are discussed. At first, an overview of the state variable modelling method which allows for the analysis of the rate-dependent deformation response of polymer resin is presented. Next, a set of one-dimensional constitutive equations based on the state variable method are presented, along with the procedures for determining the material constants. The equations are then extended to three dimensions. These equations and the composite micro-mechanics model described in an earlier section are implemented into finite element code. Then, a representative composite material AS4/polyetheretherketone (PEEK) is discussed in the implementation. A comparison between prediction results and experimental data is presented.

8.3.1 Strain Rate Effect Introduction and Review

8.3.1.1 Strain Rate Effect on Polymer Resin

Metals are known to have a rate dependent deformation response. So do polymers. Experimental studies have been performed with the goal of determining the effects of strain rate on the material properties and response of polymer matrix composites. A lot of results show that the material properties and response of polymer matrix composites do vary with strain rate, and the fibre material can affect the nature of the strain rate dependence. For graphite/epoxy and carbon/epoxy composites under tensile loading, the strain rate dependence appears to be primarily driven by the strain rate dependence of the polymer resin. These results indicate that simulating the rate dependence of the matrix correctly is important in modelling fibre reinforced polymer matrix composites.

Traditionally, in the analysis of a polymer resin's rate dependent deformation response, for very small strain, linear viscoelasticity has been used to simulate the material behaviour. In linear viscoelastic models, combinations of springs and dashpots may be used to capture the rate dependent behaviour. For cases where the strains are large enough that the response is no longer linear, nonlinear viscoelastic models have been developed. The rate dependence observed in polymer resin deformation has also been modelled empirically by scaling the yield stress as a function of strain rate.

The molecular approach is a more sophisticated technique in polymer constitutive modelling. In this method [3], the polymer deformation is assumed to be due to the motion of molecular chains over potential energy barriers. The molecular flow is due to applied stress, and the internal viscosity is assumed to decrease with increasing stress. The yield stress is defined as the point where the internal viscosity decreases to the point where the applied strain rate is equal to the plastic strain rate. Internal stresses can also be defined. These stresses represent the resistance to molecular flow that tends to drive the material back towards its original configuration. Another approach to resin deformation assumes that the deformation is due to the unwinding of molecular kinks [4]. Constitutive equations have been developed in both approaches. In these equations, the resin deformation is considered to be a function of parameters such as activation energy, activation volume, molecular radius, molecular angle of rotation and thermal constants. The deformation is assumed to be a function of state variables that represent the resistance to molecular flow caused by a variety of mechanisms. The state variable values evolve with stress, inelastic strain and inelastic strain rate.

As we described above, metals have strain rate effects. Equations have been developed for metals. With some modifications, these equations can be used for polymer resins. That is another approach to the constitutive modelling of resins [5]. Polymer resins behave differently from metals under conditions such as creep, relaxation and unloading, so the original method used for metals was modified to account for phenomena presented in polymer resins which are not encountered in metals.

From the work discussed above, several approaches have been applied to analyse the rate dependent response of polymer resins by simulating the physical deformation mechanisms. In some of them, the state variables were used to model the effect. In this study, simplified methods, or some modifications to previously existing methods, are used to perform the analysis presented here.

8.3.1.2 State Variable Modelling Overview

As described previously, the rate dependence of a polymer matrix is primarily a function of the rate dependence of the matrix constituent; and the stress–strain response of polymers is nonlinear. So, the constitutive equations which incorporate the deformation mechanisms of the material should present the nonlinear, rate dependent deformation response of the matrix material. For polymers, deformation is due to the motion of the molecular chains. At small deformation levels, there is also a resistance to the molecular flow before yield. In constitutive equations, a state variable approach can be utilised to model the mechanisms that cause material deformation. These variables can evolve as a function of external parameters such as stress, inelastic strain and the current value of the state variable. Then, the inelastic strain rate can be defined to be a function of the state variables and external variables, such as the current stresses of the matrix material.

In this study, several key assumptions were made in the constitutive model as follows:

1. Variables are defined which are meant to represent the average effects of the deformation mechanisms that are present, and the specific changes in the local details of the material microstructure are not considered.
2. There is no defined yield stress or onset of inelasticity. Alternatively, inelastic strain is assumed to be present at all values of stress. The inelastic strain is assumed to be very small compared to the elastic strain at low stress levels. When the inelastic strain increases to a higher value, the stress–strain curve will begin to present a nonlinear response.
3. A single, unified variable is utilised to represent all the inelastic strains. The effects of viscoelasticity, plasticity and creep are not separated in this type of approach, but are combined into one unified strain variable.
4. Temperature effects have been neglected in this study, and all of the results were obtained for room temperature, even though in high strain rate impact situations heating may be an important factor.
5. Small strain conditions have been assumed. As we know, in reality polymers, particularly in compression, can be subject to very large strains. And structures undergoing high strain rate impact are subject to large deformations and rotations. However, incorporating large deformation and rotation effects into the constitutive equations would add a level of complexity that is beyond the scope of this study. For the purposes of this modelling effort, therefore, all large deformation effects have been neglected. Future efforts will include modifying the constitutive equations to account for large deformation effects.
6. For the purposes of this study, stress wave effects will not be included, though stress wave effects due to dynamic loading may play a significant role in the material response when subjected to high rate loading.

In metals, the inelastic strain rate is often modelled as being proportional to the difference between the deviatoric stress and "back stress" tensors. The back stress is a resistance to slip resulting from the interaction of dislocations under a shear stress with a barrier. As dislocations pile up at a barrier, atomic forces will cause additional dislocations approaching the barrier to be repelled. This repelling force is referred to as the back stress. The back stress is in the direction opposite to the local shear stress in uniaxial loading. The net stress producing slip or inelastic strain is related to the difference between the shear stress and the back stress. An isotropic initial resistance to slip is also present in metals due to the presence of obstacles such as precipitates, grains and point defects. Similar concepts have been used in the deformation modelling of polymers. Creep strain and plastic strain can be defined as being proportional to the difference between the applied stress and an "internal stress". The internal stress is defined as evolving with increasing strain. For polymers, while the nonlinear deformation response is due to the nonlinear response of long chain molecules as opposed to the propagation of dislocations for metals, a unified inelastic strain variable can still be utilised to simulate the nonlinear behaviour. In addition, the "saturation stress" in metals and the "yield stress" in polymers (the point where the stress–strain curve becomes flat) are both defined as the stress level at which the inelastic strain rate equals the applied strain rate in constant strain rate tensile tests.

There are some limitations in this state variable modelling method. First of all, some phenomena presented in polymer resins which are not encountered in metals, such as creep, relaxation and high cycle fatigue, may not be simulated correctly in polymers with a metal-based constitutive model, even after modifications. Second, these equations will most likely only be valid for relatively ductile polymers, such as thermoplastics or toughened epoxies. To fully simulate the complete range of polymer deformation responses, a combined viscoelastic–viscoplastic model, which is very complex, would be required. However, it is not considered for this study.

8.3.2   One-Dimensional Equation and Material Constant Determination

8.3.2.1   One-Dimensional Constitutive Equation

Goldberg [6] modified the Ramaswamy and Stouffer [7] equations, which were originally developed to analyse the viscoplastic deformation of metals, to simulate the nonlinear deformation response of polymers. Here, a tensorial state variable is used to represent an "internal stress" which represents the resistance to inelastic deformation. The tensorial state variable is orientation dependent. The state variable is assumed to be equal to zero when the material is in its virgin state, and it evolves towards a maximum value at saturation. Here, the term "saturation" means yield in polymers.

A state variable based constitutive model was utilised to simulate the rate dependent, nonlinear deformation behaviour of the polymer matrix. In order to determine material constants and model correlation, one-dimensional uniaxial versions of the equations were considered first. An important point is that the flow equation, in its three-dimensional form, is based on deviatoric stresses and stress invariants, which are the primary drivers for the inelastic deformation of both polymers and metals. The total strain rate is the sum of the elastic and inelastic strain rates. Only the equations for the inelastic strain rate are presented. The total strain rate for uniaxial loading is defined as

$$\dot{\varepsilon} = \frac{\dot{\sigma}}{E} + \dot{\varepsilon}^I \tag{8.38}$$

where $\dot{\varepsilon}$ is the total strain rate, $\dot{\sigma}$ is the stress rate, $\dot{\varepsilon}^I$ is the inelastic strain rate and E is the elastic modulus of the material.

The inelastic strain rate is defined as being proportional to the exponential of the overstress, the difference between the applied stress and the internal stress.

In the uniaxial simplification of the Ramaswamy–Stouffer constitutive model, the inelastic strain rate is defined by

$$\dot{\varepsilon}^I = \frac{2}{\sqrt{3}} D_0 \, \exp\left[-\frac{1}{2} \left(\frac{Z_0}{|\sigma - \Omega|} \right)^{2n} \right] * \frac{\sigma - \Omega}{|\sigma - \Omega|} \tag{8.39}$$

where $\dot{\varepsilon}$ is the inelastic strain rate, σ is the stress and Ω is the state variable, or the internal stress which represents the resistance to molecular flow. Material constants include D_0, a scale factor which represents the maximum inelastic strain rate, n, a variable which controls the rate dependence of the deformation response and Z_0, which represents the isotropic, initial "hardness" of the material before any load is applied. The value of the state variable Ω is assumed to be zero when the material is in its virgin state. An important point to note is that Ω is a tensorial variable, not a scalar. The evolution of the state variable Ω is described by

$$\dot{\Omega} = q\Omega_m \dot{\varepsilon}^I - q\Omega|\dot{\varepsilon}^I| \tag{8.40}$$

where $\dot{\Omega}$ is the state variable rate, $\dot{\varepsilon}^I$ is the inelastic strain rate and Ω is the current value of the state variable. Material constants include q, which is the hardening rate and Ω_m, which is the maximum value of the "internal stress" at saturation. This equation is slightly different from the evolution law developed by Ramaswamy and Stouffer. For tensile loading, where the absolute value of the inelastic strain rate equals the inelastic strain rate, Equation 8.40 can be integrated to obtain

$$\Omega = \Omega_m - \Omega_m \exp(-q\varepsilon^I) \tag{8.41}$$

where ε^I is the inelastic strain and all other parameters are as defined for Equation 8.39.

8.3.2.2 Material Constant Determination

A summary of the discussion of the methods for finding the material constants for the Ramaswamy–Stouffer model will be described here. D_0 is currently assumed to be equal to a value 10^4 times the maximum applied total strain rate and is considered to be the limiting value of the inelastic strain rate. Future investigations may be conducted to investigate whether a relationship between D_0 and the shear wave speed can be determined. Such a relationship could allow the effects of stress waves to be more completely accounted for within the matrix constitutive equations.

To determine the values of n, Z_0 and Ω_m, the following method is utilised. First, the natural logarithm of both sides of Equation 8.39 is taken. The values of the inelastic strain rate, stress and state variable Ω at saturation are substituted into the resulting expression. The following equation is obtained

$$\ln\left[-2\ln\left(\frac{\sqrt{3}\dot{\varepsilon}_0}{2D_0}\right)\right] = 2n\ln(Z_0) - 2n\ln(\sigma_s - \Omega_m) \tag{8.42}$$

where σ_s equals the "saturation" stress, $\dot{\varepsilon}_0$ is the constant applied total strain rate and the remaining terms are as defined in Equations 8.39 and 8.40. To determine the value for q, Equation 8.41 is used. In a saturation situation, the value of the internal stress is assumed to approach the maximum value, resulting in the exponential term approaching zero. Assuming that saturation occurs when the following condition is satisfied

$$\exp(-q\varepsilon_s^I) = 0.01 \tag{8.43}$$

the equation can be solved for q, where ε_S^I is the inelastic strain at saturation. The inelastic strain at saturation is estimated by determining the total strain at saturation from a constant strain rate tensile curve and subtracting the elastic strain. The elastic strain is computed by dividing the saturation stress by the elastic modulus, as described by Equation 8.38. The computed inelastic strain is substituted into Equation 8.43, which is solved for q. If the inelastic strain at saturation is found to vary with strain rate, the parameter q must be computed at each strain rate and regression techniques used to determine an expression for the variation of q.

8.3.3   Three-Dimensional Constitutive Equations

In this section the one-dimensional constitutive equations (Equations 8.39–8.40) are extended to three dimensions. To account for the effect of hydrostatic stresses on the response, the effective stress term in the flow equation is appropriately modified. A discussion of the nature of the tensorial state variable is presented, and the procedure for determining the material constants for the three-dimensional version of the equations is discussed.

8.3.3.1   Original Flow Equation

The three-dimensional extension of the Ramaswamy–Stouffer flow equation given in Equation 8.38 is

$$\dot{\varepsilon}_{ij}^I = D_0 \, \exp\left[-\frac{1}{2}\left(\frac{Z_0^2}{3K_2} \right)^n \right] * \frac{S_{ij} - \Omega_{ij}}{\sqrt{K_2}} \tag{8.44}$$

where S_{ij} is the deviatoric stress component, Ω_{ij} is the component of the state variable, $\dot{\varepsilon}_{ij}^I$ is the component of inelastic strain, n and Z_0 are as defined in Equation 8.39 and K_2 is defined as

$$K_2 = \frac{1}{2}(S_{ij} - \Omega_{ij})^2 \tag{8.45}$$

8.3.3.2   Modified Equations with Shear Correction Factor

When the flow law described in Equations 8.39 and 8.40 was implemented into the composite micro-mechanics method described in an earlier section, the nonlinear deformation response and component stresses for laminates with shear dominated fibre orientation angles were predicted incorrectly. These results indicated that the nonlinear shear stresses in the matrix constituent were not being computed properly. For this work, the cause of the discrepancy was assumed to be due to the effects of the hydrostatic stresses on the inelastic response. As described earlier, hydrostatic stresses have been found to affect the yield behaviour of polymers. Furthermore, earlier work has indicated that the effective stress terms may need to be modified for the multiaxial analysis of polymers.

The effect of the hydrostatic stresses on the inelastic response of the polymer was accounted for by modifying the effective stress term K_2 in the flow law (Equations 8.43 and 8.44).

Specifically, since the shear response of the polymer required modification, the shear terms in the effective stress were adjusted to account for the effects of hydrostatic stresses. Equation 8.45 can be rewritten as

$$K_2 = \frac{1}{2}[K_{11} + K_{22} + K_{33} + 2(K_{12} + K_{13} + K_{23})] \qquad (8.46)$$

The normal terms in this expression maintain their original definition, as suggested by Equation 8.45, as follows

$$K_{11} = (S_{11} - \Omega_{11})^2 \quad K_{22} = (S_{22} - \Omega_{22})^2 \quad K_{33} = (S_{33} - \Omega_{33})^2 \qquad (8.47)$$

However, the shear terms in the effective stress definition have been modified. The primary modification to these equations is the multiplication of the shear terms in the effective stress by the parameter α. The equations become

$$K_{12} = \alpha(S_{12} - \Omega_{12})^2 \quad K_{13} = \alpha(S_{13} - \Omega_{13})^2 \quad K_{23} = \alpha(S_{23} - \Omega_{23})^2 \qquad (8.48)$$

where

$$\alpha = \left(\frac{\sigma_m}{\sqrt{J_2}}\right)^{\beta} \qquad (8.49)$$

$$\sigma_m = \frac{1}{3}(\sigma_{11} + \sigma_{22} + \sigma_{33}) \qquad (8.50)$$

$$J_2 = \frac{1}{2}S_{ij}S_{ij} \qquad (8.51)$$

In Equation 8.50, σ_m is the mean stress, J_2 is the second invariant of the deviatoric stress tensor and β is a rate independent material constant. This formulation is based on actual observed physical mechanisms. When the parameter β is set equal to zero, the value of α in Equation 8.50 is equal to 1 and Equation 8.47 is equivalent to Equation 8.48. Therefore, the modification to the constitutive equations is implemented through the use of the correlation coefficient α. The value of the parameter β was determined empirically by fitting composite data with shear dominated fibre, since only uniaxial tension is considered in this study.

8.3.3.3 Three-Dimensional Extension of Internal Stress Evolution Law

The internal stress rate, $\dot{\Omega}_{ij}$, is defined by

$$\dot{\Omega}_{ij} = \frac{2}{3}q\Omega_m\dot{\varepsilon}_{ij}^I - q\Omega_{ij}\dot{\varepsilon}_e^I \qquad (8.52)$$

where $\dot{\varepsilon}_e^I$ is the effective inelastic strain rate, defined as

$$\dot{\varepsilon}_e^I = \sqrt{\frac{2}{3}\dot{\varepsilon}_{ij}^I\dot{\varepsilon}_{ij}^I} \qquad (8.53)$$

8.3.4 Finite Element Implementation

8.3.4.1 Shell Element Simulation

The resin strain rate activities equations and the composite micro-mechanics model described in earlier sections have been implemented into the finite element code LS-DYNA. To do that, the equations must be converted into an incremental format. Specifically, Equations 8.44, 8.52 and 8.53 must be adjusted so that increments in inelastic strain, internal stress and effective inelastic strain are computed instead of their corresponding rates.

To convert Equation 8.44 into an incremental form, the rate equation is multiplied by the time increment of the current time step to compute the inelastic strain increment. The resulting equation is

$$d\varepsilon_{ij}^I = D_0 \exp\left[-\frac{1}{2}\left(\frac{Z_0^2}{3K_2}\right)^n\right]\frac{S_{ij} - \Omega_{ij}}{\sqrt{K_2}} \cdot dt \tag{8.54}$$

where all of the terms are as defined earlier. Note that the total value of the deviatoric stress components and the internal stress components are used instead of the stress increments and are the values from the previous time step. Equations 8.52 and 8.54 are modified to compute the increment in internal stress, $d\Omega_{ij}$, and the increment in effective inelastic strain, $d\varepsilon_e^I$. The following equations result

$$d\Omega_{ij} = \frac{2}{3}q\Omega_m d\varepsilon_{ij}^I - q\Omega_{ij}d\varepsilon_e^I \tag{8.55}$$

$$d\varepsilon_e^I = \sqrt{\frac{2}{3}d\varepsilon_{ij}^I d\varepsilon_{ij}^I} \tag{8.56}$$

where all the terms in the equations are as defined earlier. In Equation 8.56, the total value of the back stress from the previous time increment is used in computing the stress increment for the current time step.

The equations described above and the micro-mechanics equations described in Section 8.2 were used to compute the rate dependent, inelastic response of the matrix. Considering the resin's stain rate effect, the incremental stresses in the micro-mechanics model should be modified to a new form and the stresses in subcell B could be

$$d\sigma_{11b} = \frac{E_m}{1 - v_m^2}[(d\varepsilon_{11} - d\varepsilon_{11}^I) + (d\varepsilon_{22} - d\varepsilon_{22}^I + d\varepsilon_{33} - d\varepsilon_{33}^I) \cdot v_m] \tag{8.57}$$

$$d\sigma_{22b} = \frac{E_m}{1 - v_m^2}[(d\varepsilon_{22} - d\varepsilon_{22}^I) + (d\varepsilon_{11} - d\varepsilon_{11}^I + d\varepsilon_{33} - d\varepsilon_{33}^I) \cdot v_m] \tag{8.58}$$

$$d\sigma_{12b} = (d\varepsilon_{12} - d\varepsilon_{12}^I) \cdot g_m \tag{8.59}$$

$$d\sigma_{13b} = (d\varepsilon_{13} - d\varepsilon_{13}^I) \cdot g_m \tag{8.60}$$

$$d\sigma_{12b} = (d\varepsilon_{23} - d\varepsilon_{23}^I) \cdot g_m \tag{8.61}$$

and the stresses in subcell A could be

$$d\sigma_{22a} = E_{12m} \cdot [(d\varepsilon_{22} - d\varepsilon_{22}^I) + (W_f \cdot v_f + W_m \cdot v_m) \cdot (d\varepsilon_{11} - d\varepsilon_{11}^I + d\varepsilon_{33} - d\varepsilon_{33}^I)] \quad (8.62)$$

where

$$E_{12m} = \frac{E_1 \cdot E_2 \cdot E_m}{[E_m \cdot W_f \cdot (E_1 - E_2 \cdot v_f^2) + E_1 \cdot E_2 \cdot W_m \cdot (1 - v_m^2)]} \quad (8.63)$$

$$d\sigma_{11a} = (d\varepsilon_{11} - d\varepsilon_{11}^I) \cdot E_m + d\sigma_{22a} \cdot v_m \quad (8.64)$$

$$d\sigma_{11f} = (d\varepsilon_{11} - d\varepsilon_{11}^I) \cdot E_1 + d\sigma_{22a} \cdot v_f \quad (8.65)$$

$$d\sigma_{12a} = \frac{d\varepsilon_{12} - d\varepsilon_{12}^I}{\dfrac{W_f}{g_f} + \dfrac{W_m}{g_m}} \quad (8.66)$$

$$d\sigma_{13a} = \frac{d\varepsilon_{13} - d\varepsilon_{13}^I}{\dfrac{W_f}{g_f} + \dfrac{W_m}{g_m}} \quad (8.67)$$

$$d\sigma_{23a} = \frac{d\varepsilon_{23} - d\varepsilon_{23}^I}{\dfrac{W_f}{g_f} + \dfrac{W_m}{g_m}} \quad (8.68)$$

In all the above equations, the terms $d\varepsilon_{ij}^I$ (or the inelastic strain rate of the resin) were given in Equation 8.54. And the stresses in the whole cell could be calculated using Equations 8.57–8.68.

8.3.4.2 Solid Element Simulation

All the equations described above are used for the shell element. If a solid element is used in this model, some modifications are required. The only difference between them is the calculation of σ_{33}. In the simulation of a shell element, a plane stress assumption is made, or $\sigma_{33} = 0$. But in the case of solid element analysis, σ_{33} needs to be calculated as well as σ_{22}, which means in subcell B, it is

$$d\sigma_{11b} = \frac{E_m}{1 - v_m^2}[(d\varepsilon_{11} - d\varepsilon_{11}^I) + (d\varepsilon_{22} - d\varepsilon_{22}^I + d\varepsilon_{33} - d\varepsilon_{33}^I) \cdot v_m] \quad (8.69)$$

In subcell A

$$d\sigma_{33a} = E_{12m} \cdot [(d\varepsilon_{33} - d\varepsilon_{33}^I) + (W_f \cdot v_f + W_m \cdot v_m) \cdot (d\varepsilon_{11} - d\varepsilon_{11}^I + d\varepsilon_{22} - d\varepsilon_{22}^I)] \quad (8.70)$$

where E_{12m} is defined in Equation 8.26.

Table 8.4 Material properties of AS-4 fibre tows.

	Longitudinal Modulus (Gpa)	Trans Modulus (Gpa)	In-plane Shear Modulus (Gpa)	Out of Plane Shear Modulus (Gpa)	In-plane Poisson's Ratio	Out of Plane Poisson's Ratio
Symbol	E_L	E_T	G_{LT}	G_{TT}	v_{LT}	v_{TT}
Value	214.0	14.0	14.0	14.0	0.2	0.2

And in the whole cell, it could be

$$\sigma_{33} = W_f \cdot \sigma_{33a} + W_m \cdot \sigma_{33b} \tag{8.71}$$

8.4 Numerical Results

As a demonstration of the prediction of the developed formulation, examples will be discussed in this section. The tensile response of the AS4/polyetheretherketone (PEEK) composite material is computed. PEEK is a thermoplastic material, and AS4 is a carbon fibre. The fibre volume ratio used for the AS4/PEEK material is 0.62, a typical value for this material based on representative manufacturer information. Table 8.4 presents the material properties of AS4 fibre tows, and Table 8.5 presents the material properties of PEEK resin.

For the finite element model, both four-noded shell elements and eight-noded solid elements were used in a square mesh, with the elements on a side as shown in Figure 8.9. Each side of the model is 20 mm long. The left side of the model was clamped, and a constantly increasing specified displacement was applied to the right side of the model. These boundary conditions are chosen in order to simulate a constant strain rate tensile test. The displacements applied at each time were computed by taking the constant strain rate, multiplying it by the current time to obtain total strain and multiplying this value by the total length of the model to compute an average displacement. To create a stress–strain curve, results were read in from several elements at the centre of the model, in order to avoid edge effects.

The predicted results were compared to experimentally obtained values. During the analysis, unidirectional laminates with fibre orientations of [15°], [30°], [45°] and [90°] were considered. Comparisons between prediction result and experimental data are shown in Figures 8.10–8.13 for a strain rate of 0.1/s, using a shell element.

Table 8.5 Material properties of PEEK resin.

E_m (Gpa)	G_m (Gpa)	v	D_0 (1/s)	N	Z_0 (Mpa)	q	Ω_m (Mpa)	β
4.0	1.42	0.4	1.0E+4	0.7	630	310	52	0.45

E_m is Young's modulus, G_m is shear modulus, v_m is Poisson's ratio, D_0 is maximum strain rate factor, N is deformation response variable, Z_0 is isotropic initial hardness, q is hardening rate, Ω_m is saturation internal stress and β is a rate dependent material constant.

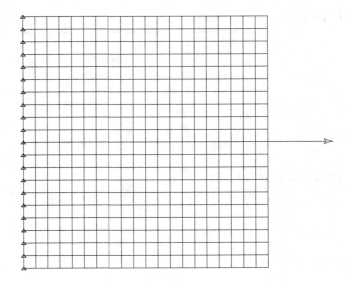

Figure 8.9 Mesh of the composite plate under tensile loading.

Plots of stress–strain curves for different strain rates, which are 0.1/s, 0.01/s and 0.001/s, using a shell element, are shown in Figures 8.14–8.17. Figures 8.18 and 8.19 present the stress–strain curves for different strain rates, which are 1/s, 0.1/s and 0.01/s, using a solid element. The fibre orientations are [30°] and [45°]. As can be seen in the figures, the analytical results match the experimental values reasonably well in general for all fibre orientation angles examined.

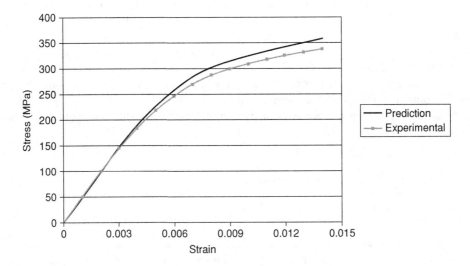

Figure 8.10 Comparison of prediction and experimental data for [15] AS4/peek at the strain rate of 0.1/s (shell element).

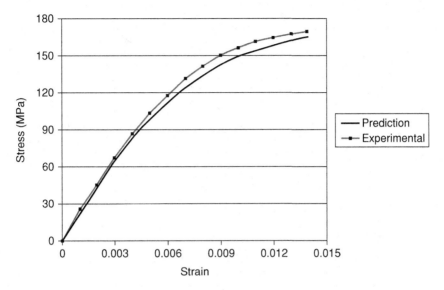

Figure 8.11   Comparison of prediction and experimental data for [30] AS4/PEEK at the strain rate of 0.1/s (shell element).

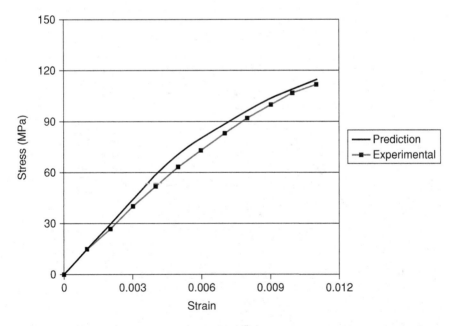

Figure 8.12   Comparison of prediction and experimental data for [45] AS4/PEEK at the strain rate of 0.1/s (shell element).

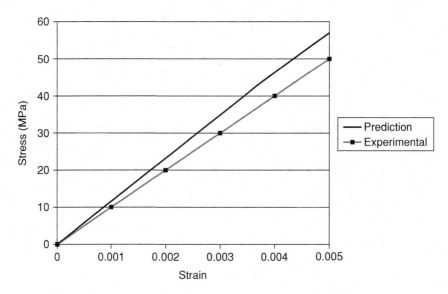

Figure 8.13 Comparison of prediction and experimental data for [90] AS4/PEEK at the strain rate of 0.1/s (shell element).

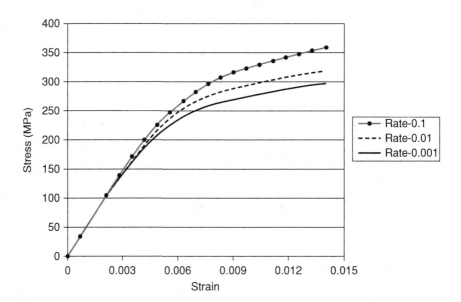

Figure 8.14 Predictions of AS4/PEEK [15] at different strain rates (shell element).

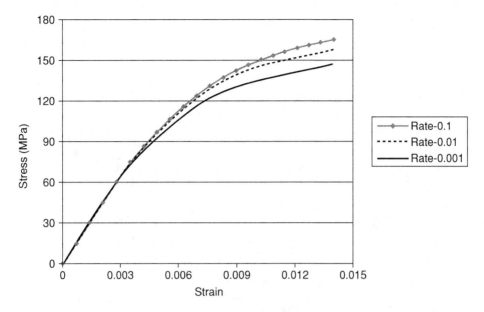

Figure 8.15 Predictions of AS4/PEEK [30] at different strain rates (shell element).

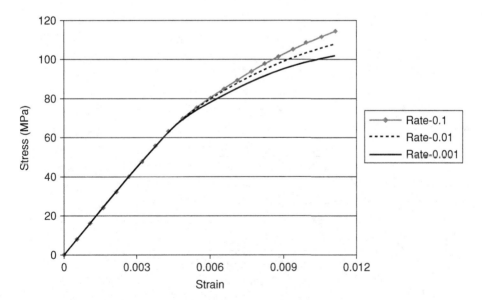

Figure 8.16 Predictions of AS4/PEEK [45] at different strain rates (shell element).

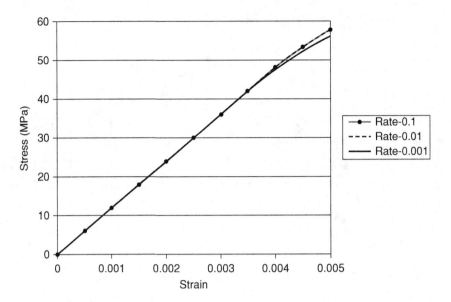

Figure 8.17   Predictions of AS4/PEEK [90] at different strain rates (shell element).

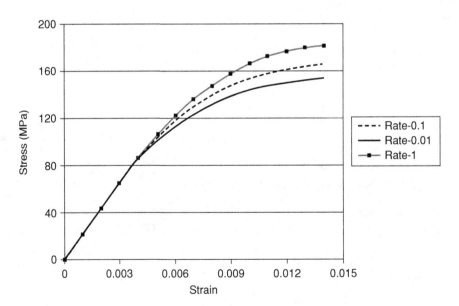

Figure 8.18   Prediction of AS4/PEEK [30] at different strain rate (solid element).

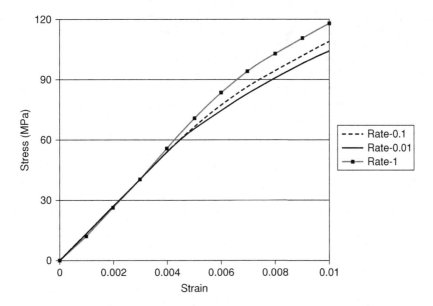

Figure 8.19   Prediction of AS4/PEEK [45] at different strain rate (solid element).

8.5   Conclusion

A micro-mechanical material model is defined for the explicit finite element method. A stress update procedure is developed and a stress/strain averaging procedure is presented for predicting the behaviour of laminated composites on the micro-level. The presented methodology is directly applied to nonlinear explicit finite element codes for structural analysis, as demonstrated in the investigation. The progressive failure in the composite material can be captured with a high level of confidence, provided that experimental characterisation is performed. This is accomplished by defining micro-failure criteria (MFC) for determination of the various failure modes, since stresses and strains in each sub-cell and each constituent is available at each time step and load increment. Validation of the implemented model is presented in this study. The model is proven to be adequate and efficient for explicit finite element simulations.

References

[1] Tabiei, A. and Chen, Q. (1998) *Micromechanics Based Composite Material Model for Impact and Crashworthiness Explicit Finite Element Simulation*, University of Cincinnati, Cincinnati, USA.
[2] Pecknold, D.A. and Rahman, S. (1994) Micromechanics-based structural analysis of thick laminated composites. *Computers and Structures*, **51**, 163–179.
[3] Ward, I.M. (1983) *Mechanical Properties of Solid Polymers*, John Wiley & Sons, Inc., New York, USA.
[4] Boyce, M.C., Parks, D.M. and Argon, A.S. (1998) Large inelastic deformation of glassy polymers. Part I: rate dependent constitutive model. *Mechanics of Materials*, **7**, 15–33.

[5] Bordonaro, C.M., (1995) Rate Dependent Mechanical Behavior of High Strength Plastics: Experiment and Modeling, PhD Dissertation, Rensselaer Polytechnic Institute, Troy, New York, USA.

[6] Goldberg, R.K. (1999) Strain Rate Dependent Deformation and Strength Modeling of a Polymer Matrix Composite Utilizing a Micromechanics Approach, PhD Dissertation, Department of Aerospace Engineering, University of Cincinnati, Cincinnati, Ohio, USA.

[7] Stouffer, D.C. and Dame, L.T. (1996) *Inelastic Deformation of Metals: Models, Mechanical Properties and Metallurgy*, John Wiley & Sons, Inc., New York, USA.

9

Design Solutions to Improve CFRP Crash-Box Impact Efficiency for Racing Applications

Simonetta Boria

School of Science and Technology, Mathematics Division, University of Camerino, Camerino, Italy

9.1 Introduction

Racing cars demonstrate maximum vehicle driving performance resulting from high tech developments in the area of lightweight materials and aerodynamic design. Since extreme racing speeds may lead to severe accidents with high amounts of energy involved, special measures were taken over time in order to ensure the drivers safety in case of high-speed crashes. Besides the driver's protective equipment (like helmet, harness or head and neck support device) and the circuit's safety features (like run-off areas and barriers), the race car itself is designed for crashworthiness and possesses special sacrificial impact structures that absorb the race car's kinetic energy and limit the deceleration acting on the human body [1–7]. In the beginning of the 1960s, Colin Chapman, chief designer of Lotus, introduced the monocoque to Formula 1 by placing thin plates around the bars of the frame. This new technology increased the stiffness of the chassis. Later in the 1970s, aluminium was mostly used for these constructions, but also these structures proved not to be resistant enough for the wings' downforce. In the early 1980s, Formula 1 with John Barnard and the McLaren team underwent the beginnings of a revolution that has become its hallmark today: the use of carbon composite materials to build a self-supporting chassis. In 1981, the McLaren drivers proved the safety and advantages of the new way of construction: John Watson twice finished second and once first in that season. Andrea De Cesaris proved the stiffness of the monocoque, with the number of crashes he made in that season. Today, most of the racing car chassis – the monocoque, suspension, wings and engine cover – is built with carbon fibre. This material has four

Advanced Composite Materials for Automotive Applications: Structural Integrity and Crashworthiness,
First Edition. Edited by Ahmed Elmarakbi.

Figure 9.1   Example of a Formula Ford car.

advantages over every other kind of material for racing car construction: it is super lightweight, super strong, super stiff and can be easily moulded into all kinds of different shapes.

This research work involves the examination of the front crash structure of a Formula Ford car (Figure 9.1) used to protect the driver in the event of a frontal impact. At the moment every manufacturer has a different method of producing their crash structures. The choice of the material to use is increasingly often dictated by weight and aerodynamic needs as well as structural performance. Recently it became preferable to make such structures in composite materials, thanks to their undoubted advantages such as high strength to weight ratio, the high level of energy absorption and the possibility to reproduce each complex geometry [8]. The excellent performance of composites, sometimes better than those of similar metallic structures, can be obtained by choosing appropriate mechanical (the stacking sequence, number of layers, type and quantity of fibres and matrix) and geometrical (the beam section shape, wall thickness, extremity joints) design parameters [9, 10].

Nowadays each manufacturer of racing cars must pass specific crash tests controlled by legislation, before seeing their vehicles in the market and in motor racing. In order to ensure the driver's safety in the case of high-speed crashes, special impact structures must be designed to absorb the race car's kinetic energy and limit the deceleration acting on the human body. In particular a Formula Ford car, before racing on circuit, must have an FIA certification regards safety and must boast, amongst the safety features, a front carbon crash structure. According to Article 15.3 of the Technical Regulation for Formula Ford with 1600 cc EcoBoost engine [11], that refers to the survival cell and frontal protection, an impact-absorbing structure must be fitted in front of the space frame. This structure must be solidly attached to it and must pass specific static and dynamic FIA tests in the presence of an FIA/ASN technical delegate in an approved testing centre. To test the attachments of the frontal impact-absorbing structure to the space frame structure, a static side load test must be performed on a vertical plane passing 400 mm in front of the front wheel axis. A constant transversal horizontal load of 30 kN must be applied to one side of the impact-absorbing structure using a pad 100 mm long and 300 mm high. The centre of the pad area must pass through the above-mentioned plane at the midpoint of the height of the structure at that section. After 30 s of load application, there must be no failure of the structure or of any attachment between the structure and the space frame. Moreover, the space frame must be capable of withstanding a frontal impact test. For the purposes of this test, the total weight of the trolley and test structure must be 560 kg and the impact velocity at least 12 m/s. The resistance of the test structure must be such that, during the impact, the average deceleration of the trolley does not exceed 25 g. Further, all structural damage must be contained within the frontal impact structure.

Designing an impact attenuator able to pass the imposed regulation is still a challenging mission due to the structural heterogeneity and the various micro- and macro-fracture mechanisms; good results can be obtained only by combining numerical analyses with experimental tests. Investigated in this research work is the best solution, in terms of energy absorption, for a Formula Ford frontal crash-box made of a sandwich structure with skins in CFRP material and a honeycomb core in aluminium. In particular, given the geometry reported by the manufacturer, it has been necessary to find the right combination of material and lamination by varying the stacking sequence and its disposition. The first analyses were conducted by joining numerical models (performed using LS-DYNA [12]) with experimental tests, through an instrumented drop tower equipment. The first aim was, in fact, to reproduce as best possible the crushing process with a finite element code using the experimental results conducted on a preliminary impact attenuator with a lower impact energy. The devices available in the laboratory did not allow, in fact, obtaining impact energies comparable with those required by the regulation. Only after reproducing the dynamic behaviour of the crash-box numerically was it possible to change the stratigraphy of the skin to find the best configuration for homologation requirements. By performing these numerical tests it was possible to have a baseline as well as a comparison between different solutions which will lead to improve driver safety in a front impact situation.

9.2   Composite Structures for Crashworthy Applications

Accidents and collisions between vehicles, where human lives are in danger, justify the study of the added value offered by new structural concepts and new materials for the protection of occupants during crash situations. This field of study, crashworthiness, deals with how materials and structures deform, fail and absorb energy in a controlled form during a crash event. Controlled form means, in this context, a deformation mode where the crushing force is kept to an approximately constant level during the collision, in such a way that a maximum amount of energy is absorbed at bearable levels of acceleration for the passengers.

Composite materials used in crashworthy applications typically consist of polymeric matrix (polyester, epoxy or thermoplastic) strengthened with fibres (typically glass or carbon). The number of structures in composite materials destined to the absorption of energy and consequent protection of people is already considerable, both in the motor sport world and in the construction of mass production vehicles [13, 14], and will certainly continue to grow in the future. This gradual replacement of metallic structures by composite ones is motivated by the higher capacity of energy absorption per unit of weight of the latter. The profit in energy absorbed per unit of weight can exceed 500%, in comparison with structures in steel or aluminium [15].

In crushing situations, composites and metals behave very differently. While metals absorb energy by plastic deformation, the process of crushing in composites is more complex (Figure 9.2). As a composite material is loaded it undergoes damage mechanisms leading to ultimate failure. Damage mechanisms occur when the fibre and/or matrix undergo deformations that produce an irreversible effect on the loading characteristics of the material. Composites are often constituted by layers and these tend to separate during the crushing, leading to

  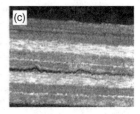

Figure 9.2 Different failure mechanisms in composites: (a) matrix cracking, (b) fibre kinking, (c) delamination.

delamination (interlaminar damage); further, in each of these layers, fibre rupture and matrix cracking might take place, either in tensile or compression (intralaminar damage). These mechanisms can be categorised into three scales: micro-, meso- and macro-scale. Micro-scale includes fibre fracture, matrix cracking and debonding at the fibre–matrix interface; meso-scale failures occur at the ply level, such as debonding between laminates; in macro-scale there is an overall failure of the laminate. Controlled testing is required to identify the material properties and monitor the onset of damage propagation.

Crash investigations on composite structures reported in the literature are mainly based on small plates submitted to bending impact and on simple bars, of circular or rectangular cross-section, of prismatic or tapered shape, submitted to axial impact. Most of the research on the energy absorption of composite materials has been limited to the axial compression of tubular structures and has been based on experimental test analysis [16–20]. The tubular devices have been shown to perform at their best when geometric, material and loading conditions are such that the axial failure of the tubes is characterised by the progression of a destructive zone of constant size at the loaded end, called the "crush zone". The design challenge is to arrange the composite material column such that the destructive zone can progress in a stable manner. The energy absorption must be as high as possible by allowing the development of a sustained high-level crushing force, with little fluctuation in amplitude as the progress zone travels along the component's axis. Experimental results showed that rectangular and square sections are less effective in energy absorption than circular ones. For tubes with structures other than circular, the crushing behaviour has been shown to be improved when the corners of polygonal thin-walled sections are rounded so as to represent segments of circular tubes. For square sections, the greater the corner radius, the higher will be the efficiency of energy absorption. Rounded corners prevent flat segments from failing by local plate buckling, with associated plate strip buckling and much lower specific energy absorption.

The failure mechanisms of composite structures during impact loading are extensive. A variety of failure mechanisms have been characterised for differing structural geometries and materials. Sigalas *et al.* [21] and Hull [22] discussed these mechanisms in detail for composite tubes and commented on the geometry and material composition influences on the structural performance. Hull's paper [23] was based on experimental work on the axial collapse of tubes (made from a variety of composite materials) by failure of the centre of the column away from the ends and by progressive crushing from one end (progressive crush failure). The only significant difference between the materials used was the arrangement and volume fraction of the glass fibres. However, it is clear that these differences resulted in large difference in the failure modes. Progressive crush failure of composite tubes is shown in Figure 9.3. The

Figure 9.3 Typical examples of progressive crush failure: (a) splaying with axial splitting; (b) fragmentation with debris [34].

crushed material can spread out on both sides of the tube wall, even if it is more common have deformation on only one side. The material on the inside of the tube is further compacted and is constrained to fold inwards, whereas the material on the outside tends to splay into a mushroom shape. Also, the behaviour of the reinforcing fibres depends on their orientation. Fibres aligned axially bend inwards and outwards with or without fracturing, depending on their flexibility and the constraints produced by other fibres.

Also, Hull [22] summarised the forces acting at various locations in the progressive crush wedge (Figure 9.4).

A detailed description is given of the effect of fibre arrangement on progressive crushing in carbon fibre-epoxy unidirectional laminated tubes, woven glass cloth–epoxy tubes, filament-wound angle-ply glass fibre–polyester tubes and in-plane random chopped glass fibre–polyester tubes. The transition between splaying and fragmentation modes is identified and related to the force-displacement response. Similar transitions have been reported with other variables, such as fibre and matrix properties, dimensions and shape of tubes, speed and temperature effects and trigger geometries.

By increasing the properties of composite material to resist these forces, the crushing strength of the tube can be influenced. Warrior *et al.* [24] investigated the influence of increasing the interlaminar properties of the composite on the crushing strength of the tube. The investigated methods of increasing the interlaminar properties of the composite included a toughened resin

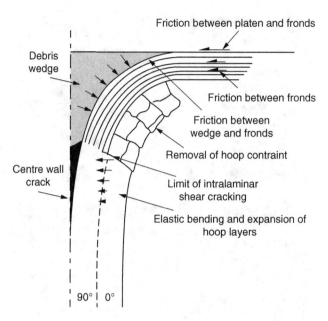

Figure 9.4 Summary of forces acting in crush zone [22]. Reproduced from Ref. [22]. Copyright 1991 Elsevier.

and interlaminar stitching. The research found that the use of a toughened resin was the best method investigated to increase the crushing strength of the tube. The use of interlaminar stitching and thermoplastic interleaves actually decreased the crushing strength due to a reduction in the in-plane properties and contact frictions.

Although significant experimental work on the axial collapse of fibre-reinforced composite shells has been carried out, studies on the theoretical modelling of the crushing process are quite limited [25–30] since the complex fracture mechanism renders difficulties when analytically modelling the collapse behaviour of composite shells. These analytical models predict some crushing parameters, such as the average crush load, the energy absorbed during collapse and the crush length, using a minimisation process on the total energy dissipated for the deformation. According to the proposed theoretical analyses the principal sources of energy dissipation are: frictional resistance between annular wedge and fronds and between fronds and platen, fronds bending owing to delamination between plies, intrawall crack propagation, axial splitting between fronds and hoop strain. Despite the simplifications adopted the comparison shows that the theoretically obtained results match the experimental ones quite well.

The design of an impact attenuator for racing applications is very similar to a square frusta. The behaviour of these structures during axial loading conditions is critical to the energy absorption and crushing strength. This was experimentally investigated by Mamalis *et al.* [16]. Square frusta structures have been observed to fail in four major modes, depending on structural dimensions, material properties and testing conditions (shown in Figure 9.5). The progressive crushing with micro-fragmentation of the composite material, associated with large amounts of crush energy, is designated as mode 1; this mode of failure is difficult to obtain in very thin- or very thick-walled shells. Brittle fracture of the component resulting in

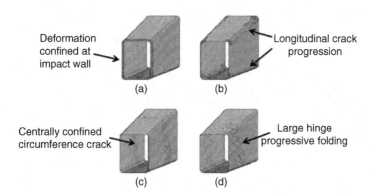

Figure 9.5 Variations in collapse modes of square frusta subjected to a frontal loading: (a) mode 1, (b) mode 2, (c) mode 3, (d) mode 4.

catastrophic failure with little energy absorption is designated as either mode 2 or mode 3, depending on the crack form. Mode 2 was observed when collapsing square frusta with 20–30° semi-apical angles with the formation of longitudinal corner cracks. Mode 3, designated as mid-length collapse mode, is characterised by the formation of circumferential fracturing of the material at a distance from the loaded end of the specimens approximately equal to the mid-height of the shell; catastrophic failure by cracking and separation of the shell into irregular shapes is involved. Tubes and frusta with small semi-apical angles and relatively small wall-thickness undergo mode 3 fracture. Progressive folding and hinging similar to the crushing behaviour of thin-walled metal and plastic tubes showing a very low energy-absorbing capacity is designed as mode 4. This mode of collapse is associated with the axial loading of very thin tubes not exceeding 15° semi-apical angles.

Typical load/displacement curves for each mode of collapse for static and dynamic loading are shown in Figure 9.6. Initially the shell behaves elastically and the load rises at a steady rate to a peak value and then drops abruptly; the magnitude of the peak is greatly affected by the shell geometry, the material characteristics and the corners' rigidity. As deformation progresses the shape of the curve depends on the mode of collapse.

An impact attenuator is designed to produce a mode 1 failure mechanism, as the structure maintains a constant crushing strength throughout the crushing process and the highest energy absorption. There are two categories for mode 1 failure: mode 1a is described as a "mushrooming" or fountain failure where the structure begins to fail by micro-cracks at the edges of the frusta and produces a split in the structure (Figure 9.7a); mode 1b is when the structure folds in one direction due to micro-cracks (Figure 9.7b) and does not display the same mushrooming effect seen in mode 1a. This effect is commonly seen in dynamic and oblique testing.

Producing a structure from composite materials can be achieved in different ways, for example by resin infusion of dry fabrics or the use of pre-impregnated plies to form laminates that are cured in an autoclave. Resin transfer moulding (RTM) infusion methods require a manufacturer to arrange the fabric in the desired shape using a mould and then infuse with a resin. The benefits of RTM include the production of a safe manufacturing environment due to enclosure of the mould and increased flexibility in defining the component shape. The disadvantages include the expense of mould manufacture and control over the component

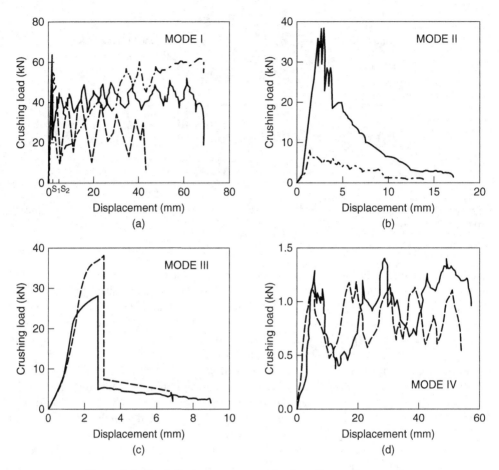

Figure 9.6   Load-displacement curves for the various collapse modes.

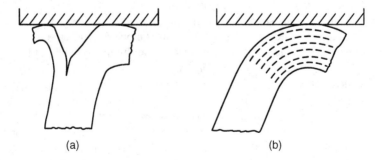

Figure 9.7   Variations in mode 1 type failure: (a) mode 1a, (b) mode 1b.

Figure 9.8   Epsilon Euskadi LMP1 front crash-box.

quality, such as resin voids, when compared with pre-impregnated composite components. RTM developed components are being introduced to an increasing number of aerospace and automotive applications, such as the bumper tube shown in Figure 9.8 used to absorb the energy from a frontal crash.

The use of pre-impregnated composites is an alternative method to resin infusion. The material requires storage at a low temperature before use and has a limited shelf life. To produce a structural component, the material is cut to size and arranged. The composite is bagged and consolidated under a vacuum; the mould is then transferred to an autoclave which subjects the composite to a high pressure/temperature cycle curing process. The method consistently produces components of a high quality with minimal voids and high fibre contents. The use of an autoclave does limit the part size and the production volume.

The choice of fibre and resin type is of greatest importance when designing a structural component. Table 9.1 compares the strength and stiffness of fibrous materials with aluminium and steel, thus highlighting the benefits of composites for low-mass applications. Hybrid fabrics, a mixture of fibre types such as aramid and carbon, can be developed to optimise properties for a specific application. Formula 1 teams with large budgets for material selection may invest in the development of hybrid fabrics to further optimise components.

Table 9.1   Tensile properties of composite materials compared with alloy materials.

Property	Aluminium 5052 grade	Mild steel	Carbon/epoxy (twill weave)	E-glass/epoxy (0°–90°)
Ultimate tensile strength (Mpa)	255	394	1084	783
Tensile modulus (Gpa)	71	208	71	40
Density (g/cm^3)	2.7	7.8	1.9	2.3

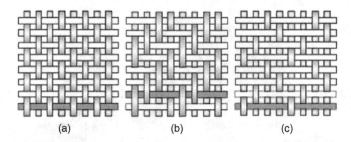

Figure 9.9   Weave types: (a) plain weave, (b) twill weave, (c) satin weave.

The orientations and weave patterns of fibres in a composite component are of great importance to the laminate properties. The simplest orientation is unidirectional (UD) where all the fibres in a single ply lie in one direction; load capacity in this direction is excellent but very poor in the transverse direction. The fibres can alternatively be weaved in a regular pattern. The interlocking of fibres in a regular pattern produces a composite ply with built-in multi-axial properties. Figure 9.9 presents three commonly used weave types and compares their properties.

For a racing impact attenuator the use of pre-impregnated materials is standard practice. The type of pre-impregnated composites used varies depending on team and application.

The addition of a honeycomb core to composite skins produces a sandwich structure with improved flexural properties (Table 9.2) with only a minor penalty in mass. As a result sandwich structures are of great interest in transport and aerospace industry. Sandwich structures consist of thin skin materials bonded to a low mass core material using an adhesive. A composite-honeycomb sandwich, as used for racing applications, uses composite laminates as skins bonded to a metallic honeycomb core, as shown in Figure 9.10.

9.3   Geometrical and Material Characterisation of the Impact Attenuator

This research work focuses on the modelling of the nosecone structure from a Formula Ford racing car during the frontal crash test. As mentioned before, the frontal impact structure

Table 9.2   Increase in flexural properties with additional core material.

	Solid metal sheet	Sandwich construction	Thicker sandwich
Relative stiffness	100%	700% (7× more rigid)	3700% (37× more rigid)
Relative strength	100%	350% (3.5× stronger)	925% (9.25× stronger)
Relative weight	100%	103% (3% increase in weight)	106% (6% increase in weight)

Figure 9.10   Sandwich configuration used.

must adhere to energy-absorbing criteria in addition to fulfilling various performance requirements [11].

The analysed crash-box (Figure 9.11) consists of a truncated pyramidal structure with an almost rectangular section. The pyramidal structure makes it possible to obtain major stability during progressive crushing, while the rectangular section has rounded edges to avoid stress concentrations [31]. The design of sacrificial structure has been completed with a trigger which consists in a smoothing (progressive reduction) of the wall thickness in order to reduce the resisting section locally. This trigger is intended both to reduce the value of the force peak and to initialise structure collapse in a stable way. For the preliminary impact attenuator, three different wall thickness zones are used, as shown in Figure 9.11. Unfortunately, for reasons of secrecy, it is not possible to report the geometric dimensions and the sequence of lamination used.

The impact attenuator is constructed from a low-mass composite-honeycomb sandwich material to maximise stiffness and energy absorption. The sandwich skins used are the combination of two different high performance pre-impregnated composite materials with carbon fibres embedded in an epoxy resin system. The honeycomb core is, instead, in aluminium to increase the structural rigidity of the component. Table 9.3 reports the main mechanical properties of the used materials, obtained using an electromechanical machine at a crosshead speed of 0.05 mm/s.

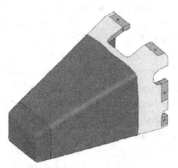

Figure 9.11   Impact attenuator with the preliminary stratigraphy.

Table 9.3 Mechanical properties of the used materials.

Property	GG200-HTA5131 plain prepreg	GG630-UTS5631 twill prepreg	1/8-5052-0.001 honeycomb
Density (kg/m^3)	1400	1500	72
Fibre volume fraction in warp and fill direction (%)	49	39	—
Young modulus in fibre direction (Mpa)	45 000	60 000	10
Young modulus in transverse direction (Mpa)	45 000	60 000	10
Poisson's ratio	0.04	0.06	0.05
Shear modulus in xy direction (Mpa)	2500	3500	10
Shear modulus in xz direction (Mpa)	3000	3000	483
Shear modulus in yz direction (Mpa)	3000	3000	214
Longitudinal tensile strength (Mpa)	500	800	1
Transverse tensile strength (Mpa)	500	800	1
Longitudinal compressive strength (Mpa)	300	500	1
Transverse compressive strength (Mpa)	300	500	1
In plane shear strength (Mpa)	60	50	1

9.4   Experimental Test

The dynamic experimental tests were performed at the Picchio S.p.A. plant in Ancarano (TE, Italy) using a drop weight test machine appropriately instrumented (Figure 9.12). The devices available in the laboratory did not allow generating impact energy greater than about 16 kJ. Therefore an impact mass of 413 kg and an initial velocity of about 8.6 m/s were adopted. During the tests every crash-box was supported at the bottom edge with four metallic angulars (Figure 9.13). It is, in fact, very important to constrain the crash-box, reproducing as closely as possible the connection with the frame. The lack of a constraint could cause catastrophic failures inconsistent with reality, implying lower energy absorption.

The acceleration of the mass and the velocity at impact were measured using an accelerometer with 180 g full scale and a photocell, respectively. The crash events were recorded using a high-speed camera Motion BLITZ EoSens mini1 able to capture up to 506 frames/s.

After the tests, the diagram representing the variation of the deceleration with the time were analysed and filtered with a CFC60 filter (Figure 9.14). CFC is the abbreviation for channel frequency class and this type represents a fourth-order Butterworth low pass phaseless digital filter with a 3 dB limit frequency of 100 Hz. The filter was used to eliminate the high frequency content introduced by vibration and noise and to give a relatively smooth trace that can be replicated. Integrating the deceleration signal once and then twice, it was possible to obtain a pattern of velocity and displacement in time (Figure 9.15). Moreover some important parameters for crash resistance characterisation have been computed and reported in Table 9.4. In particular the peak deceleration is the maximum filtered deceleration, the average deceleration is obtained from the beginning of the impact to the instant when the velocity vanishes, the absorbed energy (Figure 9.16b) corresponds to the area under the load-shortening diagram (Figure 9.16a) and the residual height is the height of the crushed impact attenuator.

From the picture of the specimen taken after the dynamic axial collapse (Figure 9.13), it can be seen that a progressive and stable failure occurs. The impact attenuator is characterised

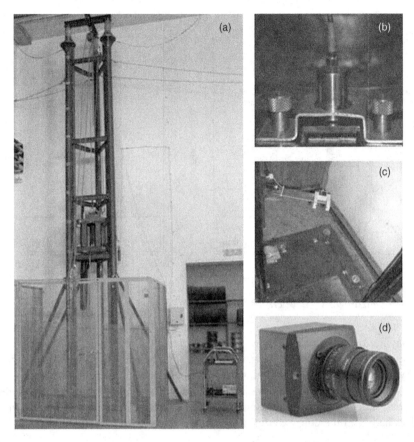

Figure 9.12   Drop tower (a), accelerometer (b), photocell (c) and high-speed recording camera (d).

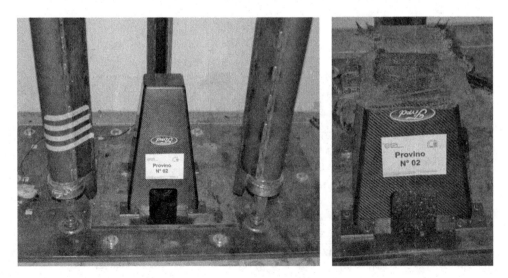

Figure 9.13   Crash-box before and after impact.

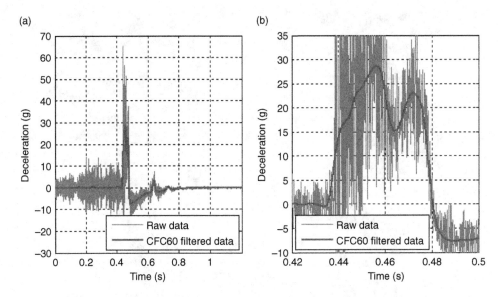

Figure 9.14   Deceleration vs time: (a) total acquisition, (b) impact phase.

by a splaying fracture mechanism, which is very desirable for its elevated energy-absorbing capabilities. It exhibited visible interlaminar separation between fabric plies and honeycomb, whereby the outer plies bent outward with large fronds of semi-intact material while the more inner plies and the core crushed or bent inward. From the final deformation it is evident how the external skin of the lower face never worked. From the early stages of impact, in fact, the skin detached from the core and did not provide continuity to the structure; traces of epoxy

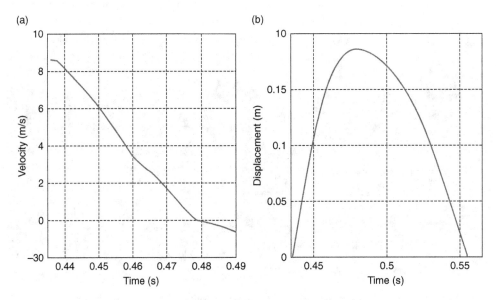

Figure 9.15   (a) Velocity vs time, (b) displacement vs time.

Table 9.4   Experimental crash-test results.

Peak deceleration (g)	Average deceleration (g)	Absorbed energy (J)	Residual height (m)	Impact time (s)
28.5	20.65	15290	0.315	0.0425

matrix present in the composite laminate show this anomaly. This aspect was also observed during numerical modelling, as will be shown later.

Analysing the experimental data obtained from acquisition systems, it can be seen how towards the end of the impact an inexplicable steep drop in the deceleration values is observed. This behaviour may be associated with some internal structural failure due to a previous lateral static test done on the same crash-box. Moreover the presence of two different slopes in the energy diagram indicates the transition from the first narrow wall thickness zone to the second one. The total crushing has not been such as to involve the third and last zone.

9.5   Finite Element Analysis and LS-DYNA

The use of finite element analysis (FEA) has been investigated to improve design process [1]. An implicit FE method is used to analyse the quasi-static crushing strength of crashworthy components. Although this is a useful tool, it is not capable of replicating the energy absorption behaviour during dynamic loading; for this an explicit FE code is necessary. In order to find the best configuration of the analysed crash-box able to absorb the total energy required by technical regulation, maintaining the average deceleration under the maximum permissible value, FEA was conducted using the non-linear dynamic code LS-DYNA [12].

The impact attenuator is modelled by four-node shell elements with the Belytschko–Tsay formulation. A multi-layered shell is used with one integration point per layer. To define a

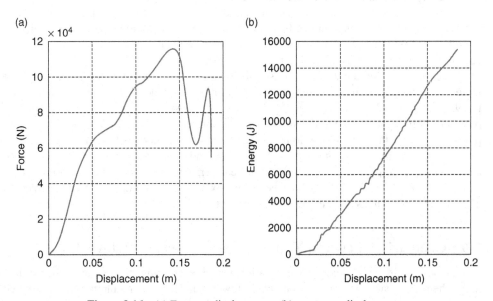

Figure 9.16   (a) Force vs displacement, (b) energy vs displacement.

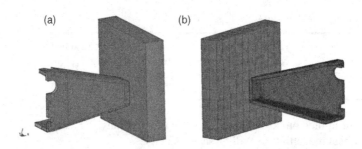

Figure 9.17   Total system: (a) CAD model, (b) FE model.

composite numerical model the card *PART_COMPOSITE is used. According to this card, the laminate has a thickness defined by the sum of each individual layer. Laminate theory is also activated with LAMSHT parameter in *CONTROL_SHELL card, to correct for the assumption of a uniform constant shear strain through the thickness of the shell. A master-surface to slave-node contact is defined between the impact surface and the nodes of the tubes. Then a self-contact algorithm, based on the penalty formulation, is defined on the impact attenuator surface to provide the friction effect between its parts during deformation and to prevent the elements penetration. Particular attention is given to the material definition of composites. The codes, in fact, contain different materials models that implement composite fabric with various failure criteria. In particular, for the LS-DYNA library the material type 55 is used for modelling impact attenuator, thanks to its ability to give a numerical behaviour near to the experimental ones [32, 33]. To avoid ductile behaviour with folding, it is important to change the element strength at some point of the collapse evolution. This is obtained thanks to a time-step failure parameter (TFAIL), which defines a limit to the element effective strain. This allows to reproduce the brittle behaviour of the material and, at the same time, to control the time-step value and therefore to reduce the analysis CPU time.

In order to assess the quality of the numerical results, initially a complete comparative analysis was developed on simple CFRP composite tubes [34]; only after having obtained a good correlation between numerical and experimental results is it possible to simulate a more complex geometry, such as the racing impact attenuator.

The total system (Figure 9.17a) is constituted by internal and external shells in a composite material, a honeycomb in aluminium and a rigid mass as 3D elements. The FE model (Figure 9.17b) has 13 178 nodes and 35 039 elements. For the symmetry of the impact attenuator only half geometry has been modelled, inserting the correct symmetry constraints in the cutting plane. As regards the initial conditions, two different impact energies have been set: initially the rigid moving mass has a finite mass of 206.5 kg and an initial velocity of about 8.6 m/s; after, in order to reproduce the homologation requirements, a mass of 280 kg and an initial velocity of 12 m/s are used. Moreover the crash-box is constrained in all degrees of freedom at its base for 10 mm, as dictated by the experimental evidence.

9.6   Comparison between Numerical and Experimental Analysis

The numerical and experimental deceleration versus shortening curves for the crash-box under the first impact conditions (total energy lower than technical regulation) are compared in Figure 9.18, while some crash parameters obtained from the experimental test are reported

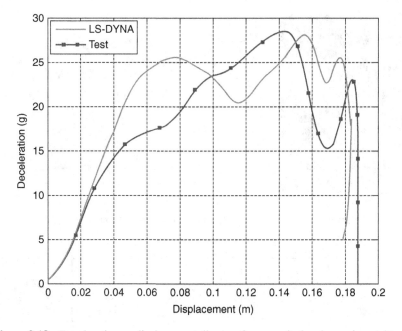

Figure 9.18   Deceleration vs displacement diagram for numerical and experimental test.

with numerical results in Table 9.5, showing also the committed error. Both the experimental data and the numerical data were filtered with the same filter, CFC60.

The difference in behaviour in the early stages of impact is due to the fact that the crash-box tested had been subjected to a static side push; the propagation of internal structural cracks resulting from the static test could be the main reason for the loss of stiffness. Despite this structural difference the numerical model is able to reproduce the crash behaviour with errors less than 5%, also reproducing the detachment of the skin outside from the lower face of the sandwich (Figure 9.19).

9.7   Investigation of the Optimal Solution

After obtaining a good correlation between the numerical and experimental results in the first case of low impact energy reached with the drop tower, attention was focused on the impact condition imposed by the technical regulation. Therefore, the same impact attenuator was numerically tested with an initial velocity of 12 m/s and a total energy of 20 160 J, obtaining a

Table 9.5   Comparison between experimental and numerical results.

	Test	LS-DYNA	Error (%)
Total crush (m)	0.185	0.182	1.6
Average deceleration (g)	20.65	21.58	4.5
Maximum deceleration (g)	28.51	28.12	1.4

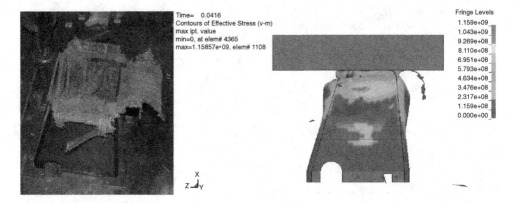

Time= 0.0416
Contours of Effective Stress (v-m)
max ipt. value
min=0, at elem# 4365
max=1.15857e+09, elem# 1108

Fringe Levels
1.159e+09
1.043e+09
9.269e+08
8.110e+08
6.951e+08
5.793e+08
4.634e+08
3.476e+08
2.317e+08
1.159e+08
0.000e+00

X
Z   Y

Figure 9.19 Comparison between experimental and numerical crush.

Version 2	Version 3	Version 4	Version 5	Version 6

Figure 9.20 Different versions of impact attenuators.

total crush equal to 0.252 m and an average and maximum deceleration of 30.58 and 42.14 g, respectively. Since it is not able to overcome the homologation test, it was decided to modify the stacking sequence and disposition of the CFRP skins. Figure 9.20 reports the different versions of the impact attenuator taken into account: versions 2 and 6 analyse the crush behaviour using four or six subdivision zones with different thicknesses, version 3 considers the crash-box made of the same lamination without thickness change, version 4 reduces the gap between the thinner laminations in the top of the crash-box, version 5 takes into account an oblique lamination into the second zone.

The numerical results can be summarised in terms of total crush and average and maximum deceleration, as reported in Table 9.6. The third and the sixth version seem to have a better behaviour in term of energy absorption; the growing progression in the lamination thickness is not so needful, because the geometry in itself leads to an increase of the section with the advance of the crushing. Modifying the orientation in the disposition of the laminate is not advantageous and it is also complex from the point of view of practical realisation.

Table 9.6 Comparison amongst different stratigraphy versions.

	Version 2	Version 3	Version 4	Version 5	Version 6
Total crush (m)	0.335	0.270	0.295	0.329	0.357
Average deceleration (g)	27.13	24.96	25.41	28.45	24.46
Maximum deceleration (g)	61.72	32.25	37.79	50.43	39.87

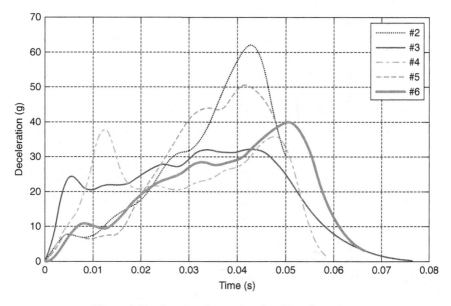

Figure 9.21   Deceleration vs time for all configurations.

In order to investigate in detail the versions taken into account, comparisons between all five structures are shown below in terms of deceleration versus time (Figure 9.21), force versus displacement (Figure 9.22) and energy versus displacement (Figure 9.23). From these diagrams it is again evident that version 2 seems to be the best configuration, with an almost constant deceleration value during the whole impact, avoiding quick variations in slope that

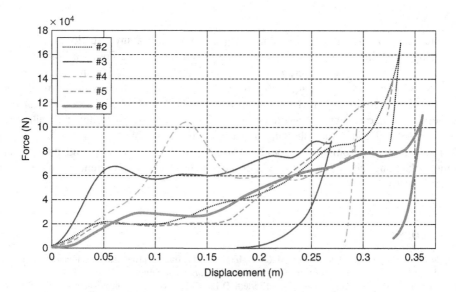

Figure 9.22   Force vs displacement for all configurations.

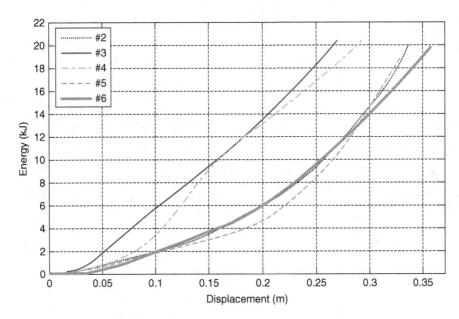

Figure 9.23   Energy vs displacement for all configurations.

can be dangerous to the human body. Moreover the average deceleration is under the limit imposed by the regulation. In view of weight reduction, choosing this configuration it is also possible to lead to a crash-box length around 350 mm.

9.8   Conclusion

The present research work describes numerical and experimental investigations on the energy-absorbing capability of a specific frontal impact attenuator, for a Formula Ford racing car, made of a CFRP sandwich structure. Initially, to calibrate the numerical model, experimental crash tests were conducted using a drop tower suitably instrumented with an impact energy lower than that required by the technical regulations. Only after obtaining a good correlation between the numerical and experimental data was the same impact attenuator numerically tested under real conditions, showing a crush behaviour unable to pass the homologation. Then, the best compromise was found regarding lamination thickness and its disposition, without changing the initial geometry, in order to obtain an impact attenuator able to verify the real test conditions.

References

[1] Savage, G., Bomphray, I. and Oxley, M. (2004) Exploiting the fracture properties of carbon fibre composites to design lightweight energy absorbing structures. *Engineering Failure Analysis*, **11**, 677–694.
[2] Bisagni, C., Di Pietro, G., Fraschini, L. and Terletti, D. (2005) Progressive crushing of fiber-reinforced composite structural components of a Formula One racing car. *Composites Structures*, **68**, 491–503.

[3] Feraboli, P., Norris, C. and McLarty, D. (2007) Design and certification of a composite thin-walled structure for energy absorption. *International Journal Vehicle Design*, **44**(3–4), 247–267.

[4] Lamb, A.J. (2007) Experimental investigation and numerical modelling of composite-honeycomb materials used in Formula 1 crash structures, PhD dissertation, Cranfield University.

[5] Heimbs, S., Strobl, F., Middendorf, P. *et al.* (2009) Crash simulation of an F1 racing car front impact structure. 7th European LS-DYNA Users Conference, Salzburg, Austria.

[6] Boria, S. (2010) Behaviour of an impact attenuator for Formula SAE car under dynamic loading. *International Journal of Vehicle Structures and Systems*, **2**(2), 45–53.

[7] Obradovic, J., Boria, S. and Belingardi, G. (2012) Lightweight design and crash analysis of composite frontal impact energy absorbing structures. *Composite Structures*, **94**, 423–430.

[8] Belingardi, G. and Chiandussi, G. (2011) Vehicle crashworthiness design-General principles and potentialities of composite material structures, in *Impact Engineering of Composite Structures* (ed. S. Abrate), Springer, Wien, pp. 193–264.

[9] Belingardi, G. and Obradovic, J. (2011) Crash analysis of composite sacrificial structure for racing car. *Mobility and Vehicles Mechanics*, **37**, 41–55.

[10] Jacob, G.C., Fellers, J.F., Simunovic, S. and Starbuck, J.M. (2002) Energy absorption in polymer composites for automotive crashworthiness. *Journal of Composite Materials*, **36**(7), 813–850.

[11] FIA (2012) Technical regulations for Formula Ford with 1600cc EcoBoost engine.

[12] LS-DYNA (2009) LS-DYNA Keywords User's Manual, Version 971 beta 4, June.

[13] Feraboli, P. and Masini, A. (2004) Development of carbon/epoxy structural components for a high performance vehicle. *Composites: Part B*, **35**, 323–330.

[14] Ickert, L. (2010) Load adapted automotive lightweight structures in composite design. 4th International CFK-Valley Stade Convention, 15–16 June, Stade, Germany.

[15] Pinho, S.T., Camanho, P.P. and De Moura, M.F. (2004) Numerical simulation of the crushing process of composite materials. *International Journal of Crashworthiness*, **9**(3), 263–276.

[16] Mamalis, A.G., Manolakos, D.E., Demosthenous, G.A. and Ioannidis, M.B. (1998) *Crashworthiness of Composite Thin-Walled Structural Components*, CRC Press, New York.

[17] Thornton, P.H. and Edwards, P.J. (1982) Energy absorption in composite tubes. *Journal of Composite Materials*, **16**, 521–545.

[18] Farley, G.L. (1983) Energy absorption of composite materials. *Journal of Composite Materials*, **17**, 267–279.

[19] Farley, G.L. (1986) Effect of specimen geometry on the energy absorption capability of composite materials. *Journal of Composite Materials*, **20**, 390–400.

[20] Farley, G.L. (1991) The effects of crushing speed on the energy-absorption capability of composite tubes. *Journal of Composite Materials*, **25**, 1314–1329.

[21] Sigalas, I., Kumosa, M. and Hull, D. (1991) Trigger mechanisms in energy absorbing glass cloth/epoxy tube. *Composites Science and Technology*, **40**, 265–287.

[22] Hull, D. (1991) A unified approach to progressive crushing of fibre-reinforced composite tubes. *Composites Science and Technology*, **40**, 377–421.

[23] Hull, D. (1983) Axial crushing of fibre-reinforced composite tubes, in *Structural Crashworthiness* (eds N. Jones and T. Wierzbicki), Butterworths, London, pp. 118–135.

[24] Warrior, N.A., Turner, T.A., Robitaille, F. and Rudd, C.D. (2004) The effect of interlaminar toughening strategies on the energy absorption of composite tubes. *Composites: Part A*, **35**, 431–437.

[25] Mamalis, A.G., Manolakos, D.E., Demosthenous, G.A. and Ioannidis, M.B. (1996) Energy absorption capability of fiberglass composite square frusta subjected to static and dynamic axial collapse. *Thin-Walled Structures*, **25**(4), 269–295.

[26] Mamalis, A.G., Manolakos, D.E., Demosthenous, G.A. and Ioannidis, M.B. (1997) Analytical modeling of the static and dynamic axial collapse of thin-walled fiberglass composite conical shells. *International Journal of Impact Engineer*, **19**(5–6), 477–492.

[27] Mamalis, A.G., Manolakos, D.E., Demosthenous, G.A. and Ioannidis, M.B. (1996) Analysis of failure mechanisms observed in axial collapse of thin-walled circular fiberglass composite tubes. *Thin-Walled Structures*, **24**, 335–352.

[28] Mamalis, A.G., Manolakos, D.E., Demosthenous, G.A. and Ioannidis, M.B. (1996) The static and dynamic axial collapse of fiberglass composite automotive frame rails. *Composite Structures*, **34**, 77–90.

[29] Gupta, N.K. and Velmurugan, R. (1999) Analysis of polyester and epoxy composite shells subjected to axial crushing. *International Journal of Crashworthiness*, **5**(3), 333–344.

[30] Zhang, X., Cheng, G. and Zhang, H. (2009) Numerical investigations on a new type of energy absorbing structure based on free inversion of tubes. *International Journal of Mechanical Sciences*, **51**(1), 64–76.

[31] Belingardi, G. and Obradovic, J. (2010) Design of the impact attenuator for a formula student racing car: numerical simulation of the impact crash test. *Journal of Serbian Society for Computational Mechanics*, **4**(1), 52–65.

[32] Mamalis, A.G., Manolakos, D.E., Ioannidis, M.B. *et al.* (2002) Axial collapse of hybrid square sandwich composite tubular components with corrugated core: numerical modeling. *Composite Structures*, **58**, 571–582.

[33] Mamalis, A.G., Manolakos, D.E., Ioannidis, M.B. and Papapostolou, D.P. (2006) The static and dynamic axial collapse of CFRP square tubes: finite element modeling. *Composite Structures*, **74**, 213–225.

[34] Boria, S. and Belingardi, G. (2012) Numerical investigation of energy absorbers in composite materials for automotive applications. *International Journal of Crashworthiness*, **17**(4), 345–356.

Part Three

Damage and Failure

10

Fracture and Failure Mechanisms for Different Loading Modes in Unidirectional Carbon Fibre/Epoxy Composites

Victoria Mollón, Jorge Bonhomme, Jaime Viña and Antonio Argüelles
Polytechnic School of Engineering, University of Oviedo, Gijón, Spain

10.1 Introduction

The mass use of composite materials in the automotive industry is an important challenge for design and manufacturing engineers, as it was in the past for the aerospace industry.

Composite materials have been used in the automotive industry for many years, although some issues of an economic and technical nature have constrained their use and expansion. In 1968, Emerson Fittipaldi drove the first competition racing car to use composites. It was built with Hexcel's [1] aluminium honeycomb. The McLaren M26 was the first Formula 1 car built using a carbon fibre/epoxy composite from Hexcel® in many areas, including the cockpit sides and front cowl, the nose cone, the airbox over the engine, the radiator air scoops, part of the floor and the fuel tank areas.

Carbon composite materials are widely used in racing cars, where the production volume is low and the team budget is high. Most racing cars employ carbon fibre composite materials for the chassis, bodywork, suspension and wings. Composite structures have extremely good impact resistance and it is thanks to the development of the composite chassis that many drivers have survived high speed crashes.

Material and processing costs play a key role in the economic viability of composite materials, particularly at high production volumes, such as those found in the automotive industry.

Advanced Composite Materials for Automotive Applications: Structural Integrity and Crashworthiness,
First Edition. Edited by Ahmed Elmarakbi.
© 2014 John Wiley & Sons, Ltd. Published 2014 by John Wiley & Sons, Ltd.

Until recently, these carbon composites have been far too costly for use in high volume mainstream applications. Many car companies are researching and developing a range of new carbon fibre materials and new manufacturing techniques in order to achieve affordable prices for high volume applications. These developments try to lighten their vehicles and improve their safety and quality standards. Nowadays the benefits of carbon composites are no longer restricted to racing cars. Road cars also benefit from these safer structures. For example, BMW uses carbon fibre reinforced plastics components on high end commercial vehicles. Nissan, General Motors, Toyota, Ford and other companies are working on the same objectives.

However, other more affordable composites such as those made out of glass fibre have long been used extensively in some parts of commercial vehicles. As an example, in 1972 the Renault 5 was the first car to use a glass fibre polyester bumper instead of the classic chromed bumpers [2]. The rear fifth door is another typical example of success in glass fibre composite applications. The development of manufacturing processes such as sheet moulding compound (SMC) and resin transfer moulding (RTM) led to a more extensive use of glass fibre composites in the automotive industry. As an example, RTM was used by Renault for the popular Espace [3]. Further, SMC has been used extensively for panels and in pickup truck cargo beds due to the reduction in weight and corrosion behaviour.

The substitution of metallic structural materials by composite materials leads to lighter, more efficient structures. Lightweight structures are of primary importance for the transport industry in order to reduce operational costs and to preserve the environment, reducing both fuel consumption and pollution. Carbon fibre/epoxy materials exhibit the best mechanical properties together with the lowest density.

Long carbon fibre/epoxy materials exhibit high specific strength (strength/density), high specific stiffness (stiffness/density) and high impact strength.

Another advantage of composite materials over metallic materials is that the former are designed by superposing plies with different fibre orientations. As the fibres are orientated in the direction of the load, the final structure is highly optimised.

It is noticeable that these materials are extensively used in the aerospace industry, where the percentage of composite materials in primary structures increases year to year. For example, Boeing 787 and Airbus 350 have introduced composites in fuselage primary structures and 50% of the total airframe in the Boeing 787 aeroplane is composed of advanced composites [4]. The use of carbon fibre/epoxy and other composites has reduced the weight by 20% compared to classical aluminium designs, thereby leading to a significant reduction in fuel consumption.

In spite of the high potential of carbon fibre/epoxy composites to reduce weight, the high performance exhibited by these materials is usually reduced by premature delamination failure due to their laminated nature.

10.2 Delamination Failure

Laminate materials are composed of plies. Each ply is formed by fibres embedded in a polymeric matrix. Fibres are responsible for laminate stiffness and strength and also act as an obstacle for crack growth by increasing fatigue strength.

The matrix holds the fibres together, allows loading transfer between fibres and protects these against mechanical abrasion and chemical or environmental degradation.

Composite failure prediction is generally a complex task. Composite laminates are composed of fibres and a matrix and each of these components presents different failure modes. There are also many other factors, such as the fibre–matrix interface strength, ply stacking sequence and environmental conditions, that have a notable influence over the final failure mode.

Delamination (the separation of two adjacent plies in composite laminates) is the most common damage observed in composite materials due to the lack of fibres to arrest crack growth in that plane, although intralaminar damage (through the thickness of the laminate, including fibre damage) could also possibly produce a fracture in the composite.

Delamination takes place in matrix-rich regions between laminate plies. It can be defined as matrix cracking between plies. The initiation of delamination can be caused by manufacturing defects, sometimes due to residual stresses induced in the matrix by machining processes, cracking after the curing cycle and so on. Cracks are sometimes formed by the coalescence of micro-cracks or voids in the interface between fibre and matrix, impacts or interlaminar stresses [5] near structural discontinuities such as free surfaces, holes, abrupt changes in thickness, adhesive joints and so on.

Figure 10.1 shows some of these discontinuities that can be the starting point for the onset and growth of delamination.

Interlaminar fracture growth can be driven by three basic loading modes [6]. Interlaminar mode I is driven by tensile stresses perpendicular to the crack plane; mode II is a sliding mode; and mode III is a tear mode (see Figure 10.2).

The delamination process is characterised by means of the energy release rate (G), which is a measurement of the energy lost in the test specimen per unit of specimen width for an infinitesimal increase in delamination length. In mathematical form

$$G = -\frac{1}{B}\frac{dU}{da} \tag{10.1}$$

In this equation B is the sample width, a is the crack length and U is the elastic energy of the system. The onset of delamination takes place when G reaches a critical value, G_c.

Several experimental and numerical models have been developed in the scientific literature to compute G_c for different loading modes (G_I, G_{II} and G_{III}) [7–17].

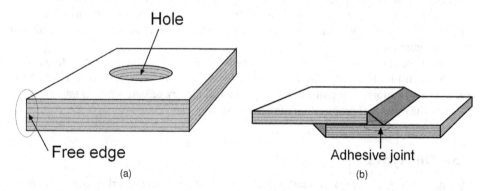

Figure 10.1   Interlaminar stresses in structural discontinuities.

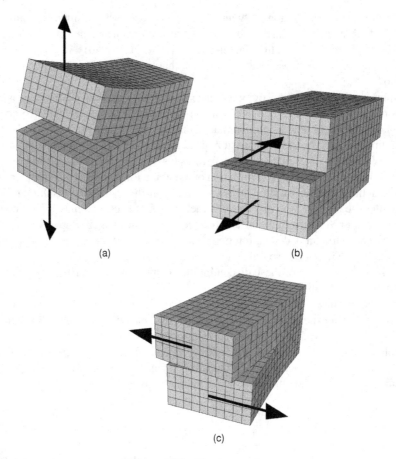

(a)   (b)

(c)

Figure 10.2   Fracture modes.

The delamination crack may propagate under the action of static or dynamic loads. It can lead to ultimate failure of the component, bringing about a reduction in stiffness and compressive load-carrying capacity of the structure. Extensive work has been carried out in recent years to predict the onset of delamination and crack growth by means of analytical methods, experimental tests and numerical procedures in different laminate structures and crack configurations.

Although scientific and technological knowledge has increased in recent decades, this mechanism still limits the potential of composite laminates to be widely used in all kinds of structural applications. Consequently, composite technology researchers continue to analyse and study the onset and propagation of delamination.

10.3   Objectives

This chapter provides an overview of fracture behaviour under interlaminar delamination failure mode in unidirectional carbon fibre reinforced epoxy composite (CFRP) laminates. The

Table 10.1 Mechanical properties of AS4/8522, AS4/3501-6 (used in the experimental programs) and M35-4/UD150 (designed specifically for the high performance car industry).

Mechanical property	Unit	M35-4/UD150	AS4/8552	AS4/3501-6
Tensile strength	MPa	2000	1703	1954
Tensile modulus	GPa	142	144	131

aim of this chapter is to study delamination toughness and the failure mechanisms operating in different loading modes (pure modes I and II and mixed mode I/II) in two different materials: AS4/8552 and AS4/3501-6 unidirectional long carbon fibre/epoxy composite laminates. These are the most widely loading modes used and well established in the scientific literature. In this work, mode III has not been considered. There are several mode III tests proposed in the scientific literature but most of them remain controversial to provide a pure mode III [18].

To do so, a broad-ranging experimental and numerical programme was developed, including fractographic observations and finite element analysis (FEA).

It should be noted that the mechanical properties of the laminates referred to in this chapter (Hexcel AS4/8552 and AS4/3506-1) are quite similar to the HexPly® M35-4 laminate, specifically developed by Hexcel for motorsport applications, including series sports cars and in general for the high performance car industry. Table 10.1 compares the mechanical properties of these three different laminates.

FEA was used to support experimental tests and to address the stress state at the crack tip under different loading modes. The stress state at the crack tip and the stress distribution were related to the surface morphology (fractographic observations) and toughness in order to gain a better understanding of the physical origin of delamination fracture toughness. These analyses are necessary to predict component and structure failures and to offer a way to optimise materials.

10.4   Experimental Programme

10.4.1   Materials and Laminate Manufacturing

Two different composites were manufactured from HexPly epoxy resin prepegs incorporating AS4 unidirectional carbon fibres. Both materials were chosen to represent a different range of toughness properties in CFRPs. One of the laminates was produced with a Hexcel 8552 modified epoxy resin in order to improve toughness properties. The commercial denomination of this composite was HexPly AS4/8552 RC34 AW196. The second composite was manufactured with a Hexcel 3501-6 epoxy resin that presented a more brittle behaviour. This composite was commercially designated as HexPly AS4/3501-6 RC37 AW190.

In order to manufacture the unidirectional laminate, 32 plies were stacked with the same fibre orientation (0°; see Figure 10.3). This direction presents the maximum stiffness and strength. Plies are usually cured together by means of pressure and temperature.

All tests were performed by means of specimens with a double beam configuration. A non-adherent insert was placed between two selected plies during lamination in order to produce an artificial delamination to initiate the crack.

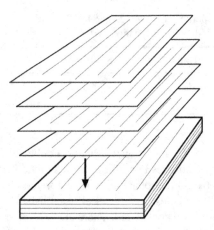

Figure 10.3   Manufacturing of unidirectional laminates.

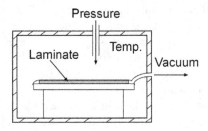

Figure 10.4   Autoclave process.

This specimen design is widely used by different testing standards due to the coincidence between fibre and crack growth directions.

In this research study, prepegs were used to manufacture the laminate. Thirty-two unidirectional plies were stacked and processed by means of an autoclave (see Figure 10.4) following the curing cycle recommended by the supplier [18]. The panels were cut into test specimens which were 6 mm thick. Figure 10.5 shows the specimen geometry.

Table 10.2 shows the fibre and matrix volume fractions of each laminate and the mechanical properties of the matrix and laminates after the curing cycle. Figure 10.6 provides a schematic

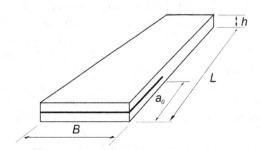

Figure 10.5   Specimens geometry. $L = 150$ mm, $a_0 = 50$ mm, $B = 25$ mm, $h = 6$ mm.

Table 10.2   Material properties.

	Property	AS4/8552	AS4/3501-6
Material description	Resin content (%)	34–38	32–45
	Fibre volume (%)	60	62
Mechanical properties of the resin	Tensile strength (MPa)	120.7	45.5
	Tensile modulus (GPa)	4.67	4.24
	Tensile elongation (%)	1.7	1.1
Mechanical properties of the laminate	E_{11} (longitudinal elastic modulus; MPa)	144 000	131 000
	E_{22} (transverse elastic modulus; MPa)	10 600	8900
	G_{12} (shear elastic modulus; MPa)	5360	5090
	σ_{11} (longitudinal tensile strength; MPa)	1703	1954
	σ_{22} (transverse tensile strength; MPa)	30.8	24.0
	σ_{12} (shear strength; MPa)	67.7	79.3

representation of the different testing specimens used to calculate the basic mechanical properties of the plies.

10.4.2   Testing Methods

A number of testing methods were developed in the scientific literature to determine the critical energy release rate (G_c) in all three pure loading modes (I, II, III) and mixed modes. Only modes I and mixed I/II tests have been raised to an international standard [7, 13]. As the mode II test is still under debate, it has been developed by means of different testing protocols. The European Structural Integrity Society (ESIS) mode II protocol [8] was followed in the present study.

In a real structure, more than one failure mode is usually present, thereby giving rise to a mixed mode mechanism. There are several experimental methods reported in the literature to determine mixed mode fracture toughness in laminated composites. The most widely used procedure is the mixed mode bending method (MMB) [9–11]. This test method allows the

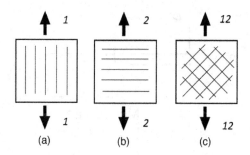

Figure 10.6   Mechanical testing modes.

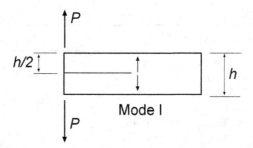

Figure 10.7 Crack in symmetry plane. Pure mode I.

calculation of G_I and G_{II}, as the test configuration controls the mode I/ mode II load percentage at the crack tip.

The asymmetric double cantilever beam (ADCB) test [12] is an interesting alternative to the MMB method. This test configuration is similar to the double cantilever beam tests (DCB). In ADCB samples, however, the crack plane lies outside the laminate midplane and so a mixed mode load state is present at the crack tip. This test is still under development (see Figures 10.7 and 10.8).

10.4.2.1   Mode I Test Method

This test was performed following the ASTM Standard D 5528-01 [7] (standard test method for mode I interlaminar fracture toughness of unidirectional fibre reinforced polymer matrix composites).

In this test, opening forces are applied to the double cantilever beam (DCB) specimens to produce mode I delamination fracture. The DCB specimen, shown in Figure 10.9, contains a non-adhesive insert at the midplane that acts as a delamination starter. The structure of the laminate was $[0°_{16}/\text{insert}/0°_{16}]$.

The load versus displacement curve recorded during the test allows the determination of the load required for the initiation and growth of the crack. This method allows the calculation of

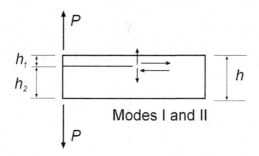

Figure 10.8 Crack in a non-symmetric plane. Mixed mode I/II.

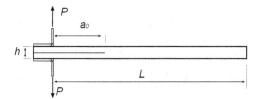

Figure 10.9 DCB test configuration.

the critical energy release rate (G_{Ic}), defined as the G needed to initiate delamination when opening loads are applied to the sample.

Specimens were loaded in displacement control at a rate of 1 mm/min on a MTS testing machine with a 5 kN load cell, recording the load–displacement response. The edge of each specimen was coated with a white water-soluble typewriter correction fluid so that the delamination could be observed more easily. A travelling optical microscope (100×) was used to measure the crack length during the test. Several criteria have been defined to determine the critical load (Figure 10.10):

- The NL criterion: the point where the load versus displacement curve deviates from linearity.
- The point at which delamination is visually observed (*VIS*).
- The point at which the compliance has increased by 5% or the load has reached a maximum value (5%/max).

The critical load, P_c, used in G_c calculations, was taken at the point of maximum load. The compliance calibration (CC) data reduction method was chosen to calculate the energy release rate, G_{Ic}, as this was the method that afforded the lowest standard deviation:

$$G_{Ic} = \frac{nP\delta}{2Ba} \qquad (10.2)$$

where P, δ, B, a and n are the critical load, load point displacement, specimen width, delamination length and a calibration parameter, respectively.

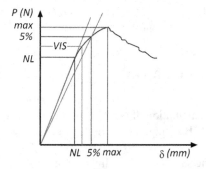

Figure 10.10 Different criteria used to define the critical load.

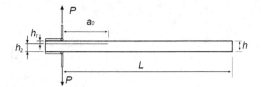

Figure 10.11 ADCB test configuration.

10.4.2.2 Mixed Mode I/II: ADCB Test

This test configuration is similar to the DCB test method. An opening load is applied to the asymmetric double cantilever beam (ADCB) specimens to produce a mixed mode delamination fracture. In ADCB samples, shown in Figure 10.11, the non-adherent insert is placed outside the laminate midplane and so a mixed mode load state is present at the crack tip (Figures 10.7 and 10.8).

This method has not been raised to an international standard. The testing fixtures needed to carry out this test are simpler than those for the MMB configuration.

The mode mixing at the crack tip varies with h_1/h_2 rate. The structure of the laminate used in this test was: six plies/insert/26 plies $[0°_6/\text{insert}/0°_{26}]$ for a total thickness of 6.18 mm. Specimens were opening mode loaded in displacement control at a speed of 0.5 mm/min on a MTS testing machine with a load cell of 5 kN.

For the ADCB specimen, in which the crack plane lies outside the laminate midplane, the calculation cannot be performed as a perfect built-in beam. Furthermore, the displacement is not symmetric with respect to the midplane. One possible approach to determine G_c is the introduction of an equivalent stiffness, (EI_{eq}).

An approach to the modified beam theory data reduction method that introduces an equivalent stiffness (EI_{eq}) was used to calculate the strain energy release rate, G_c [12]:

$$G_C = \frac{3P_C^2(a + \Delta)^2}{2BEI_{eq}} \tag{10.3}$$

where Δ and EI_{eq} can be considered as calibration parameters.

In order to calculate G_I/G and G_{II}/G (where $G = G_I + G_{II}$,), an empirical expression [12], depending on the crack position, was used. For the sample configuration employed $[0°_6/\text{insert}/0°_{26}]$, the mixed mode at the crack tip was $G_I/G = 68\%$ and $G_{II}/G = 32\%$, equivalent to a mixed-mode ratio (G_I/G_{II}) of approximately 2: 1.

10.4.2.3 Mixed Mode I/II: MMB Test

The ASTM D 6671-06 standard [13] was followed to perform MMB mixed mode I/II tests. This procedure allows the calculation of G_c, G_{Ic} and G_{IIc} for different mode I/mode II ratios. Figure 10.12 shows the MMB configuration. In this figure a_0 is the initial crack length and c the arm length.

The load is applied to the sample through hinges bonded to the end of the sample (mode I load) and by means of a rod in the centre of the sample (mode II load) in order to generate a

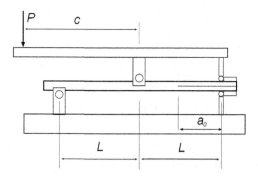

Figure 10.12   MMB test configuration.

three-point bending configuration. The mathematical expressions needed to calculate G_{Ic} and G_{IIc} are

$$G_{Ic} = \frac{12P_I^2(a + \chi h)^2}{b^2 h^3 E} \qquad (10.4)$$

$$G_{IIc} = \frac{9P_{II}^2(a + 0,42\chi h)^2}{16b^2 h^3 E} \qquad (10.5)$$

where

$$P_I = P\left(\frac{3c - L}{4L}\right) \qquad (10.6)$$

$$P_{II} = P\left(\frac{c + L}{4L}\right) \qquad (10.7)$$

$$\chi = \sqrt{\frac{E_{11}}{11G_{13}}\left[3 - 2\left(\frac{\Gamma}{1 + \Gamma}\right)^2\right]} \qquad (10.8)$$

$$\Gamma = 1.18\frac{\sqrt{E_{11}E_{22}}}{G_{13}} \qquad (10.9)$$

E_{11} and E_{22} are the Young's modulus in the direction of the fibre and transverse direction, respectively, G_{13} the shear modulus in the 1–3 plane, h the thickness of the simple arm, c the length of the loading lever, a the initial crack length and $2L$ the span of the beam.

10.4.2.4   Mode II Test Method

Mode II delamination implies the sliding of two fracture surfaces to produce the crack. The mode II test method used was the conventional end notched flexure (ENF) test. This test

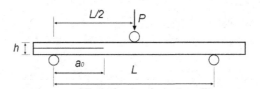

Figure 10.13   ENF test configuration.

is designed in such a way that the crack propagates along the symmetry plane of a given specimen. It was performed in flexural mode following the ESIS ENF protocol [8]. The testing configuration is shown in Figure 10.13. As previously mentioned, this test remains under debate due to unstable crack propagation and friction effects [14]. Only initiation measurements can be determined with confidence in this test.

This procedure allows the calculation of G_{IIc} for the onset of interlaminar crack growth in mode II from an initial crack, a_0, previously created in mode I ($G_{IIc\text{-}I}$) or in mode II ($G_{IIc\text{-}II}$). It shares the same specimen design as the DCB test used in mode I. The structure of the laminates was $[0°_{16}/\text{insert}/0°_{16}]$. Mode II samples were loaded in displacement control at a rate of 1 mm/min on a MTS testing machine with a 5 kN load cell.

Standard compliance calibration tests were carried out before testing in order to correlate crack length and sample compliance. The applied load and the midpoint displacement were recorded on a computer.

The experimental compliance calibration data reduction method was used to calculate the energy release rate, G_{IIc}, as this was the data reduction method that afforded the lowest standard deviation:

$$G_{IIc}^{ec} = \frac{3ma^2 P_C^2}{2B} \tag{10.10}$$

where m is the slope of the calibration line and the other parameters as in mode I.

10.5   Numerical Simulations

The simulation of delamination using the finite element method (FEM) is a useful tool to analyse fracture mechanics. The simulation of delamination using the FEM is usually performed by means of the virtual crack closure technique (VCCT) [15, 16] and the two-step extension method (TSEM) [17] with implicit solvers, or using the cohesive zone model (CZM) [19–21], generally with explicit solvers.

In the present study, FE simulations were performed by means of the two-step extension method (TSEM) using an implicit solver. Numerical simulations were compared with experimental results performed by means of DCB (mode I), ENF (mode II) and MMB and ADCB (mixed mode I/II) tests in order to support the experimental results and to gain a more in-depth knowledge of the delamination process. This technique was also used to study the stress distribution ahead of the crack tip.

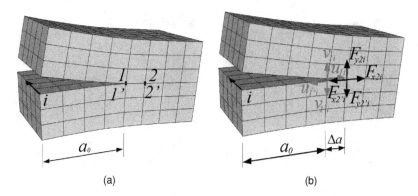

Figure 10.14   VCCT method.

10.5.1   Virtual Crack Closure Technique

The virtual crack closure technique (VCCT) is a method is based on two basic assumptions when a given crack extends a length Δa:

- The released energy is identical to the energy required to close the crack.
- When Δa is small enough, the stress state at the crack tip does not change significantly.

The latter assumption is a simplification that allows the calculation of the energy release rate with data gathered in only one calculation step. Figure 10.14 shows the FEM model. The crack path is modelled using pairs of coincident nodes with identical coordinates. Coincident nodes are initially joined. They have all degrees of freedoms (DOFs) coupled together in the unloaded model. When the imposed load or displacement reaches a critical value (P_c, δ_c), the coupled DOFs of the nodes at the crack tip are released. The nodes are then allowed to move and the crack extends one element length, Δa (Figure 10.14a and b).

The critical energy release rate (G_{Ic} and G_{IIc}) is then calculated in a second step by means of the force at the crack tip and the displacement of the released nodes. The above procedure can be analytically described as follows:

$$G_I = \frac{1}{2B\Delta a} \sum_{i=1}^{n} F_{y2i}(v_{1i} - v_{1'i}) \tag{10.11}$$

$$G_{II} = \frac{1}{2B\Delta a} \sum_{i=1}^{n} F_{x2i}(u_{1i} - u_{1'i}) \tag{10.12}$$

Where F_{x2i}, F_{y2i} are forces at the crack tip (nodes 2–2′), u_{1i}, v_{1i} are horizontal and vertical displacements of the released nodes (nodes 1–1′), B is the specimen width and Δa is the crack increment. The suffix i takes into account the extension to a 3D system, where i nodes are placed along the crack front.

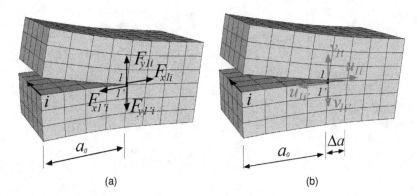

(a)   (b)

Figure 10.15   TSE method.

10.5.2   *Two-Step Extension Method*

The two-step extension method (TSEM) is similar to the VCCT, but only the first hypothesis is assumed. The crack path is once again modelled using pairs of coincident nodes. However, the forces at the crack tip are now calculated in the first step (Figure 10.15a). The imposed displacement in the sample is then held and the coupled DOFs of the nodes at the crack tip are released in the second step (Figure 10.15b).

The energy release rate is then calculated by means of the load calculated in the first step and the displacement from the second step using the following expressions:

$$G_I = \frac{1}{2B\Delta a} \sum_{i=1}^{n} F_{y1i}(v_{1i} - v_{1'i})$$
(10.13)

$$G_{II} = \frac{1}{2B\Delta a} \sum_{i=1}^{n} F_{x1i}(u_{1i} - u_{1'i})$$
(10.14)

For elements with midside nodes, as described by Krueger [15], two nodes must be released at the same time in order to avoid kinematic incompatibilities.

10.5.3   *Cohesive Zone Model*

Compared to the VCCT and TSEM, the cohesive zone model (CZM) method presents the advantage of being able to predict the onset and propagation of a crack without the need to implement a pre-existing crack [22].

The combined effects of the damage processes are defined by means of CZMs. Using these CZMs, it is possible to accurately represent the onset and growth of an initial crack.

In a cohesive element, the initial crack extension is associated with a maximum stress, while a maximum crack length is associated with zero bond strength. Two laws are defined to characterise mode I and mode II loads. The simplest relationship between stress and displacement for pure mode I and II can be implemented by means of a triangular law (bilinear law; see Figure 10.16).

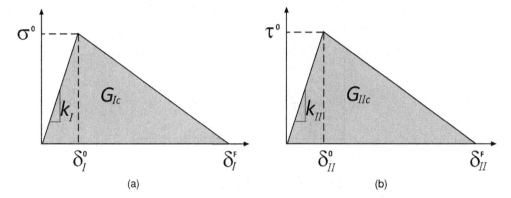

Figure 10.16   Bilinear law: (a) normal mode, (b) shear mode.

The fracture toughness of the bond between plies is a material input (G_{Ic} and G_{IIc}). The energy release rate is equal to the integral of the element traction (normal or shear) versus the crack opening:

$$G_{Ic} = \int_0^{\delta_I^F} \sigma d\delta \tag{10.15}$$

$$G_{IIc} = \int_0^{\delta_{II}^F} \tau d\delta \tag{10.16}$$

The initial slope of both curves is the penalty stiffness (k). This value is usually high in order to better reproduce the crack behaviour. Its value is usually considered to be between 10^4 and 10^7 N/mm [23, 24]. In pure modes I or II, when the normal or shear traction reaches the critical value (σ^0 or τ^0), the stiffness is progressively reduced to zero, as can be seen in Figure 10.16.

It is well known that explicit calculations with CZM show spurious oscillations of computed forces when the applied load reaches the critical load (delamination initiation and growth) [25, 26]. This problem is caused by instabilities resulting from the mathematical discontinuity at maximum load shown by bilinear laws. This problem can be controlled or avoided by means of certain techniques. For example, using artificial damping with additional energy dissipations, as proposed by Gao and Bower [20]. Another solution consists in implementing a smooth law, thus avoiding the mathematical discontinuity (Figure 10.17).

Other authors have developed cohesive models specially designed to overcome this problem, such as that of Hu et al. [21]. These authors developed a model with a pre-softening zone ahead of the existing softening zone where the initial stiffness and the interface strengths of the cohesive elements are gradually reduced as the relative displacements at these points increase. This model is known as the adaptive cohesive model (ACM). In this work none of these techniques have been used, as only the onset of crack growth was studied.

For pure modes I and II, the onset of delamination can be determined by comparing G_I or G_{II} with their critical values (G_{Ic} or G_{IIc}).

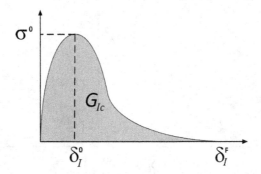

Figure 10.17   Smooth nonlinear law.

A number of theories have been developed to determine the onset and growth of cracks in a mixed load state, where delamination may occur before any of the energy release rate values reach their respective critical values. The power law [27–29] and the Benzeggagh and Kenane [30] law are two which have been widely used by the scientific community.

10.6   Fractography

A number of predictive methods have been developed in recent decades to predict delamination in composite materials. However, these models base their formulations on real (experimental) damage studies. The use of fractography to analyse the delaminated surface is thus a very useful technique to comprehend the mechanisms that drive delamination growth and lead to final structural failure.

Fractography provides important information about the causes of failure, the location of the origin of failure and the path of crack progression. In some materials, fractography also allows the determination of the environmental conditions and the stress state at the onset of crack growth [29].

In contrast to isotropic materials, fracture can take place in a number of different planes in laminate materials. Fractographic observations can help determine the specific mechanism (interlaminar, intralaminar or translaminar) that may take place during crack propagation [30–32]. The delamination fracture surfaces were examined to deduce information about damage in each tested material using a Jeol scanning electron microscope (SEM). SEM photomicrographs were taken just beyond the delamination insert and in regions far from it.

Fractographic morphologies, failure modes and crack tip stress state were related in the present study. For this purpose, the stress state at the crack front was modelled by FE analysis, as described below.

10.7   Results and Discussion

10.7.1   Experimental Results

Figures 10.18–10.20 show the different testing fixtures and configurations used for mode I (DCB), mixed mode I/II (MMB) and mode II (ENF) testing, respectively.

Figure 10.18   Mode I test fixture.

Figure 10.19   MMB test fixture.

Figure 10.20   ENF test fixture.

Table 10.3   Experimental critical energy release rate, G_c (J/m^2), for different loading modes.

Loading mode	AS4/8552	AS4/3501-6
Pure I (100% G_I) DCB	274.2 (±29.1)	90.6 (±18.3)
Mixed (68% G_I, 32%G_{II}) ADBC	297.5 (±44.8)	205.9 (±38.2)
Mixed (69% G_I, 31%G_{II}) MMB	294.2 (±60.1)	174.6 (±16.8)
Mixed (60% G_I, 40%G_{II}) MMB	—	369.3 (±44.3)
Pure II (100% G_{II}) ENF	738.0 (±103.8)	943.4 (±67.2)

The critical energy release rate (G_c) of each material under the different loading modes obtained in the experimental tests is presented in Table 10.3. In this work, the AS4/3501-6 material was tested in two different mode mixing ratios: $G_{II}/G = 0.31$ and 0.4, while the AS4/8552 laminate was tested only at $G_{II}/G = 0.31$.

As expected, the AS4/8552 material in mode I and mixed I/II was tougher than the AS4/3501-6 laminate because of the behaviour of the matrix. However, the critical energy release rate, G_c, was higher in the brittle epoxy matrix material, AS4/3501-6, in pure mode II.

This behaviour could be associated with a higher cusp or "hackle morphology" formation, especially in brittle systems (see Section 10.7.3). The rougher the surface, the more energy will be consumed. "Hackles" are usually associated with mode II fractures [33, 34] in brittle composite systems. In two studies, Hibbs and Bradley [35] also observed that "hackles" were absent from ductile resin systems such as AS4/Dow-Q6 due to extensive yielding and ductile fracture. It seems that "hackles" are the result of coalescence of micro-cracks, rather than the result of matrix yielding and plastic deformation.

As can be seen in Table 10.3, the critical energy release rate increases with increasing mode mixing. Similar results were found by other authors [36]. As also found by other authors, the scattering of the results increases as the mode II component increases [37], that is the standard deviation increases.

10.7.2   Numerical Results

Every sample tested in the experimental programme was modelled by means of an Ansys® software package. The critical load obtained in each test was implemented as an input in the FE models. The two-step extension method was used to obtain G_c.

2D models were prepared in order to simplify the calculation process. In previous work, it was demonstrated that 3D and 2D models furnished similar results [38]. The element used was plane 42 with the plane strain option. This element is defined by four nodes with two degrees of freedom at each node: translations in the nodal x and y directions (u_x and v_y). The crack plane was modelled by means of coincident nodes. The degrees of freedom of these nodes were coupled together ahead of the crack tip. These nodes were released in the second calculation step in order to extend the crack one element length.

Figure 10.21 shows the mesh used to model symmetric specimens (DCB, ENF and MMB). This mesh is regular and composed of standard elements in its entirety. Stress and deformation fields around the crack tip generally have high gradients. The mesh around the crack tip should be refined enough to find accurate results. Ansys tutorials recommend a collapsed

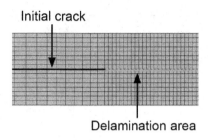

Figure 10.21   Mesh for FE models.

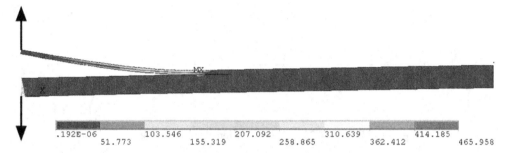

Figure 10.22   ADCB FEM model.

eight-node quadratic solid element to model the crack tip. In previous work, the optimal element length needed to find accurate results was studied [38]. The element length was set to 0.33 mm at the crack tip ($\Delta a/a_0 = 0.0067$). It was likewise demonstrated in the same work that the use of collapsed elements furnished G_c results similar to the standard elements.

As recommended by Ansys, eight-node collapsed elements would improve the stress solution near the crack tip. However, in the energy calculation it was found that standard elements afford sufficiently accurate results. Figure 10.22 shows the deformed shape and stress distribution for the ADCB specimen. Table 10.4 shows the obtained FEM results.

A good agreement was found between experimental and numerical results (Tables 10.3 and 10.4), as the results obtained in both procedures show errors below 10%.

Table 10.4   Numerical results G_c (J/m^2), for different loading modes (FEM).

Loading mode	AS4/8552	AS4/3501-6
Pure I (100% G_I) DCB	259.4 (±47.8)	95.0 (±20.3)
Mixed (68% G_I, 32% G_{II}) ADBC	276.7 (±44.5)	227.4 (±45.4)
Mixed (69% G_I, 31% G_{II}) MMB	260.7 (±53.3)	170.0 (±16.1)
Mixed (60% G_I, 40% G_{II}) MMB	—	341.5 (±41.2)
Pure II (100% G_{II}) ENF	791.3 (±146,9)	951.5 (±65.5)

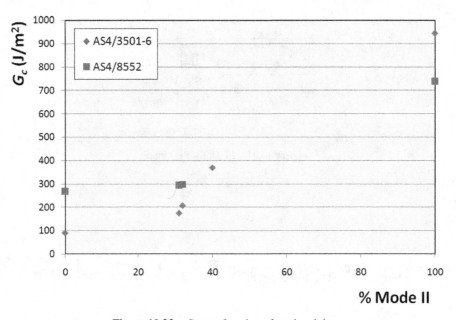

Figure 10.23 G_c as a function of mode mixing.

10.7.3   Fractographic Analysis

The delamination toughness and associated fracture morphology of unidirectional carbon fibre/epoxy composites varies with mode mixing [39]. Figure 10.23 shows the delamination toughness for different mode ratios in both the AS4/8552 and AS4/3501-6 materials.

Figures 10.24–10.26 show the SEM observations of the delaminated surfaces. Clean fibres and many imprints were observed in both materials in the different loading states, where fibres

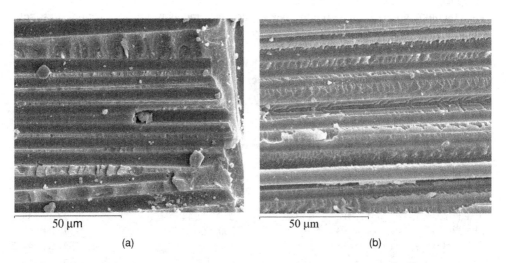

Figure 10.24   DCB: 100% mode I, 0% mode II. (a) AS4/8552. (b) AS4/3501-6.

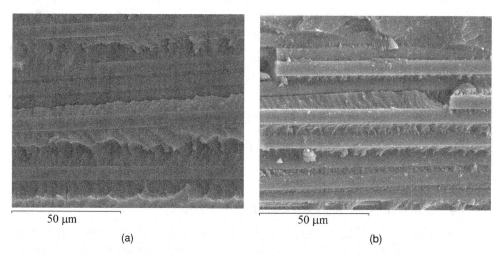

50 μm   50 μm

(a)   (b)

Figure 10.25   MMB: 69% mode I, 31% mode II. (a) AS4/8552. (b) AS4/3501-6.

have been pulled away from the matrix, indicating interfacial failure (crack progression across the fibre/matrix interfaces).

As can be seen in Figure 10.24a, b, the fracture surfaces under pure mode I are generally flat and smooth, indicating a brittle cleavage fracture which would explain the low mode I fracture toughness. Therefore, toughness is controlled by cohesive fracture or matrix cleavage and fibre bridging. Composites hence exhibit the lowest toughness. This topographical feature is typical of mode I and can also be observed in Figure 10.24b.

Figure 10.25a, b shows 2: 1 mixed-mode ratio fracture surfaces. Here, mode I is still predominant. The micrographs show inclined lines with respect to the axis of the fibres. This morphology can be related to a plastic micro-creep [40] of the matrix. The confluence of

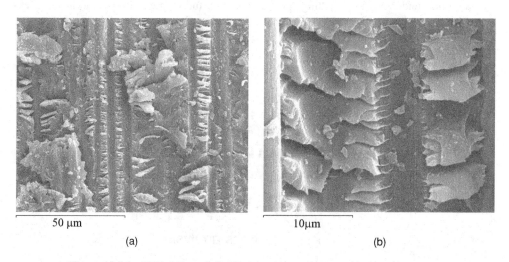

50 μm   10μm

(a)   (b)

Figure 10.26   ENF: 0% mode I, 100% mode II. (a) AS4/8552. (b) AS4/3501-6.

these lines, known as "river markings" [41] or "corrugated roof" [35] structure, indicates the direction of crack propagation These "river markings" are the result of the interaction of adjacent fracture planes. This structure is expected to be found in the MMB test because of the higher contribution of mode I to G ($G_I/G = 69\%$).

Figure 10.26a, b show mode II fracture surfaces. These surfaces are coarser and the typical features of brittle composite materials in mode II delamination, that is "hackle markings", are present on both mode II surfaces of the AS4/8552 and AS4/3501-6 composites. This texture is associated with regions of shear stress and is characterised by cusps or "hackle" formation. Morris [42] studied the slope of the shear cusps, which is particularly useful for failure analysis of composite structures.

10.7.4   Stress State at the Crack Front

A finite element analysis was developed to model the stress state ahead of the crack to better understand the fracture micro-mechanisms acting when the sample is tested under the different loading modes (I, II and I/II) and to explain how the different morphologies are formed. The aim of this analysis was to study the influence of the stress state on the formation and growth of micro-cracks and on delamination fracture toughness.

To analyse the principal tensile stresses at pure I and II modes and mixed mode I/II, a simple FE model including fibres and matrix was developed (Figure 10.27). This model was loaded in mode I, mode II and mixed mode I/II ($G_{Ic}/G_c = 68\%$). The elastic modulus of the AS4 carbon fibre and matrix implemented in the model were $E = 231.0$ GPa and $E = 4.67$ GPa according to Hexcel data sheets [43,44].

Figures 10.28–10.30 show the results obtained in the FEM analysis. These figures show how the principal tensile stress orientation changes with the mode mixing at the crack front.

Under pure mode I, fracture initiates as micro-cracks at the fibre/matrix interface. The plane of these micro-cracks coincides approximately with the delamination plane because the first principal tensile stresses are perpendicular to this plane (see Figure 10.31a, b).

Therefore, under opening loading (pure I mode) to the laminate plane, the local cracks extend towards one another and coalesce to form the crack morphology. As can be seen in Figure 10.31, relatively little fractured surface is generated and the corresponding toughness is low.

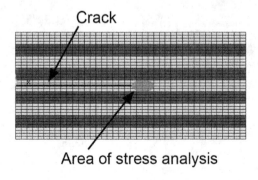

Figure 10.27   Fibre-matrix FE model used to study the principal tensile stresses.

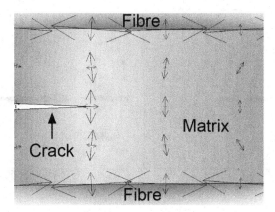

Figure 10.28   Pure mode I. Principal tensile stresses at the crack tip.

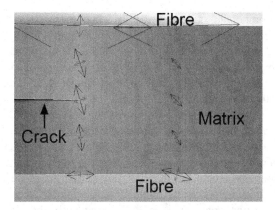

Figure 10.29   Mixed mode ratio of 2: 1 (68% mode I + 32% mode II). Principal tensile stresses at the crack tip.

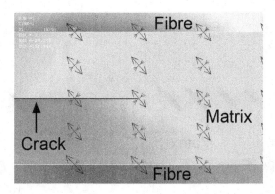

Figure 10.30   Pure mode II. Principal tensile stresses at the crack tip.

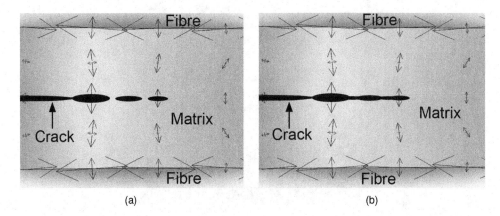

(a)   (b)

Figure 10.31   Pure mode I. (a) Micro-crack formation. (b) Micro-crack coalescence.

As the mode II component increases, that is when shear stresses are introduced (see Figure 10.29), the first principal tensile stresses turn, forming an angle with respect to the delamination plane. Under pure mode II loading (Figure 10.30), this angle is inclined 45° to this plane, as found by other authors [45]. As a result of the inclination of the principal stresses, the micro-cracks are reoriented, so that they develop perpendicular to the new direction of the stresses. This is illustrated in Figure 10.32.

Micro-cracks consequently coalesce, leading to the formation of cusps [46] or hackles (Figure 10.33), a characteristic morphology associated with mode II delamination. Macroscopically, these cusps increase surface roughness and the total fractured area.

The cusps accordingly reorient, forming an angle to the fracture plane. This angle is related to the mode II component percentage; that is it depends on the mode mixing ratio (G_I/G_{II}). As the mode II percentage increases, the angle or inclination increases, the matrix roughness increases, the total fracture area increases and consequently the toughness increases (see Table 10.3). Under a 2: 1 mixed mode ratio testing, the inclination of the cusps is also low and no clear change in the failure mechanism is observed.

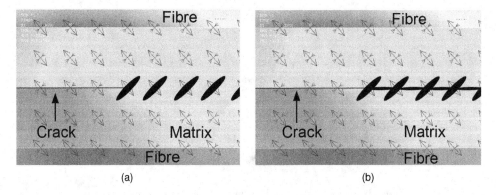

(a)   (b)

Figure 10.32   Pure mode II. (a) Micro-crack formation. (b) Micro-crack coalescence.

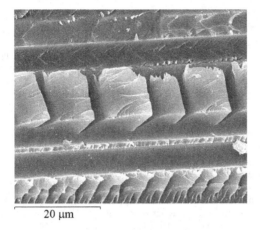

20 μm

Figure 10.33   Pure mode II. Hackle morphology.

10.8   Conclusion

In this chapter, the onset and growth of delamination in unidirectional composite materials has been studied from different points of view, employing experimental and numerical procedures and fractographic observation of the delaminated surfaces.

A good agreement was found between experimental and numerical results (Tables 10.3 and 10.4), as the results obtained in both procedures show errors below 10%.

Fractographic observations and numerical simulation of the areas ahead of the crack tip have shown that the principal stress orientation has a strong influence on the formation of micro-cracks and the final mechanisms that lead to material failure and the resulting surface features.

As found by other authors, the total energy release rate G_c increases with increasing mode II loading. This observation could be related to the rough surface (rougher than that for mode I) typically observed in mode II, the creation of which would consume more energy.

Finally, laminates manufactured with the modified matrix afforded higher G_c values in mode I and mixed mode I/II, while the laminate manufactured with the unmodified matrix gave higher G_c results in mode II. This finding is in agreement with the fractographic observations, as the unmodified matrix showed a rougher surface (more energy consumption) in mode II than that of the modified matrix. The "hackle" structure produces a rougher fracture surface that involves more energy consumption during the delamination process.

References

[1] Hexcel Composites Company (1999) Hexcel Composites Company, www.hexcel.com (Accessed 18th February 2012).

[2] Bunsell, A.R. and Renard, J. (2005) *Fundamentals of Fibre Reinforced Composite Materials*, 1st edn, Taylor and Francis, London, ISBN 10:0750306890.

[3] Owen, M.J., Middleton, V. and Jones, I.A. (2000) *Integrated Design and Manufacture Using Fibre-Reinforced Polymeric Composites*, Woodhead Publishing Ltd, London, ISBN 10:1855734532.

[4] Gardiner, G. (2011) Primary structure repair: the quest of quality. *High-Performance Composites*, **73**, 41–52.
[5] Green, E. (1999) *Marine Composites*, 2nd edn, Eric Greene Associates Inc., Annapolis, Maryland. ISBN 0-9673692-0-7.
[6] Irwin, G.R. (1958) *Fracture. Handbuck der Physik*, Springer, Heidelberg, vol. **6**, pp. 551–590.
[7] ASTM (2007) D5528-01e3. Standard Test Method for Mode I Interlaminar Fracture Toughness of Unidirectional Fiber-Reinforced Polymer Matrix Composites.
[8] Moore, D.R., Pavan, A. and Williams, J.G. (2001) *Fracture Mechanics Test Methods for Polymers, Adhesives and Composites*, ESIS Publication No. 28, Elsevier, New York, pp. 271–359.
[9] Crews, J.H. Jr and Reeder, J.R. (1988) *A Mixed-Mode Bending Apparatus for Delamination Testing*. NASA Technical Memorandum 100662, NASA Langley Research Centre, Hampton, VA.
[10] Reeder, J.R. and Crews, J.H. Jr (1990) The mixed-mode bending method for delamination testing. *AIAA Journal*, **28**(7), 1270–1276.
[11] Reeder, J.R. and Crews, J.H. Jr (1992) Redesign of the mixed-mode bending delamination test to reduce nonlinear effects. *Journal of Composites Technology and Research*, **14**(1), 12–19.
[12] Mollón, V., Bonhomme, J., Viña, J. and Argüelles, A. (2010) Mixed mode fracture toughness: an empirical formulation for G_I/G_{II} determination in asymmetric DCB specimens. *Engineering Structures*, **32**, 3699–3703.
[13] ASTM (2006) D6671/D6671M. Standard Test Method for Mixed Mode I-Mode II Interlaminar Fracture Toughness of Unidirectional Fiber Reinforced Polymer Matrix Composites.
[14] Davies, P., Blackman, B.R.K. and Brunner, A.J. (1998) Standard test methods for delamination resistance of composite materials: current status. *Applied Composite Materials*, **5**, 345–364.
[15] Krueger, R. (2004) Virtual crack closure technique. History, approach and applications. *Applied Mechanics Reviews*, **57**, 109–143.
[16] Rybicki, E.F. and Kanninen, M.F. (1977) Finite element calculation of stress intensity factors by a modified crack closure integral. *Engineering Fracture Mechanics*, **9**(4), 931–938.
[17] Bonhomme, J., Argüelles, A., Viña, J. and Viña, I. (2009) Computational models for mode I composite fracture failure: the virtual crack closure technique versus the two-step extension method. *Meccanica*, **45**, 297–304.
[18] Hexcel (2011) http://www.hexcel.com/Resources/prepreg-data-sheets (Accessed 8th February 2012).
[19] Needleman, A. (1999) An analysis of intersonic crack growth under shear loading. *Journal of Applied Mechanics*, **66**, 847–857.
[20] Gao, Y.F. and Bower, A.F. (2004) A simple technique for avoiding convergence problems in finite element simulations of crack nucleation and growth on cohesive interfaces. *Modelling and Simulation in Materials Science and Engineering*, **12**, 453–463.
[21] Hu, N., Zemba, Y., Okabe, T. *et al.* (2008) A new cohesive model for simulating delamination propagation in composite laminates under transverse loads. *Mechanics of Materials*, **40**, 920–935.
[22] Xie, D. and Waas, A.M. (2006) Discrete cohesive zone model for mixed-mode fracture using finite element analysis. *Engineering Fracture Mechanics*, **73**, 1783–1796.
[23] Camanho, P.P., Dávila, C.G. and de Moura, M.F. (2003) Numerical simulation of mixed-mode progressive delamination in composite materials. *Journal of Composite Materials*, **37**(16), 1415–1438.
[24] Zou, Z., Reid, S.R., Li, S. and Soden, P.D. (2002) Modelling interlaminar and intralaminar damage in filament wound pipes under quasi-static indentation. *Journal of Composite Materials*, **36**, 477–499.
[25] Mi, Y., Crisfield, M.A. and Davis, G.A.O. (1998) Progressive delamination using interface element. *Journal of Composite Materials*, **32**, 1246–1272.
[26] Gonçalves, J.P.M., de Moura, M.F.S.F., de Castro, P.M.S.T. and Marques, A.T. (2000) Interface element including point-to-surface constraints for three-dimensional problems with damage propagation. *Engineering Computations*, **17**(1), 28–47.
[27] Camanho, P.P. and Matthews, F.L. (1999) Delamination onset prediction in mechanically fastened joints in composite laminates. *Journal of Composite Materials*, **33**(10), 906–927.
[28] Dávila, C.G. and Johnson, E.R. (1993) Analysis of delamination initiation in post buckled dropped-ply laminates. *AIAA Journal*, **31**(4), 721–727.
[29] Cui, W., Wisnom, M.R. and Jones, M. (1992) A comparison of failure criteria to predict delamination of unidirectional glass/epoxy specimens waisted through the thickness. *Composites*, **23**(3), 158–166.
[30] Benzeggagh, M.L. and Kenane, M. (1996) Measurement of mixed-mode delamination fracture toughness of unidirectional glass/epoxy composites with mixed-mode bending apparatus. *Composites Science and Technology*, **56**, 439–449.

[31] Parrington, R.J. (2002) Fractography of metals and plastics. *Journal of Failure Analysis and Prevention*, **2**, 16–19.
[32] Baas, S.J.W. (1994) Garteur AG14: Fractography of Composites. Garteur TP No 083.
[33] Richards-Frandsen, R. and Naerheim, Y. (1983) Fracture morphology of graphite /epoxy composites. *Journal Composite Materials*, **17**, 105–113.
[34] Friedrich, K. (1989) *Fractographic Analysis of Polymer Composites. Application of Fracture Mechanics to Composite Materials* (ed. K. Friedrich), Elsevier, Amsterdam, pp. 425–487.
[35] Hibbs, M.F. and Bradley, W.L. (1987) Correlations between Micromechanical Failure Processes and the Delamination Toughness of Graphite /Epoxy Systems. ASTM STP 948, pp. 68–97.
[36] Mathews, M.J. and Swanson, S.R. (2007) Characterization of the interlaminar fracture toughness of a laminated carbon/epoxy composite. *Composites Science and Technology*, **67**, 1489–1498.
[37] Ducept, F., Gamby, D. and Davies, P. (1999) A mixed-mode failure criterion derived from tests on symmetric and asymmetric specimens. *Composites Science and Technology*, **59**(4), 609–619.
[38] Bonhomme, J., Argüelles, A., Viña, J. and Viña, I. (2009) Numerical and experimental validation of computational models for mode I composite fracture failure. *Computational Materials Science*, **45**, 993–998.
[39] Greenhalgh, E.S. (1998) Characterisation of Mixed-Mode Delamination Growth in Carbon-Fibre Composites, Ph.D. Thesis, Imperial College, UK.
[40] Purslow, D. (1986) Matrix fractography of fibre reinforced epoxy composites. *Composites*, **17**, 289–303.
[41] Smith, B.W. and Grove, R.A. (1987) Determination of Crack Propagation Directions in Graphite /Epoxy Structures. Fractography of Modern Engineering Materials. ASTM STP 948.
[42] Morris, G.E. (1979) Determining Fracture Directions And Fracture Origins On Failed Graphite /Epoxy Surfaces. Nondestructive evaluation and flaw criticality for composite materials, ASTM STP 696, pp. 274–297.
[43] Hexcel (2011) http://www.hexcel.com/resources/datasheets/carbon-fiber-data-sheets/as4.pdf (Accessed 8th February 2012).
[44] Hexcel (2011) http://www.hexcel.com/Resources/DataSheets/Prepreg-Data-Sheets/8552_eu.pdf (Accessed 8th February 2012).
[45] Fleck, N.A. (1991) Brittle fracture due to an array of microcracks. *Proceedings of the Royal Society of London, Series A (Mathematical and Physical Sciences)*, **432**, 55–76.
[46] Greenhalgh, E.S. (2009) *Failure Analysis and Fractography of Polymers Composites*, Woodhead Publishing, ISBN 10:1845692179.

11

Numerical Simulation of Damages in FRP Laminated Structures under Transverse Quasi-Static or Low-Velocity Impact Loads

Ning Hu[1], Ahmed Elmarakbi[2], Alamusi[1], Yaolu Liu[1], Hisao Fukunaga[3], Satoshi Atobe[3] and Tomonori Watanabe[1]

[1]*Department of Mechanical Engineering, Chiba University, Yayoi-cho 1-33, Inage-ku, Chiba 263-8522, Japan*
[2]*Department of Computing, Engineering and Technology, Faculty of Applied Sciences, University of Sunderland, Sunderland, SR6 0DD, UK*
[3]*Department of Aerospace Engineering, Tohoku University, Aramaki-Aza-Aoba 6-6-01, Aoba-ku, Sendai 980-8579, Japan*

11.1   Introduction

It is well known that fibre reinforced plastic (FRP) composites possess high specific strength (strength to density ratio) and high specific stiffness (strength to density ratio) and are widely applied to the fields of automobiles, aerospace, sports, constructions, civil engineering and on on. Further, it is a well recognised phenomenon that this kind of material is very susceptible to the transverse quasi-static or low-velocity impact loads due to the various possible types of damage, for example fibre breakage, matrix cracking and delamination, which lead to a significant decrease of material strength and stiffness. This chapter addresses the numerical modelling and simulations of the occurrence and propagation of these damages in FRP laminated structures under transverse quasi-static or low-velocity impact loadings. The focus of this chapter is on the key issue of numerical modelling of delamination using cohesive

Advanced Composite Materials for Automotive Applications: Structural Integrity and Crashworthiness,
First Edition. Edited by Ahmed Elmarakbi.
© 2014 John Wiley & Sons, Ltd. Published 2014 by John Wiley & Sons, Ltd.

elements, which is a conventional difficulty due to numerical instability in the simulation process. To overcome this numerical instability, several recent achievements of effective numerical approaches are proposed and reported by the present authors in this chapter. These numerical approaches are employed to successfully tackle this complex problem of delamination propagation and some other typical damage, for example, fibre breakage and matrix cracking in FRP laminated plates or shells under quasi-static and low-velocity impact loadings with additional experimental verifications.

In recent years, the automotive industry has been widely viewed as being the industry in which the greatest volume of advanced FRP materials will be used in the future. Because this industry is mature and highly competitive, the principal motivation for introducing composites is cost savings. The most visible example in recent years is the Pontiac Fiero, which has all-composite exterior panels. With the introduction of various kinds of short fibre or long fibre reinforced plastic materials, the automotive industry can realise the following advantages: (i) weight savings of 20% or more, which leads to the large amount of saving in fuels, (ii) improvement of corrosion resistance due to composites, which leads to a longer life span of vehicles (predicted 20 years) and (iii) improvement of passenger's amenity and production cost-cutting.

Also, for another important viewpoint of operation safety of drivers, from some recent limited research outcomes by experiments and numerical simulations, it has been clarified that the impact resistance of composite components might be higher than that provided by metallic components. However, for the practical application of FRP materials in the automotive industry, little knowledge exists on how FRP responds in automotive applications during transverse quasi-static or impact loading conditions. It is crucial to understand and quantify the basic deformation and failure mechanisms active in fibre materials during complex transverse loading conditions through powerful computational modelling, besides experiments.

It is well known that very complicated damage phenomena occur in composite laminated structures under transverse loads. Understanding the mechanisms of the occurrence and propagation of this damage is crucial for properly designing this kind of composite structures. Generally, there are two main categories of the various types of damage in composite laminates under transverse loads. The first category consists of various kinds of in-plane damage, such as fibre failure and transverse matrix cracking and on on. The second category includes interface damage, that is, delaminations between multiple laminae. Clearly understanding the damage mechanism of only one damage pattern in the composites mentioned above is a tough task. Moreover, the interaction between the different types of damage makes this problem more difficult. So far, a lot of studies have been conducted to experimentally or numerically investigate the damage phenomena of composite laminates under transverse loads [1–10]. In the following, only work in the field of theoretical models and numerical simulations is briefly described.

First, for various types of damage, such as in-plane damage and delamination, some authors have proposed various stress-based criteria. For example, Chang and Chang [1] proposed the tension–shear failure criteria for matrix cracking. In this model, the damage conditions relate to single ply failure, which are not unique, but have a great degree of commonality with other widely accepted criteria. Hou et al. [2] coupled the Chang–Chang model [1] with the delamination criterion of Brewer and Lagace [3]. This model thoroughly considers the influences of various stress components on damage. In their latter work, Hou et al. [4] pointed out that the above criteria [2] for some in-plane failure modes, such as fibre failure and matrix cracking, can predict failure quite well and remain unchanged. The predicted delamination, however, was underestimated as it was prohibited due to out of plane compression. In reality, delamination is able to develop in regions where the interlaminar shear is high and the out of plane compression is relatively low. They further modified the delamination criterion by taking into account the interaction between out of plane compression and interlaminar shear [4]. From many previous studies (e.g. [5]) it has been concluded that the stress-based criteria are enough to predict the initiation of various types of damage, especially for in-plane damage, such as fibre failure and matrix cracking. However, there has been a lot of debate on using stress-based criteria to predict damage propagation, especially for delamination extension. As pointed out by Davies and Zhang [5], in the stress-based criteria for delamination, the scale effects would not be exhibited as in a fracture model. Therefore, it is not accurate to use the stress-based criteria to predict the delamination size or propagation. It certainly requires an energy release algorithm based on fracture mechanics.

To understand the mechanism of delamination occurring on the interfaces of different layers, besides the above stress-based criteria, some methods based on fracture mechanics have also been proposed. For instance, in Refs. [6–8], the strain energy released rate of the mixed-mode at the delamination front is directly evaluated and used to predict the delamination propagation. This kind of method cannot deal with the initiation of delamination; therefore, some initial pre-existing small delamination areas are needed. Further, cohesive interface models, which can tackle the initiation and propagation of delamination simultaneously, are widely used to simulate interface damage propagation due to their inherent simplicity and efficiency [9–15]. However, using cohesive finite elements to simulate interface damage, such as delamination, leads to a problem of strong numerical instability [9–15], especially when the interface is strong and comparatively coarse meshes are utilised. This problem is caused by a well known elastic snap-back instability, which occurs just after the stress reaches the peak strength of the interface. Traditionally, this problem can be controlled using some direct techniques. For instance, a very fine mesh can alleviate this numerical instability; however, it leads to a very high computational cost. Also, very low interface strength and the initial interface stiffness can partially remove this convergence problem; however, this leads to a lower slope of loading history in the loading stage before the occurrence of damage. Further, various generally orientated methodologies can be used to remove this numerical instability, for example, Riks method [16] to follow the equilibrium path after instability, as well as the artificial damping method with additional energy dissipation proposed by Gao and Bower [15] in the implicit time integration scheme (Newmark-β method).

Although there has been a lot of work in the numerical simulation of damage propagation in composite laminates, there are only a few studies considering the various damage patterns simultaneously. Some researchers (e.g. Ref. [17]) employed stress-based criteria entirely to deal with in-plane damage and delamination. As stated previously, there may be some problems when using stress-based criteria to tackle delamination, especially large delamination. Also, for some simple structures, such as beam-like laminates, Geubelle and Baylor [10] used a 2D plane-stress cohesive element to study damage in thin composite beams subjected to low-velocity impact. In their work, cohesive elements were introduced along the boundaries of the inner layers and inside the transverse plies to simulate the spontaneous initiation and propagation of transverse matrix cracks and delamination. However, this kind of method is difficult to extend into three-dimensional (3D) complex composite structures due to extremely high computational cost. As a matter of fact, up to now, there have been only a few reports [1] for the numerical simulation of complex damage propagations in 3D composite structures, such as shells. In the authors' previous work [7, 8], for cross-ply CFRP plates, the in-plane damage and delamination are treated independently. There, the in-plane damage is dealt with using stress-based criteria [2] while the delamination is evaluated using the strain energy release rate based on the Mindlin plate element and the implicit time integration scheme (Newmark-β method). For the application of 3D cohesive elements, in the authors' recent work [18], a move-limit technique to restrict the movement of nodes located in the cohesive zone to avoid the numerical instability is proposed by using two assumed rigid walls. This technique can lead to very stable numerical simulations of delamination propagation in 3D problems, which, however, introduces the artificial external work done by the contact forces of rigid walls during the delamination propagation. Moreover, this method is only suited for quasi-static problems. To overcome the above drawbacks, we have further developed an adaptive cohesive model [19] to stably and accurately simulate the delamination propagations in composite laminates under quasi-static and low-velocity impact transverse loads using comparatively *coarse meshes*. In this model, a *pre-softening zone* ahead of the existing traditional softening zone is proposed. In this pre-softening zone, the initial stiffness and the interface strengths at the integration points of cohesive elements are gradually reduced as the corresponding effective relative displacements at these points increase. However, the onset displacement corresponding to the onset damage is not changed in this model. Moreover, the fracture toughness of materials for determining the final displacement of complete decohesion is kept constant. This model is a revolutionary technique which can be effectively and easily implemented in various commercial codes. This technique has been further extended into a rate-dependent cohesive model in our following work [20] with its implementation in LS-DYNA commercial software, which verifies its effectiveness once again. The above mentioned three studies [18–20] are mainly based on the explicit time integration scheme for quasi-static and low-velocity impact problems with transverse loadings. Moreover, 3D brick finite elements, which can deal with the possible large deformation during low-velocity impacts (e.g. [21]) are used for the laminate portion.

With the previous background in mind, the objective of this chapter is to describe the numerical simulations of various types of damage in complex FRP laminated structures under transverse quasi-static or low-velocity impact loads. Two categories of damage patterns in composite structures under transverse loads are tackled independently. The first category includes the various types of in-plane damage, that is, fibre failure, transverse matrix cracking and transverse matrix crushing. A kind of stress-based criterion [2] is adopted to deal with this kind of damage at the integration points within an individual 3D element. The strategy for updating the in-plane stiffness due to various types of in-plane damages suggested. For the second category of damage, that is, delamination between multiple laminae, a bi-linear cohesive interface model [9] is adopted. However, to stabilise the finite element simulations of delamination propagation in composite laminates when using the cohesive interface model, we focus on the description of our recent several outcomes about the improvement on the traditional cohesive model, which include: (i) implementation of artificial damping technique in the explicit time integration scheme, which is similar to that used in the implicit time integration scheme [15], (ii) move-limit technique [18] and (iii) a new adaptive cohesive model [19] and its extension into rate-dependent problems [20]. Besides the description of theoretical contents, a typical experiment, that is, the double cantilever beam (DCB) problem is employed to characterise the properties of the above techniques. Moreover, a low-velocity impact example is used to show the effectiveness of the adaptive cohesive model.

11.2   Theory

11.2.1   Theory of Finite Element Method

The explicit time integration scheme is widely used to simulate the interface crack propagation [10–12], because unlike the implicit time integration scheme, the Newton–Raphson iteration scheme is not needed. In this section, the explicit central difference algorithm is adopted to simulate the delamination propagation. The motion equation of laminated plates at the nth time step can be simply described as

$$\mathbf{M}\ddot{\mathbf{u}}_n + \mathbf{C}\dot{\mathbf{u}}_n + \mathbf{p}_n = \mathbf{f}_n \tag{11.1}$$

where $\mathbf{M}$ and $\mathbf{C}$ are mass and damping matrices, and $\mathbf{p}_n$ and $\mathbf{f}_n$ are the internal nodal forces provided by finite elements due to deformation and external forces, respectively.

The displacement vector can be evaluated as

$$\mathbf{u}_{n+1} = \left(\mathbf{M} + \frac{\Delta t}{2}\mathbf{C}\right)^{-1}\left[(\Delta t)^2(-\mathbf{p}_n + \mathbf{f}_n) + 2\mathbf{M}\mathbf{u}_n - \left(\mathbf{M} + \frac{\Delta t}{2}\mathbf{C}\right)\mathbf{u}_{n-1}\right] \tag{11.2}$$

where Δt is the time increment, and the internal forces can be predicted as

$$\mathbf{p}_n = \sum_{i=1}^{NE}\mathbf{k}_n^i\left(\mathbf{u}_n^i\right)\mathbf{u}_n^i \tag{11.3}$$

where NE is the total number of elements in the structures, including the cohesive elements at the interfaces and the structural elements. $\mathbf{k}_n^i(\mathbf{u}_n^i)$ is the static effective stiffness matrix of the ith element. For the cohesive elements, the static effective stiffness is a function of the deformation state when considering the damage in analysis. The finite elements for analysing the laminate portion and the time integration scheme will be described in detail later.

11.2.2   Damage Models

Various forms of damage in FRP laminates are caused by transverse loads and are categorised into two types, that is, various in-plane forms of damage and delamination or interface damage. These two types of damage are simultaneously modelled by using the different approaches, which are described as follows.

11.2.2.1   In-Plane Damage

For in-plane damage, the stress criteria proposed by Hou *et al.* [2] are adopted. As shown by the stress state in Figure 11.1, the following several in-plane failure modes exist.

For fibre failure

$$\text{If } e_f = \left(\frac{\sigma_{11}}{X_T}\right)^2 + \left(\frac{\sigma_{12}^2 + \sigma_{13}^2}{S_f^2}\right)^2 \geq 1 \tag{11.4a}$$

$\sigma_{11}, \sigma_{22}, \sigma_{33}, \sigma_{12}, \sigma_{23}$ and σ_{13} are set to be zero.

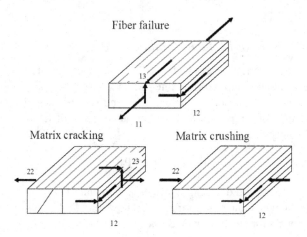

Figure 11.1   In-plane damage model.

For matrix cracking (when $\sigma_{22} \geq 0$)

$$\text{If } e_{mc} = \left(\frac{\sigma_{22}}{Y_T}\right)^2 + \left(\frac{\sigma_{12}}{S_{12}}\right)^2 + \left(\frac{\sigma_{23}}{S_{m23}}\right)^2 \geq 1 \tag{11.4b}$$

σ_{22} and σ_{12} are set to be zero.

For matrix crushing (when $\sigma_{22} < 0$)

$$\text{If } e_{mr} = \frac{1}{4}\left(\frac{\sigma_{22}}{S_{12}}\right)^2 + \frac{Y_c\sigma_{22}}{4S_{12}^2} - \frac{\sigma_{22}}{Y_c} + \left(\frac{\sigma_{12}}{S_{12}}\right)^2 \geq 1 \tag{11.4c}$$

σ_{22} is set to be zero.

In the above, X_T is the fibre directional (1-axis) strength, Y_T is the transversal (2-axis) tensile strength, Y_C is the transversal (2-axis) compressive strength, S_{12} is the laminate plane (1–2 plane) shearing strength, S_f is the shearing strength causing fibre failure and S_{m23} is the cross-sectional (2–3 plane) shearing strength causing matrix cracking.

The above criteria are used to check the failure state of each Gauss integration point within brick elements, where $3 \times 3 \times 3$ Gauss integration scheme is used. An important feature for in-plane damage is that, within one time step, although the stress states at some points may satisfy the above criteria, these points may not certainly fail. In practical computations, first we evaluate the highest e_f, e_{mc} and e_{mr} as

$$e_f^{\max} = \max_{i=1}^{TNINT}\{e_f^i\}, \quad e_{mc}^{\max} = \max_{i=1}^{TNINT}\{e_{mc}^i\}, \quad e_{mr}^{\max} = \max_{i=1}^{TNINT}\{e_{mr}^i\} \tag{11.5}$$

where *TNINT* is the total number of all integration points within all 3D structural elements. Then, we only pick up the most dangerous points from all integration points in structural elements according to the following conditions

$$e^i_f > 1.0 \quad \text{and} \quad \left(e^i_f - 1.0\right)/\left(e^{\max}_f - 1.0\right) > 1.0 - DE \quad i = 1 \ldots TNINT \qquad (11.6a)$$

$$e^i_{mc} > 1.0 \quad \text{and} \quad \left(e^i_{mc} - 1.0\right)/\left(e^{\max}_{mc} - 1.0\right) > 1.0 - DE \quad i = 1 \ldots TNINT \qquad (11.6b)$$

$$e^i_{mr} > 1.0 \quad \text{and} \quad \left(e^i_{mr} - 1.0\right)/\left(e^{\max}_{mr} - 1.0\right) > 1.0 - DE \quad i = 1 \ldots TNINT \qquad (11.6c)$$

where DE is a constant which, based on experience, is set to be 0.3 in our computations. Here, DE represents the ratio of the most dangerous points in all integration points. When DE is 0.3, for instance for Equation 11.6a, it means that only those points satisfying $e^i_f/e^{\max}_f > 0.7$ are in the state of fibre failure.

The above consideration is based on two aspects: (i) it can avoid a nonphysical sudden structural crash leading to the stop of computations and (ii) after the occurrence of damage at the most dangerous points, the stiffnesses at these locations are updated and the stresses within the whole structure will be redistributed. Therefore, the stress states at other integration points can be more physically evaluated. For the explicit time algorithm, the time step size is very small, usually lower than 1×10^{-7} s, therefore, this above method is theoretically and physically reliable.

Further, according to Maxwell's principle $v_{ij}/E_j = v_{ji}/E_i$, stresses listed in Equation 11.4 and the constitutive relation of the single layer in laminates, we introduce an equivalent strategy for updating elastic constants in Table 11.1. If the stress state of one Gauss integration point satisfies the failure criteria, the material properties at this point should be modified according to Table 11.1. Parameters e_t, e_c and e_v in Table 11.1 are very small and determined by pilot calculation based on comparison with some experimental data. For instance, e_t and e_c describe the final damage-resisting capability. e_c can be usually taken as the Young's modulus of polymer matrix.

Table 11.1 Updating scheme for material properties of in-plane damages.

	Fibre failure	Matrix cracking	Matrix crushing
When Equation (11.6) is satisfied in the current step, it updates the strategy of elastic constants at the Gauss points of 3D brick elements	$E_i = e_t, i = 1,2$ $v_{ij} = e_v, i,j = 1,2$ $G_{ij} = e_t/2, i,j = 1,2$	$E_2 = e_t,$ $v_{12} = e_v,$ $G_{12} = e_t/2$	$E_2 = e_t,$ $v_{12} = e_v$
In the following time steps, the material constants are determined from the sign of normal stress σ_i ($i = 1,2$)	$\sigma_i \geq 0$ $E_i = e_t, v_{ij} = e_v,$ $G_{ij} = e_t/2, i,j = 1,2$	$\sigma_2 \geq 0$ $E_2 = e_t, v_{12} = e_v,$ $G_{12} = e_t/2$	$\sigma_2 \geq 0$ $E_2 = e_t,$ $v_{12} = e_v$
	$\sigma_i < 0$ $E_i = e_c, v_{ij} = e_v,$ $G_{ij} = e_c/2, i,j = 1,2$	$\sigma_2 < 0$ $E_2 = e_c, v_{12} = e_v,$ $G_{12} = e_c/2$	$\sigma_2 < 0$ $E_2 = e_c,$ $v_{12} = e_v$

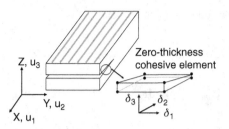

Figure 11.2   Cohesive interface element.

11.2.2.2   Theory of Traditional Cohesive Element for Modelling Delamination

To analyse the delamination propagation at interfaces in laminates, a lot of cohesive models have been proposed in many previous studies. Here a zero-thickness *rate-independent* cohesive element with eight nodes in Figure 11.2 is adopted [9] to simulate the resin-rich layer connecting the several laminae of a composite laminate. It should be noted that most of the contents in this chapter will be based on this rate-independent cohesive model, with the exception of the extension of our "adaptive cohesive model" into a rate-dependent one as stated later.

The constitutive equation of zero-thickness cohesive elements is established in terms of relative displacements and tractions across the interface. The relative displacements for an element with a general orientation in 3D space are defined in Figure 11.2.

In this figure, at each integration point of cohesive element, the relative displacements $\boldsymbol{\delta}_s = \{\delta_1, \delta_2, \delta_3\}^T$ in local coordinates are obtained from the displacement vector $\mathbf{u} = \{u_1, u_2, u_3\}^T$ in the global coordinates as

$$\boldsymbol{\delta}_s = \mathbf{B}_s \mathbf{u} \tag{11.7}$$

The constitutive relationship of the cohesive element, $\mathbf{D}_s$, at each integration point relates the tractions, $\boldsymbol{\tau}_s$, to the relative displacements $\boldsymbol{\delta}_s$ as

$$\boldsymbol{\tau}_s = \mathbf{D}_s \boldsymbol{\delta}_s \tag{11.8}$$

The stiffness matrix of the cohesive element can be obtained from the principle of virtual work as follows

$$\mathbf{k}_s = \int_{\Gamma} \mathbf{B}_s^T \mathbf{D}_s \mathbf{B}_s d\Gamma \tag{11.9}$$

The 4×4 Newton–Cotes closed integration scheme, which can overcome the locking caused by the strong initial interface stiffness [9], is adopted in this work to evaluate the stiffness matrix of cohesive element.

Here, it is a fundamental task to build up an appropriate constitutive equation in the formulation of the cohesive element for accurate simulations of the interlaminar cracking process. It is considered that there is a process zone or cohesive zone ahead of the delamination tip, which physically represents the coalescence of crazes in the resin rich layer located at the delamination tip and reflects the way by which the material loses load carrying capacity. As

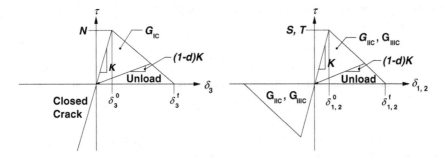

Figure 11.3   Constitutive law of traditional cohesive interface element.

shown in Figure 11.3 for a bi-linear model [9] in the cases of typical pure mode I, II or III, after the interfacial normal or shear tractions attain their respective interlaminar tensile or shear strengths at an integration point of the cohesive element, the stiffness of the cohesive element at this point is gradually reduced to zero. The softening onset displacements are obtained as

$$\delta_3^0 = N/K, \quad \delta_2^0 = S/K, \quad \delta_1^0 = T/K \tag{11.10}$$

where N, S and T are the interlaminar tensile and shear strengths, respectively, and K is the initial stiffness of interface.

As shown in Figure 11.3, the area under the traction-relative displacement curves is the respective (mode I, II, III) fracture toughness (G_{IC}, G_{IIC}, G_{IIIC}) which is used to define the final relative displacements corresponding to complete decohesion (i.e. δ_3^f, δ_2^f and δ_1^f) as

$$\int_0^{\delta_3^f} \tau_3 d\delta_3 = G_{IC}, \quad \int_0^{\delta_2^f} \tau_2 d\delta_2 = G_{IIC}, \quad \int_0^{\delta_1^f} \tau_1 d\delta_1 = G_{IIIC} \tag{11.11}$$

The final displacements for the state of complete decohesion are then obtained as

$$\delta_3^f = 2G_{IC}/N, \quad \delta_2^f = 2G_{IIC}/S, \quad \delta_1^f = 2G_{IIIC}/T \tag{11.12}$$

For the mixed-mode, the current effective relative displacement δ_m is defined as

$$\delta_m = \sqrt{\delta_1^2 + \delta_2^2 + \langle\delta_3\rangle^2} = \sqrt{\delta_{shear}^2 + \langle\delta_3\rangle^2} \tag{11.13}$$

where δ_{shear} represents the norm of the vector defining the tangential relative displacements of the element, and the MacCauley bracket (i.e. $< >$) is defined as

$$\langle x \rangle = \begin{cases} 0 \Leftarrow x \le 0 \\ 0 \Leftarrow x > 0 \end{cases} \tag{11.14}$$

Assuming $S = T$, the single-mode relative displacements at softening onset are defined from Equation 11.10 as follows

$$\delta_3^0 = N/K, \quad \delta_1^0 = \delta_2^0 = \delta_{shear}^0 = S/K \tag{11.15}$$

When the opening displacement δ_3 is greater than zero, the mode mixity ratio β is

$$\beta = \frac{\delta_{shear}}{\delta_3} \tag{11.16}$$

The softening onset displacement of mixed-mode, that is, δ_m^0, is then defined as

$$\delta_m^0 \begin{cases} \delta_3^0 \delta_1^0 \sqrt{\dfrac{1 + \beta^2}{\left(\delta_1^0\right)^2 + \left(\beta \delta_3^0\right)^2}} \Leftarrow \delta_3 > 0 \\ \delta_{shear}^0 \Leftarrow \delta_3 \leq 0 \end{cases} \tag{11.17}$$

The final displacement of mixed-mode corresponding to the state of complete decohesion is obtained from the well known Benzeggagh–Kenane (B–K) model as follows

$$\delta_m^f \begin{cases} \dfrac{2}{K\delta_m^0} \left[G_{IC} + (G_{IIC} - G_{IC}) \left(\dfrac{\beta^2}{1+\beta^2}\right)^\eta \right] \Leftarrow \delta_3 > 0 \\ \sqrt{\left(\delta_1^f\right)^2 + \left(\delta_2^f\right)^2} \Leftarrow \delta_3 \leq 0 \end{cases} \tag{11.18}$$

η is chosen through the comparison with experimental results (usually it ranges from 1.3 to 1.8). Further, δ_m^{max} is defined to be the maximum effective relative displacement of one integration point within a cohesive element in the loading history. Using the maximum value of the effective relative displacement rather than the current value δ_m prevents healing of the interface. Finally, the constitutive matrix $\mathbf{D}_s$ in Equation 11.8 for mixed-mode is evaluated by the penalty parameter, that is, the initial stiffness of interface K, the damage evolution function d and the softening onset and final displacements of mixed-mode, that is, δ_m^0 and δ_m^f, respectively, as

$$D_{sr} \begin{cases} \bar{\delta}_{sr} K \Leftarrow \delta_m^{max} \leq \delta_m^0 & \text{intact} \\ \bar{\delta}_{sr}[(1-d)K] \Leftarrow \delta_m^0 < \delta_m^0 \text{ and } \delta_3 > 0 & \\ \bar{\delta}_{sr}[(1-d)K + Kd\bar{\delta}_{s3}] \Leftarrow \delta_m^0 < \delta_m^{max} < \delta_m^f \text{ and } \delta_3 \leq 0 & \text{softening} \\ 0 \Leftarrow \delta_m^{max} \geq \delta_m^f \text{ and } \delta_3 > 0 & \text{intact} \\ \bar{\delta}_{s3}\bar{\delta}_{sr} K \Leftarrow \delta_m^{max} \geq \delta_m^f \text{ and } \delta_3 \leq 0 & \text{complete decohesion} \end{cases} \tag{11.19a}$$

$$d = \frac{\delta_m^f \left(\delta_m^{max} - \delta_m^0\right)}{\delta_m^{max} \left(\delta_m^f - \delta_m^0\right)}, \quad d \in [0, 1] \tag{11.19b}$$

where $\bar{\delta}_{sr}$ is the Kronecker delta and d represents the extent of stiffness loss with an increase of effective relative displacement. It should be noticed that the above equation avoids the interpenetration of the crack faces of the cohesive element in the state of compression for softening and complete decohesion states.

11.3   Techniques for Overcoming Numerical Instability in Simulation of Delamination Propagation

From many previous studies [9–15], it was found that computations that use cohesive zones to model crack nucleation often experience convergence difficulties at the point where the crack first nucleates. These problems are known to arise from an elastic snap-back instability, which occurs just after the stress reaches the peak strength of the interface. As stated previously, various approaches can be used to resolve these convergence problems; for instance, the Riks method [16] may be used to follow the unstable branch of the solution during the snap-back. Convergence is guaranteed only if there exists a static equilibrium path that connects the states of the solid at the start and end of a time increment. Under a fixed remote loading, the work done by the remote boundaries during debonding is positive or zero. Consequently, if the strain energy release rate during debonding exceeds the rate of work done against the cohesive zone tractions, a static equilibrium path cannot exist. Procedures such as the Riks method [16] resolve this issue by providing a way for the remote boundaries to do negative work on the solid. In general, these schemes require some effort to implement. Recently, Gao and Bower [15] have proposed a simple technique to introduce a kind of artificial damping to eliminate this numerical instability.

In this section, three techniques proposed by the present authors are described to alleviate the numerical instability occurring in the simulations of delamination propagation, which are termed as: "artificial damping technique" [15], "move-limit technique" [18] and "adaptive cohesive model" [19], respectively. Their characteristics, application ranges and effectiveness for avoiding the numerical instability will be explored in detail. Note that the applications of these three techniques for the rate-independent cohesive model are mainly investigated. Further, the extension of only the "adaptive cohesive model" to a rate-dependent cohesive model [20] is also introduced in this section.

11.3.1   Artificial Damping Technique

First, we extend the damping technique proposed by Gao and Bower [15] into the bi-linear cohesive model for composites and the explicit time integration scheme. Gao and Bower [15] incorporated a kind of damping force into the Xu–Needleman interfacial cohesive model and the implicit time integration scheme; and they found that, in this case, the numerical computation became more stable. For the bi-linear cohesive model, as shown in Equation 11.19, we modify it as shown in the following equation where damping terms are different from those proposed by Gao and Bower [15].

$$
D_{ij}
\begin{cases}
\bar{\delta}_{ij} K \Leftarrow \delta_m^{\max} \leq \delta_m^0 \\[2ex]
\bar{\delta}_{ij} \left[(1-d)K + \xi \frac{\dot{\delta}_i}{\delta_i} \right] \Leftarrow \delta_m^0 < \delta_m^{\max} < \delta_m^f, \delta_3 > 0 \\[2ex]
\bar{\delta}_{i3}\bar{\delta}_{3j} K \Leftarrow \delta_m^0 < \delta_m^{\max} < \delta_m^f, \delta_3 \leq 0 \\[2ex]
\bar{\delta}_{ij} \xi \frac{\dot{\delta}_i}{\delta_i} \Leftarrow \delta_m^{\max} \geq \delta_m^f, \delta_3 > 0 \\[2ex]
\bar{\delta}_{i3}\bar{\delta}_{3j} K \Leftarrow \delta_m^{\max} \geq \delta_m^f, \delta_3 \leq 0
\end{cases}
\tag{11.20}
$$

where ξ is the damping parameter, which is defined as $\xi = \zeta N \delta_m^0$ by referring to [15].

Note that there are two features in the above cohesive model:

- One is that, even when the relative opening displacement is larger than δ_m^f, there are still damping forces. Although they can be set to be zero in Equation 11.20, the effect of stabilisation of the method becomes weaker.
- The other is that D_{11}, D_{22} or D_{33} may be different due to the probably different relative velocities in the different directions.

Although the additional viscosity makes the solution rate dependent and introduces additional energy dissipation into the computations, the solution can converge more easily. In problems involving unstable crack nucleation or growth, this additional dissipation can be regarded as approximately equivalent to the energy that would be radiated from the crack in elastic waves during the instability [15]. Further, we incorporated this model into an explicit time integration scheme.

11.3.2 Move-Limit Technique Enforced on Cohesive Zone

From the physical understanding of the numerical instability problem, when the instability occurs, it means that there are sudden changes in the displacements of the nodes located in the cohesive region due to the sudden release of internal loads caused by the zero stiffness of the damaged cohesive elements. These sudden changes in displacements lead to unstable loads in the remote loading points. Here, the displacement increments of nodes are restricted in the cohesive zone at the time steps when the delamination occurs and propagates. Taking the example of the mode I case as shown in Figure 11.4, we insert moving rigid walls to restrict the absolute value of displacement increments of nodes located in the cohesive zone. For instance, for the nodes located on the upper layer within the cohesive zone, there are two rigid walls represented by bold lines. When the nodes cross over these two rigid walls, they will automatically be drawn back to the positions of the rigid walls. For nodes located on the lower

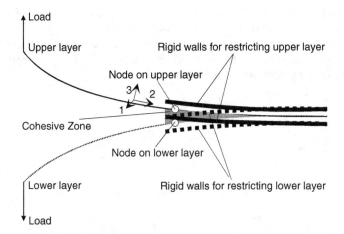

Figure 11.4 Schematic of move-limit model for restricting displacement increments in cohesive zone.

layer, there are still the same two rigid walls to restrict their increments, which are represented by two dotted bold lines in Figure 11.4. This idea is similar to the concept of move-limit used in the sequential linear programming technique in optimisation problems, which restricts the variation of design variables; therefore, it is termed a "move-limit" technique.

For the description of this contact-like problem, as the forward increment Lagrange multiplier method [22], the incremental equation of motion is

$$\mathbf{M}\ddot{\mathbf{u}}_n + \mathbf{p}_n + \mathbf{G}_{n+1}^T \boldsymbol{\lambda}_n = \mathbf{f}_n \tag{11.21}$$

where n represents the number of time step. $\mathbf{G}$ and $\boldsymbol{\lambda}$ are the constraint matrix and Lagrange multiplier vector, respectively. For the ith direction of the node k defined in the local coordinates within the cohesive zone in Figure 11.4, the constraint conditions of contact are expressed as

$$\left(u_{n+1}^{ki} - u_n^{ki} - Q^i\right) = 0, \quad \text{if } u_{n+1}^{ki} - u_n^{ki} > Q^i \tag{11.22a}$$

$$\left(u_{n+1}^{ki} - u_n^{ki} + Q^i\right) = 0, \quad \text{if } u_{n+1}^{ki} - u_n^{ki} \leq -Q^i \tag{11.22b}$$

where u^{ki} is the displacement component of the node k in the ith direction in the cohesive zone. Q^i represent the half widths of band area defined by two rigid walls in three local axes directions, and they are positive values, as obtained later. The above constraint conditions can be finally summarised as

$$\mathbf{G}_{n+1}^T (\mathbf{u}_{n+1} - \mathbf{u}_n + \mathbf{Q}) = 0 \tag{11.23}$$

To solve Equation 11.23, the well known multi-step central difference method is employed, which is given by

$$\mathbf{u}_n = \frac{1}{2\Delta t}(\mathbf{u}_{n+1} - \mathbf{u}_{n-1}), \quad \ddot{\mathbf{u}}_n = \frac{1}{\Delta t^2}(\mathbf{u}_{n+1} - 2\mathbf{u}_n + \mathbf{u}_{n-1}) \tag{11.24}$$

Direct substitution from the above equations into Equation 11.21 leads to

$$\mathbf{u}_{n+1} + \mathbf{u}_{n+1}^* + \mathbf{u}_{n+1}^C \tag{11.25}$$

where $\mathbf{u}_{n+1}^* = \Delta t^2 \mathbf{M}^{-1}[\mathbf{f}_n - \mathbf{p}_n] + 2\mathbf{u}_n - \mathbf{u}_{n-1}$, which is completely the same as that of the central difference scheme without a consideration of contact effects.

Also, the modification of displacement due to contact conditions is expressed as [22]

$$\mathbf{u}_{n+1}^C = -\Delta t^2 \mathbf{M}^{-1} \mathbf{G}_{n+1}^T \boldsymbol{\lambda}_n \tag{11.26}$$

The contact forces provided by the rigid walls can be predicted as

$$\boldsymbol{\lambda}_n = [\Delta t^2 \mathbf{G}_{n+1} \mathbf{M}^{-1} \mathbf{G}_{n+1}^T]^{-1} \mathbf{G}_{n+1}(\mathbf{u}_{n+1}^* - \mathbf{u}_n^* + \mathbf{Q}) \tag{11.27}$$

Naturally, due to the introduction of external contact forces into the system, the work done by the contact forces will influence the original system. The following simple formulation is used to define the bandwidth between two rigid walls.

$$Q^i = \alpha V_L \Delta t \qquad (11.28)$$

where V_L is the prescribed loading velocity at the remote loading points, α is a parameter depending on the location and initial stiffness of the cohesive zone, as well as some structural parameters. Through computational trials, generally, α can be set as: $0.5 \leq \alpha \leq 3$.

To discuss this problem in more detail, we define an approximate formulation to determine α dynamically during the computation. It is expected that the introduced constraint conditions will remove the numerical instability; however, it does not interfere with normal crack propagations, due to the excessive external work provided by these contact conditions. The computational procedure will be stable if the external work is larger than the energy dissipated by the cohesive elements. For a specified time increment, there is

$$\sum_{i=1}^{NF} F_i V_L \Delta t \geq \sum_{i=1}^{NEC} \left(\frac{1}{2} \Delta u_1^2 K + \frac{1}{2} \Delta u_2^2 K + \frac{1}{2} \Delta u_3^2 K \right) A_i \qquad (11.29)$$

where NF is the total number of external forces, A_i is the area of the ith cohesive element and NEC is the number of cohesive elements with drastic crack propagation. Originally, the above equation represents a clear physical meaning, that is, when the dissipated energy by crack propagation is larger than the external work, the superfluous energy will be transferred into system dynamic energy, which causes the physical instability. However, in the practical implementation of the cohesive model, by seeing Equation 11.29, when the element size is very large, the dissipated energy by drastic propagation of many large cohesive elements can be easily larger than the practical external work, which causes the numerical instability. This is because the linear distribution of relative displacement within one present cohesive element cannot realistically reflect the real distribution of a displacement field at the crack front. For small cohesive elements, the energy can be dissipated gradually, which leads to stable propagation simulations.

In Equation 11.29, if we consider that the relative displacements in the three directions are identical for simplicity: $\Delta u = \Delta u_1 = \Delta u_2 = \Delta u_3$. Assuming that the absolute displacement increments of the upper and lower layers are identical but of different signs, the relative displacement increment can be simply written as $\Delta u = 2\Delta u_c$, where u_c is the absolute value of the displacement increments at the upper or lower layer. With the assumption of $u_c = \alpha V_L \Delta t$ in Equation 11.28, from the above equations, we can define α as

$$\alpha = \left(\sqrt{\sum_{i=1}^{NF} F_i V_L \Delta t \Big/ \sum_{i=1}^{NEC} 6 K A_i} \right) \Big/ V_L \Delta t \qquad (11.30)$$

Also, in Equation 11.29, *NEC* is set to be the number of those cohesive elements of comparatively higher maximum total relative displacement which satisfy the following condition

$$\left(\sum_{i=1}^{NINT} \delta_m^{\max^i} \right) \bigg/ NINT > \varepsilon \qquad (11.31)$$

where *NINT* is the number of integration points within one cohesive element and ε is a very small pre-set value.

Also, during the computation, *NEC* does not include those elements, which are completely in a state of decohesion. It should be noted that, in the above prediction of α, we neglect the influence of increments of structural strain energy and kinetic energy within this time incremental step. Therefore, it is a rough formulation for predicting α, which defines the upper limit of the bandwidth. However, as shown in the numerical examples, we find that the above formulation can yield a reasonable value of α.

11.3.3   Adaptive Cohesive Model

The above described two techniques, that is, the "artificial damping" and "move-limit" techniques introduce artificial external work to stabilise the computational process. As shown later, although these two methods can remove the numerical instability when using comparatively coarse meshes, there is usually an error in the peak load of the load–displacement curve. The numerical peak load, for example, for a double cantilever beam (DCB) problem is usually higher than the real one, as observed by [13, 18]. Moreover, basically, the "move-limit" technique is only suited for quasi-static problems since it is difficult to control the movement of two rigid walls in complex dynamic 3D problems. Based on the above facts, we proposed a new cohesive model named the "adaptive cohesive model" (ACM), which is effective and simple for various applications [19]. This simple model was further updated to a rate-dependent version in [20].

11.3.3.1   Rate-Independent Adaptive Cohesive Model

To remove numerical instability when using coarse meshes, as shown in Figure 11.5, ahead of the traditional softening zone, we insert an assumed transition area called a *pre-softening zone*. In this zone, the initial stiffness and interface strength of the integration points in cohesive elements are gradually reduced as $\delta_m^{\max}$ increases. In Figure 11.5, the stiffness K and the interface strength, for example, N for mode I, are linearly updated with the increase of $\delta_m^{\max}$ as follows

$$N_i = \frac{\delta_m^{\max}}{\delta_m^0}(N_{\min} - N_0) + N_0, \quad (N_0 > N_{\min}) \quad \text{and} \quad \left(\alpha\delta_m^0 < \delta_m^{\max} < \delta_m^0\right) \qquad (11.32a)$$

$$K_i = \frac{\delta_m^{\max}}{\delta_m^0}(K_{\min} - K_0) + K_0, \quad (K_0 > K_{\min}) \quad \text{and} \quad \left(\alpha\delta_m^0 < \delta_m^{\max} < \delta_m^0\right) \qquad (11.32b)$$

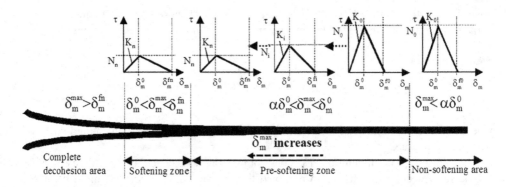

Figure 11.5   Schematic view of adaptive cohesive model ($\alpha = 0$–1.0).

where N_0 is the initial interface strength, N_{min} the lower limit of interface strength, K_0 the initial stiffness and K_{min} is the lower limit of stiffness. Note that Equation 11.32a also holds for S and T used in modes II and III. By choosing the proper ratio between the lower limits of strength and stiffness (e.g. N_{min} and K_{min}) from Equation 11.10, the following relations can be realised easily.

$$\delta_3^0 = \frac{N_0}{K_0} = \frac{N_i(\delta_m^{max})}{K_i(\delta_m^{max})} = \frac{N_{min}}{K_{min}}, \quad \delta_2^0 = \delta_1^0 = \frac{S_0}{K_0} = \frac{S_i(\delta_m^{max})}{K_i(\delta_m^{max})} = \frac{S_{min}}{K_{min}} \quad \text{(for } S = T\text{)} \quad (11.33)$$

Therefore, from Equation 11.17, the onset displacement δ_m^0 in the pre-softening zone is the same as that in the traditional cohesive model, which does change in the updating process of the interface stiffness and strength.

Moreover, to keep the constant fracture toughness G_c when reducing K, N and S in Equation 11.32, the final displacement δ_m^{fi} is adjusted correspondingly according to Equations 11.12 and 11.18, which is schematically shown in Figure 11.6. Once the integration point enters into the *real softening process*, that is, $\delta_m^{max} > \delta_m^0$, the current values of strength and stiffness, that is, N_n and K_n in Figures 11.5 and 11.6, will be constantly used in the subsequent computations. It should be noted that α in Equation 11.32 is a parameter to define the size of the pre-softening zone. When $\alpha = 1$, the present ACM model degenerates into the traditional cohesive model.

In the above adaptive model, N_0 can be taken as the real interface strength. Also, it is crucial to define N_{min} in Equation 11.33 from the consideration of computational stabilisation and accuracy. Mi et al. [14] concluded that several elements in the softening zone are needed to realise stable numerical simulations of the interface crack propagation. Here, by referring to this statement, for instance, for the case of mode I, the size of the softening zone R was defined by Geubelle and Baylor [15] as

$$R = N_c R_n = \frac{\pi}{2} \frac{E_{polymer}}{1 - \nu^2} \frac{G_{IC}}{N_{min}^2} \quad (11.34)$$

where N_c is the number of elements in the softening zone, ranging from 2 to 5 from our numerical experiences, R_n is the element size and $E_{polymer}$ and ν are the Young's modulus

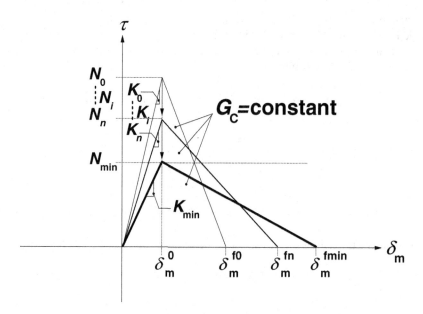

Figure 11.6   Constitutive law of adaptive cohesive model.

and the Poisson ratio of the polymer matrix, respectively. Finally, N_{min}, which depends on the element size, can be calculated as

$$N_{min} = \sqrt{\frac{\pi}{2} \frac{E_{polymer}}{1 - v^2} \frac{G_{IC}}{N_c R_n}} \qquad (11.35)$$

For the mixed mode, the similar formulation can be set up by simply replacing G_{IC} by G_c, which is equal to $G_{IC} + (G_{IIC} + G_{IC}) \left(\frac{\beta_2}{1 + \beta_2}\right)^\eta$ from Equation 11.18.

11.3.3.2   Rate-Dependent Adaptive Cohesive Model

In our previous work [20], the above adaptive technique was further extended into a rate-dependent cohesive model, for stably and accurately simulating delamination propagations in composite laminates under transverse loads. This rate-dependent adaptive cohesive model is basically similar to the previous rate-independent cohesive model; however, its traction includes a cohesive zone viscosity parameter (η) to vary the degree of rate dependence and to adjust the peak or maximum traction. Here, this rate-dependent cohesive model is formulated and implemented in LS-DYNA [21] as a user defined material (UMAT). LS-DYNA is one of the explicit FE codes most widely used by the automotive and aerospace industries. It has a large library of material options; however, continuous cohesive elements are not available within the code. The formulation of this model is fully 3D and can simulate mixed-mode delamination. Here, we describe this model that can be used to capture delamination onset and growth under quasi-static and dynamic loading conditions. For simplicity, we only consider the mode I case and the bilinear model in Figure 11.3.

Similar to the content in Section 11.3.3.1, the adaptive interfacial rate-dependent constitutive response for a mode I case is implemented as follows:

1. In the pre-softening zone, $\alpha \delta_3^o < \delta_3^{\max} < \delta_3^o$, the constitutive equation is given by

$$\tau = (N + \eta \dot{\delta}_3) \frac{\delta_3}{\delta_3^o} \tag{11.36}$$

and

$$N = K \delta_3^o \tag{11.37}$$

where τ is the traction (Figure 11.6) and K is the penalty stiffness and can be written as

$$K = \begin{cases} K_o & \delta_3 \leq 0 \\ K_i & \delta_3^{\max} < \delta_3^o \\ K_n & \delta_3^o \leq \delta_3^{\max} < \delta_3^f \end{cases} \tag{11.38}$$

where δ_3 is the relative displacement in the interface between the top and bottom surfaces (in this study, it equals the normal relative displacement for mode I), δ_3^o is the onset displacement and it remains constant in the simulation and can be determined by Equation 11.33. For each increment and for time $t + 1$, δ_3 is updated. The updated stiffness and interface strength are determined using Equation 11.32. In our computations, we set $\alpha = 0$ in Equation 11.32. From our numerical experiences, the size of the pre-softening zone has some influence on the initial stiffness of loading–displacement curves, but not so significant. The reason is that, for the region always far from the crack tip, the interface decrease or update according to Equation 11.32 is not obvious since $\delta_3^{\max}$ is very small.

The energy release rate for mode I, G_{IC}, also remains constant. Therefore, the final displacements associated to the complete decohesion δ_3^{fi} are adjusted as shown in Figure 11.6 as

$$\delta_3^{fi} = \frac{2G_{IC}}{N_i} \tag{11.39}$$

Once the maximum relative displacement of an element located at the crack front satisfies the condition $\delta_3^{\max} > \delta_3^o$, this element enters into the real softening process. As shown in Figure 11.5, the real softening process denotes a stiffness decreasing process caused by accumulated damage. Then, the current strength N_n and stiffness K_n, which are almost equal to $N_{\min}$ and $K_{\min}$, respectively, will be used in the softening zone.

2. In the softening zone, $\delta_3^o \leq \delta_3^{\max} < \delta_3^f$, the constitutive equation is given by

$$\tau = (1 - d)(N + \eta \dot{\delta}_3) \frac{\delta_3}{\delta_3^o} \tag{11.40}$$

where d is the damage variable, as defined in Equation 11.19b. The above adaptive cohesive modes in Sections 11.3.3.1 and 11.3.3.2 are of the engineering meaning when using coarse meshes for complex composite structures. They are, in fact, an "artificial" means for achieving a stable numerical simulation process. A reasonable explanation is that all numerical techniques are artificial, if their accuracy strongly depends on their mesh sizes, especially at the front of a crack tip. To remove the factitious errors in the simulation results caused by the coarse mesh sizes in the numerical techniques, we artificially adjust some material properties in order to partially alleviate or remove the numerical errors. Otherwise, we have to resort to very fine meshes, which may be computationally impractical for very complex problems due to the capabilities of most current computers. Of course, the modified material parameters should be those which do not have a dominant influence on the physical phenomena. For example, the interface strength usually controls the initiation of interface cracks. However, it is not crucial for determining the crack propagation process and final crack size from the viewpoint of fracture mechanics. Moreover, there has been almost no clear rule to exactly determine the interface stiffness, which is a parameter determined with a high degree of freedom in practical cases. Therefore, the effect of the modifications of interface strength and stiffness can be very small since the practically used onset displacement δ_m^0 for delamination initiation remains constant in our model. The parameters which dominate the fracture phenomena should be unchanged. For instance, in our model, the fracture toughness dominating the behaviour of interface damage is kept constant.

11.4   Numerical Examples

To verify our proposed schemes for stabilising the numerical simulation of delamination propagation, except for the rate-dependent adaptive model implemented in LS-DYNA in Section 11.3.3.2, a powerful 3D eight-noded brick element proposed by the authors [21] is adopted to model the laminate portions. This element is further extended into the large deformation analysis in [18]. For the rate-dependent adaptive cohesive model in Section 11.3.3.2, the integrated S/R eight-noded solid brick element in the LY-DYNA element library is used. Also, for quasi-static and dynamic problems, the explicit time algorithm based on the central difference method is employed.

11.4.1   DCB Problem

A DCB test specimen of a $(0°)_{24}$, T300/977-2 carbon fibre-reinforced epoxy laminate, containing a thin insert at the mid-plane near the loaded end, is simulated [9].

As shown in Figure 11.7, this specimen is 150 mm long, 20 mm wide, with two 1.98 mm thick plies and with an initial crack length of 55 mm. The material properties are: $E_{11} = 150$ GPa, $E_{22} = E_{33} = 11.0$ GPa, $G_{12} = G_{13} = 6.0$ GPa, $G_{23} = 3.7$ GPa, $v_{12} = v_{13} = 0.25$, $v_{23} = 0.45$ and $\rho = 1445$ kg/m^3. It is a static experiment [9]. After several computational trials, the loading speed is chosen as 10 mm/s in our computations which, from our numerical experiences, can produce stable results without inertia effects. The tensile strength N, mode I fracture toughness G_{IC} and the initial stiffness of cohesive zone K are defined in Table 11.2. Three cases are considered in Table 11.2, where cases 1 and 2 employ the assumed material properties for discussing the characteristics of the two methods, while case 3 employs the

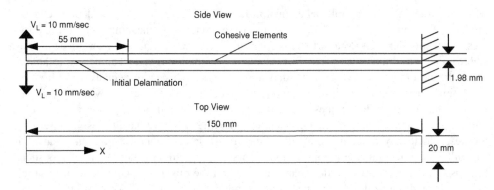

Figure 11.7   Geometry of a [0°/0°] DCB problem.

material properties provided in [9]. For these three cases, the relationship between the stress and the opening displacement for the bi-linear cohesive model is shown in Figure 11.8.

11.4.1.1   Standard Numerical Simulations

First, when using the conventional cohesive technique, the relationships between the load and the opening displacement at the loading point are shown in Figure 11.9. Three experimental results [9] are also plotted for comparison. The time step of the explicit time integration scheme is determined as: $\Delta t = 2/\omega_{max}$, where ω_{max} is the highest natural circular frequency of the original structure. From Figure 11.9, for case 3, when using 1 and 2 mm mesh sizes in the cohesive zone, the computation is stopped due to the sudden drop of the external loads. It is not strange, for the linear cohesive model used in this work, the rough size R of the cohesive zone can be predicted as [15]: $R = \frac{\pi}{2} \frac{E}{1-v^2} \frac{G_{IC}}{N^2}$. For those cohesive elements with non-zero thickness, it is suggested that $K = E/t$, where t is the thickness of resin layer. In our cases, the cohesive element is of zero thickness. Therefore, it is difficult to directly set up the proper relationship between the initial stiffnesses K and E of the interface layer. Based on the following properties of resin of PMMA ($E = 3.24$ GPa, $G_{IC} = 0.268$ kJ/m², $v = 0.35$, $N = 45$ MPa), R is predicted as 0.7676 mm. Mi *et al.* [14] concluded that at least it needs several elements in the softening cohesive zone R to obtain a stable numerical simulation of interface crack propagation.

Therefore, in this case, if we employ three cohesive elements within R, the maximum cohesive element size should be approximately equal to 0.256 mm. We have confirmed that, when the mesh size is 0.25 mm, a stable simulation can be obtained. For case 2 of 2 mm mesh size in the cohesive zone, although the computation is not stopped, numerical spurious

Table 11.2   Material properties for DCB specimen.

Case No.	N (MPa)	K (N/mm³)	G_{IC} (kJ/m²)	δ_m^0 (mm)	δ_m^f (mm)
1	10	2.0×10^3	0.378	0.0050	0.0756
2	20	2.0×10^3	0.378	0.0100	0.0378
3	45	3.0×10^4	0.268	0.0015	0.0119

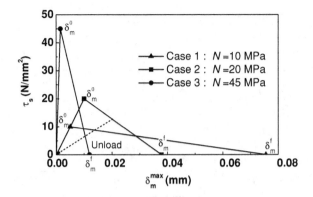

Figure 11.8   Constitutive law of cohesive element for three cases.

oscillations are significant in Figure 11.9. For case 1, by lowering the tensile strength N and initial stiffness K, we can get stable results in the stage of delamination propagation, as shown in Figure 11.9.

However, in this case, due to the much lower tensile strength and initial stiffness, the slope of the loading–displacement curve before the occurrence of damage is lower than that of the experimental results in Figure 11.9.

To investigate the influence of mesh size on numerical stability, we studied case 2 particularly. The curves of load–opening displacement are shown in Figure 11.10. When using 1 mm mesh size in the cohesive zone, we can get very stable results. However, with a mesh size of 2 mm, spurious oscillations become more obvious. By increasing the mesh size up to 3 mm, the numerical results drastically oscillate, which finally leads to the termination of computation.

From the above results, we can conclude that, with an increase in the interface strength, the initial interface stiffness and the mesh size of cohesive element, the numerical instability caused by the elastic snap-back becomes more significant, since in this case the energy released by the cohesive element increases. By decreasing the interface strength and initial stiffness,

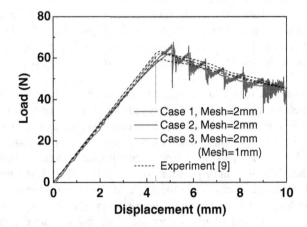

Figure 11.9   Numerical results (cases 1–3, mesh = 2 mm).

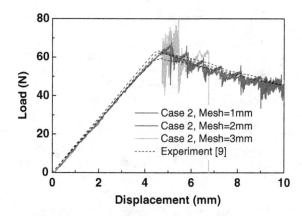

Figure 11.10   Numerical results (case 2, mesh = 1–3 mm).

it is anticipated to stabilise the numerical computation process, which, however, makes the initial slope of the load-opening displacement curve before the happening of damages lower. To employ the real interface strength and initial interface stiffness, it is usually hard to obtain stable results when using a comparatively coarse mesh.

11.4.1.2   Artificial Damping Technique

In this section, we introduce artificial damping as shown in Equation 11.20. The results are shown in Figure 11.11(a) for case 2 with a mesh size of 2 mm. The damping parameter ζ ($\times 10^{-2}$ s/mm^2) is set to be 0.5, 1.5 and 10.0, respectively. Compared with Figure 11.9, the computation becomes more stable with the introduction of damping. The high spurious oscillation can be removed. However, for the higher damping parameter, that is, the case of $\zeta = 10.0$, the external load decreases more slowly in the stage of delamination propagation. The reason can be found from Equation 11.20; even when $\delta_m^{max} \geq \delta_m^f$, that is, a state of complete decohesion, there are still damping forces. Inversely, if we remove these damping forces when $\delta_m^{max} \geq \delta_m^f$, the results will oscillate more obviously. Figure 11.11(b) shows the results of case 3 with a mesh size of 2 mm in the cohesive zone. In this case, the computation is not terminated; however, the spurious numerical oscillations after the peak load become more remarkable.

From the above results, for a comparatively higher interface strength, a higher initial interface stiffness and coarse meshes, it can be concluded that the numerical simulation becomes more stable with the introduction of artificial damping. However, as shown in Figure 11.11(b), it is hard to completely remove the spurious oscillations after the peak load by using this technique. Naturally, it is anticipated to increase the damping parameter to stabilise the numerical computation further. However, in this case, as shown in Figure 11.11(a), too high a damping parameter may lead to a higher load in the stage of delamination propagation. Therefore, when using this technique, it is important to choose the proper damping parameter, which can stabilise the numerical simulation but does not introduce unnecessary energy dissipation into the original system, resulting in a higher load in the stage of delamination propagation.

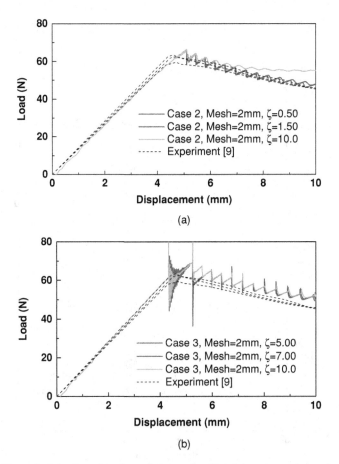

Figure 11.11   Numerical results using the technique of artificial damping: (a) case 2, mesh = 2 mm; (b) case 3, mesh = 2 mm.

11.4.1.3   Move-Limit Technique

To study the effects of the constraints of the displacement increments on the nodes located in the cohesive zone, in case 2 with 2 mm mesh size, the results are shown in Figure 11.12(a) for $\alpha = 0.7$, 1.4 and as a result of Equation 11.30, respectively. For this pure mode I problem, $6KA_i$ in Equation 11.30 is replaced by $2KA_i$ since we only consider the relative displacements in one direction. In our computation for this problem, when ε is set to be 10^{-15} in Equation 11.31, the maximum α computed by Equation 11.30 is around 2.2. From Figure 11.12(a), it can be found that the numerical instability can be removed effectively by setting the move-limit on the nodes in the cohesive zone. The spurious oscillations occurring in Figure 11.10 can also be controlled effectively. Further, there is no obvious difference in the results when using three kinds of α, therefore it is suitable to employ Equation 11.30 to predict α.

In case 2 with 3 mm mesh size, the results are plotted in Figure 11.12(b). Compared with Figure 11.12(a), it can be found that there are bigger oscillations in the load–opening

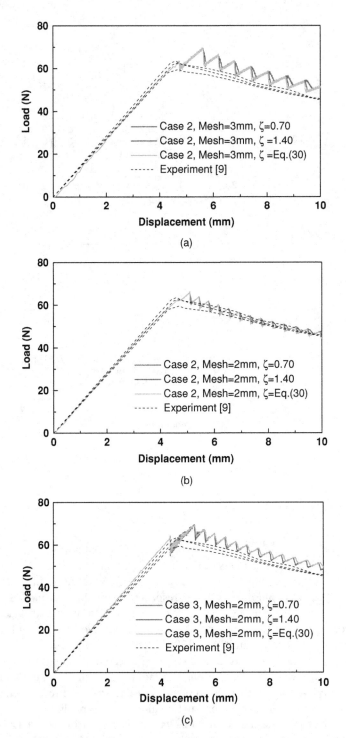

Figure 11.12 Effects of the move-limit technique: (a) case 2, mesh = 2 mm; (b) case 2, mesh = 3 mm; (c) case 3, mesh = 2 mm.

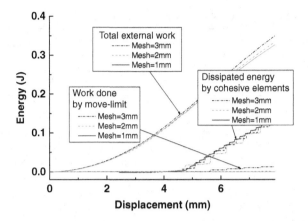

Figure 11.13   Influence of the move-limit technique on system energy (case 2).

displacement curve although the computation is not terminated, as shown in Figure 11.10. The peak load is a little higher than the experimental one, which is caused by the larger size for this linear cohesive element. Also, compared with Figure 11.11(a) for the artificial damping of $\zeta = 10.0$, the phenomenon that the load is higher in the stage of delamination propagation does not appear. Figure 11.12(c) shows the results of case 3 with 2 mm mesh size. By comparing with Figure 11.11(b), it can be seen that the large numerical oscillations occurring in Figure 11.11(b) have been removed clearly in Figure 11.12(c), although there is a discrepancy with the experimental results.

To investigate the influence of the move-limit, we calculate the work done by the contact forces of rigid walls during the delamination propagation. For case 2, the total external work done by the applied external forces, the dissipated energy by the cohesive elements and the work done by the contact forces of rigid walls during the delamination propagation are shown in Figure 11.13. With 1 mm mesh size, the computation is very stable even without a move-limit, as can be seen from Figure 11.10. Therefore, there is no extra work induced by the contact forces of rigid walls. In the cases of 2 and 3 mm mesh sizes, the extra work done by the contact forces increases. However, comparing the total external work done by the applied forces to the dissipated energy by cohesive elements, this extra work is sufficiently small. Although the move-limit technique is very effective to avoid a numerical instability, there are other problems for this method. As shown in Figure 11.12(b) and (c), the peak load is higher than the real when using a comparatively large mesh size. Moreover, it is difficult to apply for this method for dynamic problems. These problems can be solved by using the ACM model.

11.4.1.4   Rate-Independent ACM

For the analysis using ACM, we focus on case 3 in Table 11.2, which is the most unstable case. The mesh size and predicted $N_{\min}$ using Equation 11.35 are listed in Table 11.3 when $E_{\text{polymer}} = 3.0$ GPa and $N_c = 3$ (three elements in the softening zone). Also, $\alpha = 0$ in Figure 11.5 is constantly used in all examples.

The various results are shown in Figure 11.14 for two kinds of cohesive mesh sizes. First, a comparison of the results of a traditional cohesive element (cases 1 and 3 in Table 11.2),

Table 11.3 Mesh size and predicted N_{min} in Equation (11.35).

Mesh size R_n (mm)	Initial $N_0 \rightarrow N_{min}$ (MPa)
1.0	$45.0 \rightarrow 22.5$
2.0	$45.0 \rightarrow 15.0$

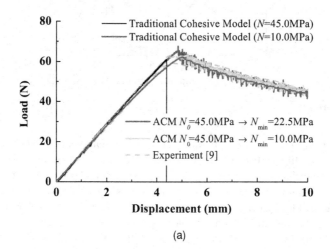

(a)

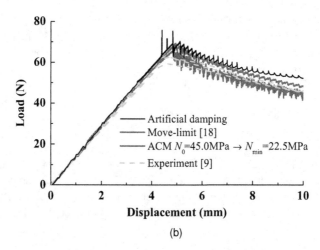

(b)

Figure 11.14 Comparison of different results of DCB problem. (a) Results of traditional cohesive element, ACM and experiments (mesh size $R_n = 1$ mm). (b) Results of stabilising techniques (artificial damping; move-limit), ACM and experiments (mesh size $R_n = 1$ mm). (c) Results of traditional cohesive element, ACM and experiments (mesh size $R_n = 2$ mm). (d) Results of stabilising techniques (artificial damping; move-limit), ACM and experiments (mesh size $R_n = 2$ mm).

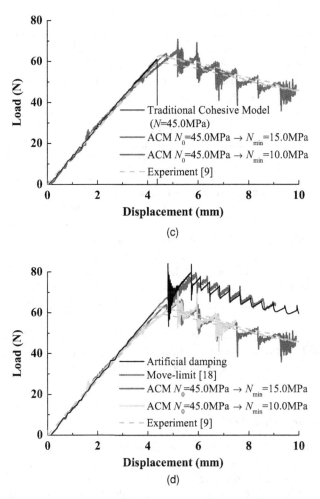

Figure 11.14   (*Continued*)

ACM and experiment [9] is shown in Figure 11.14(a) for a mesh size of 1 mm. From this, we can find that, when the practical interface strength is used in the traditional cohesive model, that is, 45.0 MPa, the result has a sudden stop due to very strong numerical instability. With a decrease of interface strength to 10.0 MPa in the traditional model, the result is very stable. However, the slope of the loading curve before the peak load is obviously lower than those of experimental ones [9]. For the results of ACM, when N_{min} is 22.5 MPa predicted by Equation 11.35, very good results can be obtained by comparing to the experimental ones.

However, when N_{min} is taken as 10.0 MPa in ACM, the same result as that of the traditional cohesive model of the same interface strength is obtained. In Figure 11.14(b), the results of artificial damping and the move-limit technique [18] are plotted. From this figure, we can find that both techniques work stably; however, the peak loads predicted by these techniques are slightly higher than the experimental ones.

ACM can overcome this problem. When the mesh size is 2 mm, from Figure 11.14(c), it can be found that the traditional cohesive element cannot track the loading–displacement history due to a sudden stop. The results of ACM for two values of $N_{\min}$ are good although the oscillation is more significant compared with those of 1 mm mesh size. Also, the slope of the loading curve of ACM using 15 MPa as $N_{\min}$, which is predicted by Equation 11.35, is closer to the experimental results compared with that of ACM using 10 MPa as $N_{\min}$. In Figure 11.14(d), we can find that the results of the artificial damping and the move-limit [18] yield much higher peak loads than the experimental ones. By comparing with Figure 11.14(b) for the case of 1 mm mesh size, we can find that, with the increase of mesh size, the error in peak loads increases too in these two methods. Naturally, this phenomenon is not caused by artificial damping and the move-limit technique. The reason is from the employed linear cohesive elements. For a cohesive element located at the crack tip, the distribution of the relative displacements within one element is linear. If the elemental size is too large, this distribution of the relative displacements cannot reflect the real one in the crack tip area, which leads to a higher external peak load. The error of this higher peak load was also identified in [13]. However, in Figure 11.14(d), the errors in the peak load of ACM are much smaller than those predicted by artificial damping and the move-limit technique due to the proper decrease of interface stiffness. From the above discussions, we can find that ACM can yield very good results from the aspects of the peak load and the slope of loading curve if $N_{\min}$ is properly defined.

A comparison of results obtained from the different mesh sizes is also performed, which shows that the different mesh sizes result in almost the same loading curves. The delamination propagation speeds are almost the same for various mesh sizes.

11.4.1.5   Rate-Dependent ACM

In [20], being different from the previous quasi-static problem in Figure 11.7, to test the effectiveness of our rate-dependent ACM, a high loading speed V_L, that is, 650 mm/s is used for a DCB made of an isotropic fibre-reinforced laminate. This problem was analysed by Moshier [23]. The width and half-thickness of the DCB specimen are 25 and 1.5 mm, respectively, which are different from those in Figure 11.7. The initial crack length in this case is 34 mm. Young's modulus, density and Poisson's ratio of carbon fibre-reinforced epoxy material are given as $E = 115$ GPa, $\rho = 1566$ Kg/m^3, and $\upsilon = 0.27$, respectively. The properties of the DCB specimen interface are given as follows: $G_{IC} = 0.7$ kJ/m^2, $K_o = 1 \times 10^5$ N/ mm^3, $K_{\min} = 0.333 \times 10^5$ N/mm^3, $N_o = 50$ MPa and $N_{\min} = 16.67$ MPa.

In the adaptive rate-dependent cohesive zone model with 1 mm mesh size in the span direction, two different values of viscosity parameter are used in the simulations: $\eta = 0.01$ and 1.0 N·s/mm^3, respectively. Note that η is a material parameter depending on the deformation rate, which appears in Equations 11.36 and 11.40. By observing Equation 11.36, η determines the ratio between viscosity stress $\eta \dot{\delta}_m$ and interface strength N. Therefore, $\eta = 0.01$ N·s/mm^3 and $\eta = 1.0$ N·s/mm^3 correspond to a low and a high rate dependence, respectively.

A plot of a reaction force as a function of the applied end displacement of the DCB specimen using cohesive elements with viscosity value of 0.01 N·s/mm^3 is shown in Figure 11.15. It is clearly shown from Figure 11.15 that the bilinear formulation results in a severe instability once the crack starts propagating. However, the adaptive constitutive law is able to model

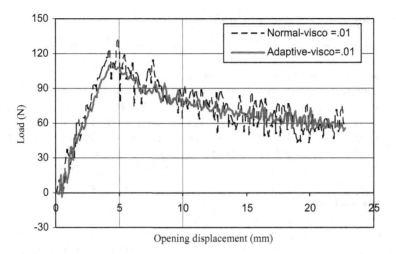

Figure 11.15 Load–displacement curves obtained using both bilinear and adaptive formulations ($\eta = 0.01$).

smooth, progressive crack propagation. It is worth mentioning that the bilinear formulation might bring smooth results by decreasing the element size. The load–displacement curves obtained from the numerical simulation of both the bilinear and adaptive cohesive models using a viscosity parameter of 1.0 N·s/mm^3 is presented in Figure 11.16. It can be seen that, again, the adaptive constitutive law is able to model the smooth, progressive crack propagation while the bilinear formulation results in a severe instability once the crack starts propagating. The average maximum load obtained using the adaptive rate dependent model is 110 N, whereas the average maximum load predicted form the bilinear model is 120 N.

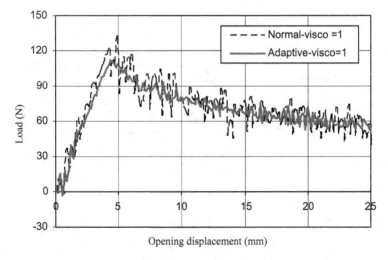

Figure 11.16 Load–displacement curves obtained using both bilinear and adaptive formulations ($\eta = 1$).

11.4.2   Low-Velocity Impact Problem

The above proposed techniques for stabilising the numerical simulations of delamination propagation have been applied for various plate and shell problems under transverse quasi-static loadings with experimental verifications [18, 19].

In this section, we describe an example of CFRP laminates under low-velocity impact that is used to verify the proposed technique for modelling the delamination propagation.

First, in our numerical model, the following material properties of lamina of CF/epoxy are used: $E_1 = 135.0$ GPa, $E_2 = E_3 = 10.0$ GPa, $G_{12} = G_{13} = 5.50$ GPa, $G_{23} = 4.50$ GPa, $v_{12} = 0.0183$, $v_{13} = 0.45$, $v_{23} = 0.25$, $\rho = 1489$ kg/m^3. Also, the properties for damage simulations are listed as follows: $N_0 = 85.0$ MPa, $S_0 = T_0 = 106.0$ MPa, $N_{min} = 76.5$ MPa, $S_{min} = T_{min} = 95.4$ MPa, $G_{IC} = 0.5$ kJ/m^2, $G_{IIC} = G_{IIIC} = 1.0$ kJ/m^2. Note that only N_{min}, S_{min} and T_{min} are assumed parameters and the others are experimentally obtained parameters. To deal with the possible the in-plane damages, the damage model in Section 11.2.2 and strength parameters of CFRP material [18] are adopted. Also, to calculate the impact force, the modified Hertz contact law [24] is employed to deal with the contact between the ball and the laminate.

For the integrity of the content, the experiment is briefly described. The low-velocity impact experiments were performed by the present authors using a weight-drop impact test machine (Dynatup 9250HD). The experimental setup is shown in Figure 11.17. The specimens were prepared according to the SACMA standard of CAI test. As shown in Figure 11.18, a quasi-isotropic CFRP laminated plate of 32 plies as $[(45°/0°/-45°/90°)_4]$s is put on the bottom

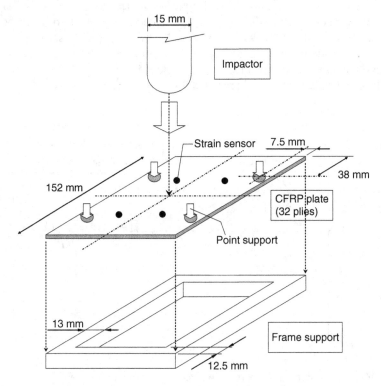

Figure 11.17   Schematic view of impact test.

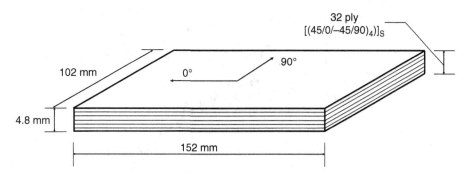

Figure 11.18   Specimen of 32-ply quasi-isotropic CFRP plates.

frame with four fixed points. This plate is impacted by an impacting body of a lower semi-spherical shape and mass of 4.6 kg.

We performed eight tests for four impact energy levels, that is, 3.0, 4.8, 6.0 and 7.2 J. Two tests were carried out for each energy level. When the impact energy is 3.0 J, there is no impact-induced damage. It is interesting to note that, for an impact energy of 4.8 J, the damage occurs in one specimen, but does not occur in the other one. Therefore, the impact energy of 4.8 J can be thought of as the threshold of impact energy, which induces possible damage in CFRP laminates. When the impact energy is higher than 4.8 J, for example, 6.0 and 7.2 J, there is obvious impact-induced damage in four specimens. The ultrasonic results of specimens after impact for these two energy levels will be shown later. From the ultrasonic results, it can be found that the damage area in the impacted side is larger than that of the side opposite to impact.

First, in our numerical model, the following material properties of lamina of CF/epoxy are used: $E_1 = 135.0$ GPa, $E_2 = E_3 = 10.0$ GPa, $G_{12} = G_{13} = 5.50$ GPa, $G_{23} = 4.50$ GPa, $v_{12} = 0.0183$, $v_{13} = 0.45$, $v_{23} = 0.25$, $\rho = 1489$ kg/m^3. Also, the properties for damage simulation are listed as follows: $N_0 = 85.0$ MPa, $S_0 = T_0 = 106.0$ MPa, $N_{min} = 76.5$ MPa, $S_{min} = T_{min} = 95.4$ MPa, $G_{IC} = 0.5$ kJ/m^2, $G_{IIC} = G_{IIIC} = 1.0$ kJ/m^2. In this numerical model, the total number of elements including the cohesive elements is 24 696. The contact force between the ball and CFRP laminates is simulated by a distributed load applied on a 4×4 mm central square area of plate since it is observed that there is an approximate circular unrecoverable indentation area of radius of around 2.0–2.5 mm on specimens after impact.

For the cases of 6.0 and 7.2 J, the comparisons of the impact force histories are demonstrated in Figure 11.19(a) and (b), respectively. From these figures, it can be found that the numerical results agree with the experimental ones very well. One can observe that the repeatability of two experimental results is acceptable, considering the shape of these curves and force levels for the damaged samples. For the features of impact force history, after the peak load, there is a sudden drop in the force history. After this drop, the impact force decreases gradually.

From Figure 11.19(b), we can find that the impact force of 7.2 J is basically similar to that of 6.0 J in Figure 11.19(a). However, after the sudden drop from the peak load, the impact load of 7.2 J does not decrease as immediately as that of 6.0 J. In contrast, there is a platform where the impact load keeps almost constant. After this platform, the impact load decreases gradually. The numerical result reproduces this feature very well compared with the experimental one in Figure 11.19(b).

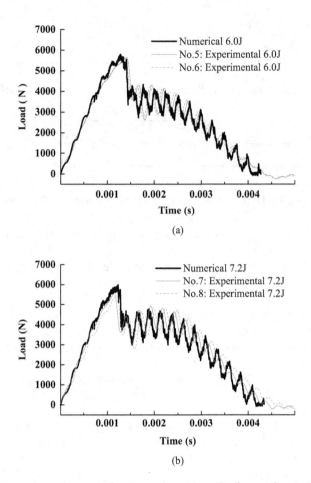

Figure 11.19 Comparison of numerical and experimental results (impact force): (a) 6.0 J, (b) 7.2 J.

Further, the delamination configurations on the side of impact and the side opposite to impact, which are obtained from numerical computations and ultrasonic inspections of specimens, respectively, are shown in Figures 11.20 and 11.21. In these figures, for the numerically obtained delamination on the side of impact, the delamination shapes between the first and 10th interfaces are plotted. Meanwhile, for the delamination on the side opposite to impact, the delamination shapes between the 21st and 30th interfaces are plotted. From these figures, we can find that the numerically obtained delamination shapes agree with the experimental ones very well, although the maximum sizes of delamination are slightly smaller than those of the experimental ones at the side opposite to impact. Also, compared with the experimental results, the numerically obtained delamination shape is more unsymmetrical. As to the computational instability problem, after investigating various cases, it is very interesting to find that the computational process tends to be more unstable as the impacting speed of the ball decreases at the same impact energy level. Naturally, the present ACM can still be employed to avoid this instability. For higher impacting speeds, the numerical instability is not obvious.

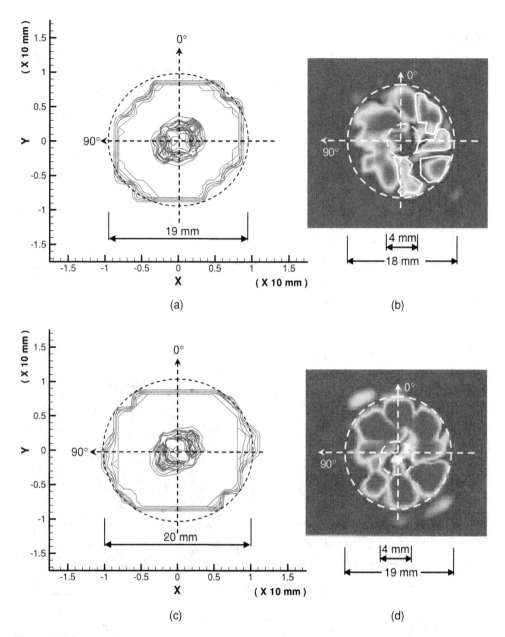

Figure 11.20 Comparison of delamination obtained from FEM and experiments at the side of impact on specimen: (a) numerical (6.0 J), (b) experimental (6.0 J), (c) numerical (7.2 J), (d) experimental (7.2 J).

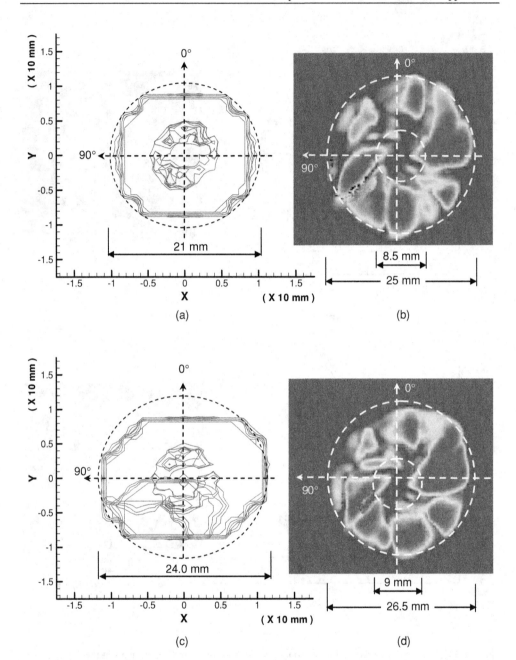

Figure 11.21 Comparison of delamination obtained from numerical computations and experiments at the opposite side of impact on specimen: (a) numerical (6.0 J), (b) experimental (6.0 J), (c) numerical (7.2 J), (d) experimental (7.2 J).

11.5   Conclusion

In this chapter, we have described some recent outcomes on the numerical simulations of damage propagation in FRP laminated structures under quasi-static and low-velocity impact transverse loads. Two categories of damage patterns in composite structures under transverse loads are tackled independently, that is, the various types of in-plane damage and delamination. Especially, for the delamination propagation between multiple laminae based on the cohesive interface model, we have described three kinds of techniques to improve the stability and accuracy and to decrease the computational cost of the traditional cohesive model, which include: (i) artificial damping technique for the explicit time integration scheme, (ii) move-limit technique, (iii) a new adaptive cohesive model and its extension into rate-dependent problems. Besides the description of theoretical contents, a typical experiment, that is, a DCB problem is employed to characterise the properties of the above techniques. Moreover, a low-velocity impact example is used to show the effectiveness of the adaptive cohesive model. From the obtained results, it can be found that the adaptive cohesive model for rate-independent or rate-dependent problems may be the most effective one compared with other techniques from the aspects of simplicity, computational accuracy, computational efficiency and universality for various problems.

References

[1] Chang, F.K. and Chang, K.A. (1987) A progressive damage model for laminated composite containing stress concentrations. *Journal of Composite Materials*, **21**, 834–855.

[2] Hou, J.P., Petrinic, N., Ruiz, C. and Hallet, S.R. (2000) Prediction of impact damage in composite plates. *Composites Science and Technology*, **60**, 273–281.

[3] Brewer, J.C. and Lagace, P.A. (1988) Quadratic stress criterion for initiation of delamination. *Journal of Composite Materials*, **22**, 1141–1155.

[4] Hou, J.P., Petrinic, N. and Ruiz, C. (2001) A delamination criterion for laminated composites under low-velocity impact. *Composites Science and Technology*, **61**, 2069–2074.

[5] Davies, G.A.O. and Zhang, X. (1995) Impact damage prediction in carbon composite structures. *International Journal of Impact Engineering*, **16**(1), 149–170.

[6] Zheng, S. and Sun, C.T. (1995) A double-plate finite-element model for the impact-induced delamination problem. *Composites Science and Technology*, **53**, 111–118.

[7] Li, C.F., Hu, N., Yin, Y.J. et al. (2002) Low-velocity impact-induced damage of continuous fibre-reinforced composite laminates. Part I. An FEM numerical model. *Composites Part A: Applied Science and Manufacturing*, **33**, 1055–1062.

[8] Li, C.F., Hu, N., Chen, J.G. et al. (2002) Low-velocity impact-induced damage of continuous fibre-reinforced composite laminates. Part II. Verification and numerical investigation. *Composites Part A: Applied Science and Manufacturing*, **33**, 1063–1072.

[9] Camanho, P.P. and Davila, C.G. (2002) Mixed-Mode Decohesion Finite Elements for the Simulation of Delamination in Composite Materials. NASA/TM-2002-211737.

[10] Geubelle, P.H. and Baylor, J.S. (1998) Impact-induced delamination of composites: a 2D simulation. *Composites Part B: Engineering*, **29**, 589–602.

[11] Reddy, E.D. Jr, Mello, F.J. and Guess, T.R. (1997) Modeling the initiation and growth of delaminations in composite structures. *Journal of Composite Materials*, **31**(8), 812–831.

[12] Segurado, J. and Llorca, J. (2004) A new three-dimensional interface finite element to simulate fracture in composites. *International Journal of Solids and Structures*, **41**, 2977–2993.

[13] Goncalves, J.P.M., De Moura, M.F.S.F., De Castro, P.M.S.T. and Marques, A.T. (2000) Interface element including point-to-surface constraints for three-dimensional problems with damage propagation. *Engineering Computations*, **17**(1), 28–47.

[14] Mi, Y., Crisfield, M.A. and Davis, G.A.O. (1998) Progressive delamination using interface element. *Journal of Composite Materials*, **32**(14), 1246–1272.

[15] Gao, Y.F. and Bower, A.F. (2004) A simple technique for avoiding convergence problems in finite element simulations of crack nucleation and growth on cohesive interfaces. *Modelling and Simulation in Materials Science and Engineering*, **12**, 453–463.

[16] Riks, E. (1979) An incremental approach to the solution of snapping and buckling problems. *International Journal of Solids and Structures*, **15**, 529–551.

[17] Guan, Z.D. and Yang, C.D. (2002) Low-velocity impact and damage process of composite laminates. *Journal of Composite Materials*, **36**(7), 851–871.

[18] Hu, N., Zemba, Y., Fukunaga, H. et al. (2007) Stable numerical simulations of propagations of complex damages in composite structures under transverse loads. *Composites Science and Technology*, **67**, 752–765.

[19] Hu, N., Zemba, Y., Okabe, T. et al. (2008) A new cohesive model for simulating delamination propagation in composite laminates under transverse loads. *Mechanics of Materials*, **40**, 920–935.

[20] Elmarakbi, A., Hu, N. and Fukunaga, H. (2009) Finite element simulation of delamination growth in composite materials using LS-DYNA. *Composites Science and Technology*, **69**, 2383–2391.

[21] Cao, Y. P., Hu, N., Lu, J. et al. (2002) A 3D brick element based on Hu-Washizu variational principle for mesh distorsion. *International Journal for Numerical Methods in Engineering*, **53**, 2529–2548.

[22] Carpenter, N.J., Taylor, R.L. and Katona, M.G. (1997) Lagrange constraints for transient finite element surface contact. *International Journal for Numerical Methods in Engineering*, **32**, 103–128.

[23] Moshier, M. (2006) Ram Load Simulation of Wing Skin-Spar Joints: New Rate-Dependent Cohesive Model. RHAMM Technologies LLC, USA, Report No. R-05-01.

[24] Tan, T.M. and Sun, C.T. (1985) Use of statistical indentation laws in the impact analysis of laminated composite plates. *Journal of Applied Mechanics*, **52**, 5–12.

12

Building Delamination Fracture Envelope under Mode I/Mode II Loading for FRP Composite Materials

Othman Al-Khudairi, Homayoun Hadavinia, Eoin Lewis, Barnaby Osborne and Lee S. Bryars
Material Research Centre, SEC Faculty, Kingston University, London, UK

12.1 Introduction

Fibre reinforced polymer (FRP) composite materials are in common use in many ground and space vehicles, particularly in automotive, aircraft and marine structures, in racing yachts and in sport equipment and bicycles since they allow a lighter structure, which increases efficiency by reducing weight, fuel consumption and weight-based maintenance. FRP composite materials also have a better fatigue performance relative to metals. Further; their resistance to corrosion is superior to metals. These factors reduce scheduled maintenance and increase lifetime. FRP composite materials have also excellent damping property, higher specific stiffness and strength ratios relative to those of metallic materials.

However, one of the main sources of damage in laminated composite structures is delamination, separation of the plies in the low resistance thin resin-rich interface between adjacent layers, particularly under compressive loading, impact or free-edge stresses. This is more of a problem when there is a lack of any reinforcement in the thickness direction. Laminated composite materials are prone to damage during trimming, finishing and machining activities, which can lead to edge delamination. When subjected to transverse loading, contact stresses and impact loading they are also susceptible to delamination, fibre pullout, fibre fracture and local matrix cracking. Other causes of delamination are the existence of contaminated fibres during the manufacturing process, insufficient wetting of fibres, curing shrinkage of the resin

Advanced Composite Materials for Automotive Applications: Structural Integrity and Crashworthiness,
First Edition. Edited by Ahmed Elmarakbi.
© 2014 John Wiley & Sons, Ltd. Published 2014 by John Wiley & Sons, Ltd.

and out of plane impact. The existence of high stress gradients near geometric discontinuities in composite structures such as holes, cut-outs, flanges, ply drop-offs, stiffener terminations, bonded and bolted joints promote delamination initiation.

Delamination will cause failures in a typically weak matrix of FRP materials, which can lead to local buckling of the laminate and degradation of the load-carrying capacity of the structures. The initiation and rapid propagation of a crack will cause an abrupt change in both sectional properties and load paths within the affected damaged area. Impact often causes internal cracking and delamination in the resin-rich zone between the plies for lower energy levels, while high impact energies cause penetration and excessive local shear damage. When a low velocity impact happens, the matrix material becomes overstressed resulting in micro-cracking which leads to redistribution of the load and the concentration of energy and stress at the inter-ply regions where large differences in material stiffness exist. Real-life examples of low velocity impact include in-service loads such as a dropped tool or the impact of debris from the runway on an aircraft constructed from a laminated composite. Compressive loads are sources of continuous growth on the damaged area, with a corresponding decrease of residual strength and the subsequent risk of structural failure under service loads. The presence and growth of delamination in laminates also significantly reduces the overall buckling strength of a structure while delamination grows rapidly in the post-buckling region. Delamination is an important energy absorption mechanism and it reduces the load-carrying capacity in bending and the fatigue life of the structures.

There is nondestructive evidence that shows a large extent of delaminations occurring between individual plies under impact loading. In the design of a ground or space vehicle, the need to protect its occupants from serious injury or death in the case of impact and accident is of prime importance and this should be addressed when fibre reinforced plastics (FRP) materials are used in their design. The failure behaviour of FRP materials under impact loading is a complex process and a detailed analysis of various mode of failure is necessary. The response of laminated composite structures to foreign object impact at various velocities has been a subject of intense research in recent years. Impact generally causes a global structural response and often results in internal cracking and delamination in the resin rich zone between the actual plies for lower energy levels while high impact energies cause penetration and excessive local shear damage. Real life examples of low velocity impact include in-service loads such as a dropped tool or crash of car made from laminated composite to side rails. Compressive load will especially cause continuous delamination growth of damaged area when subjected to impact loads, with the corresponding decrease of their residual strength and the subsequent risk of structural failure under service loads.

The objective of this chapter is to explain the experimental methodology for characterising the delamination fracture envelope of F RP composite materials. The double cantilever beam (DCB), three-point end notched flexure (3ENF) and mixed-mode bending (MMB) tests will be examined to establish the full delamination failure envelope under various mode mixity ratios, G_{II}/G_T. Experimental results will show the detailed procedure for finding the failure envelope for any FRP materials.

12.2 Experimental Studies

There are a number of publications in the literature on delamination fracture toughness testing using the DCB test. Amongst them, Refs. [1–6] showed that a DCB test is an accurate method to obtain GIC delamination toughness (where GIC is the opening mode I interlaminar fracture toughness). This test is standardised in ASTM D5528-01 (2007) [7]. For analysis of the experimental data, modified beam theory (MBT), simple beam theory (SBT), compliance calibration method (CCM), modified compliance calibration (MCC) and experimental compliance method (ECM) are used.

Experiments were carried out according to the ASTM 2007 standard test method for mode I interlaminar fracture toughness of unidirectional fibre-reinforced polymer matrix composites [7]. The loading tests on all experiments used a Zwick/Roell Z050 loading rig (manufactured by Zwick GmbH). This load cell has a loading capacity of 50 kN. The data capture and export from the loading rig were carried out using the Zwick/Roell TestXpert software which comes as part of the apparatus installation. For the fracture toughness testing in the DCB, 3ENF and MMB tests, it is necessary to observe the fracture front propagation and to capture the data at key stages. To observe this, a travelling microscope was positioned alongside the test specimen and manually migrated along the fracture front as the crack grew.

A preformed starter crack or delamination initiator at the mid-plane of the specimen was used. It is typically formed using a very thin non-stick film such as Teflon (the standard specifies a thickness not greater than 13 μm) which is added during layup of the composite laminates. The film material depends on the material curing temperature.

All specimens used in testing were produced using a unidirectional GFRP prepreg which was stored at a low temperature to increase its working shelf-life. The oven was initially heated to 90 °C for ramping up the aluminium plates and the FRP material to the necessary curing temperature. After 30 min, the temperature was increased to 120 °C and it was held for 1.5 h at this temperature. All test specimens were produced from [09//09] lay-up.

The end notched flexure (ENF) test was used to quantify the interlaminar mode II shear fracture toughness, GIIC, of laminated composites. The experiment consists of a three-point bending test on an end notched GFRP composite laminated beam. The compliance calibration method (CCM), modified beam theory (MBT), direct beam theory (DBT), corrected beam theory with effectivecrack length (CBTE) and compliance based beam method (CBBM) were used for analysis of the tests. ENF mode II fracture toughness was first projected by Refs. [8, 9]. Further works were carried out by Refs. [10–14]. ENF testing is standardised by the ASTM D-30 Committee to produce ASTM standardisation. In ENF testing [15] there is a major difficulty in designing test specimens for pure mode II. The difficulty lies in preventing any crack opening without introducing excessive friction between the two crack faces.

The mixed-mode bending (MMB) test is used for characterisation of the delamination under mixed mode loading. MMB test results were analysed using beam theory to calculate mode I and mode II fracture toughness and to obtain the energy release rates GI and GII under different mode mixity. MMB testing was developed through the works of [16–22]. In 2006, ASTM released a standard test method for MMB fracture toughness of unidirectional fibre reinforced polymer [23].

The objective of the current chapter is to introduce the procedure for producing the fracture envelope under various loading conditions for FRP laminates. For analysis of delamination, the mechanical properties of the FRP are required. The mechanical properties such as Young's modulus and shear modulus can be found from tensile testing of 0°, 90° and ±45° shear testing and the volume fraction of the composite can be obtained from burnout tests. For our case, the mechanical properties of the GFRP composite are summarised in Table 12.1.

Table 12.1   Mechanical properties of GFRP.

E_{11} (GPa)	E_{22} (GPa)	G_{12} (GPa)	ν_{12}	ν_{21}	Fibre volume fraction (%)
38.7 ± 2.6	12.7 ± 2.0	5.0 ± 1	0.249 ± 0.016	0.094 ± 0.014	61.7 ± 4.5

12.3   Mode I Delamination Testing: Double Cantilever Bending Test Analysis and Results

ASTM [7] states that the simple beam theory (SBT) for the strain energy release rate of a flawlessly built-in double cantilever beam, where it is clamped at the delamination front, can be found from

$$G_{IC} = \frac{3P\delta}{2ba} \tag{12.1}$$

There is some inaccuracy in using Equation 12.1 as the beam is supported by an elastic foundation and experiences large deformation of the bending arm. Taking this into account, the modified beam theory (MBT) is derived

$$G_{IC} = \frac{3P\delta}{2b(a + |\Delta|)} \tag{12.2}$$

The mode I interlaminar fracture based on the compliance calibration method (CCM) can be found from Equation 12.3.

$$G_{IC} = \frac{nP\delta}{2ba} \tag{12.3}$$

Also based on ASTM [7] and from the plot of the delamination crack length normalised by the thickness of specimen (a/h) versus cube root of compliance $C^{1/3}$, the slope A_1 can be determined. Then the interlaminar fracture toughness can be found from

$$G_{IC} = \frac{3P^2 C^{\frac{2}{3}}}{2A_1 bh} \tag{12.4}$$

where

$$A_1 = \frac{a/h}{C^{1/3}} \tag{12.5}$$

Using the Irwin–Kies equation

$$G_{IC} = \frac{P^2}{2b} \frac{dC}{da} \tag{12.5b}$$

G_{IC} can be directly found using a power law for compliance calibration [5]. In Equation 12.6 the constants n and k are both experimentally determined. G_{IC} is found by differentiating Equation 12.6 and substituting in Irwin–Kies results in Equation 12.7. The constants F and N are correction factors for arm shortening and end block stiffening.

$$C = ka^n \tag{12.6}$$

$$G_{IC} = \frac{nP\delta}{2ba}\frac{F}{N} \tag{12.7}$$

The DCB tests were performed on a Zwick tensile testing machine (as shown in Figure 12.1) at a constant displacement rate of 2 mm/min. The testing was carried in dry laboratory conditions at a constant temperature of $21 \pm 1\,°C$. The specimens were measured for their length, width and thickness, with three measured values taken to produce an average value for each. The load-opening displacement tests results for four specimens are shown in Figure 12.2. Samples of the crack front in DCB test are shown in Figure 12.3. Table 12.2 summarises the results of DCB tests from various methods. It can be seen that the G_{IC} value obtained by SBT is overestimating and MBT and CCM are very close and a correct representation of mode I delamination fracture toughness.

12.4 Mode II Delamination Testing: End Notched Flexure Test Analysis and Results

The experimental set-up of the three-point end-notched flexure (3ENF) test and the specimen dimensions are shown in Figure 12.4. The analysis of 3ENF by the compliance calibration method (CCM) was developed by Russell [8] and Russell and Street [9]. The classical data reduction schemes to acquire the mode II critical fracture energy (pure) G_{II} are

$$C = C_0 + ma^3 \tag{12.8}$$

$$\frac{dC}{da} = 3ma^2 \tag{12.9}$$

$$G_{II} = \frac{P^2}{2b}\frac{dC}{da} = \frac{3ma^2P^2}{2b} \tag{12.10}$$

Based on the direct beam theory (DBT), Murri and O'Brien [10] derived Equation 12.11 that uses crack length within the calculation to obtain G_{IIC}. However, measuring the crack length is not so easy in mode II as the position of crack tip is not well defined due to the formation of shear bands.

$$G_{IIC} = \frac{9P\delta a^2}{2b(2L^3 + 3a^3)} = \frac{9P^2a^2C}{2b(2L^3 + 3a^3)} \tag{12.11}$$

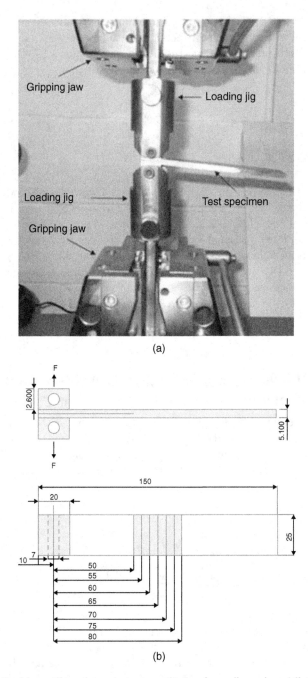

Figure 12.1   (a) Double cantilever beam test set up. (b) Specimen dimensions (all dimensions in mm).

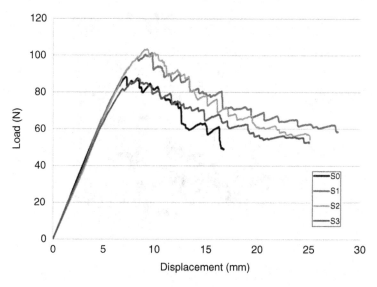

Figure 12.2   Experimental results of load-opening displacement for DCB tests.

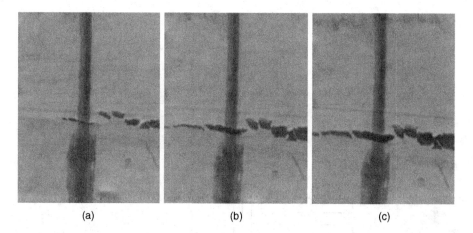

Figure 12.3   Crack front in DCB test: (a) crack begins to form, (b) multiple bridging, (c) growing of crack.

Table 12.2   Results of interlaminar fracture toughness in mode I testing DCB.

Method of analysis	MBT	CCM	MCC	ECM	SBT
G_{IC} (J/m^2)	733 ± 20	729 ± 29	740 ± 20	804 ± 39	963 ± 22

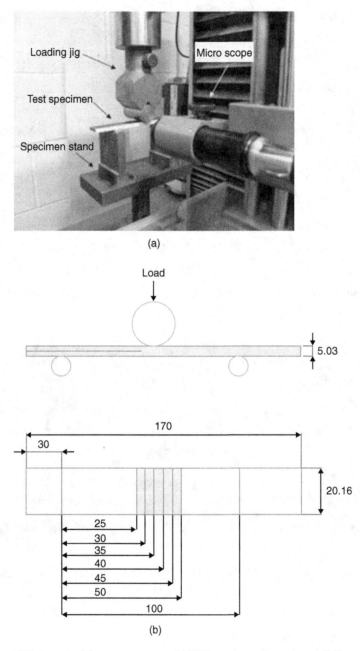

Figure 12.4   (a) End-notched flexure test set-up. (b) ENF specimen dimensions (all dimensions in mm).

Williams [12] and Wang and Williams [13] derived a modified beam theory (MBT) for ENF analysis. In MBT, Equation 12.15, the crack correction length Δ_{II} is calculated from Equation 12.14.

$$\Delta_I = h \sqrt{\frac{E_{11}}{11G_{13}} \left[3 - 2\left(\frac{\Gamma}{1+\Gamma}\right)^2\right]} \tag{12.12}$$

$$\Gamma = 1.18 \frac{\sqrt{E_{11}E_{22}}}{G_{13}} \tag{12.13}$$

$$\Delta_{II} = 0.42\Delta_I \tag{12.14}$$

$$G_{IIC} = \frac{9(a + \Delta_{II})^2 P^2}{16b^2 h^3 E_1} \tag{12.15}$$

Murri and O'Brien [10] used the corrected beam theory with effective crack length (CBTE) stating that, when using certain methods such as MBT, DBT or CCM to obtain G_{IIC}, accurate crack length measurements during propagation are required, which is not an easy task. Because of this issue, significant errors can occur during the fracture characterisation under pure mode II loading. CBTE is one of the few methods that do not require crack length to obtain G_{IIC}.

$$C = \frac{\delta}{P} \tag{12.16}$$

$$C_C = C - \frac{3L}{10G_{13}bh} \tag{12.17}$$

$$a_e = \sqrt[3]{\frac{8E_f bh^3 C_C}{3} - \frac{2L^3}{3}} \tag{12.18}$$

$$G_{IIC} = \frac{9P^2 a_e^2}{16b^2 E_f h^3} \tag{12.19}$$

Murri and O'Brien [10] also used the compliance based beam method (CBBM) to calculate G_{IIC}. In CBBM also, the crack length is not required to obtain G_{IIC}. Equation 12.21 is used to calculate the flexural modulus, E_f, from the three-point bending test of composites.

$$C_{0C} = C_0 - \frac{3L}{10G_{13}bh} \tag{12.20}$$

$$E_f = \frac{3a_0^3 + 2L^3}{8C_{0C}bh^3} \tag{12.21}$$

$$C_C = C - \frac{3L}{10G_{13}bh} \tag{12.22}$$

$$a_e = \left[\frac{C_C}{C_{0C}}a_0^3 + \frac{2}{3}\left(\frac{C_C}{C_{0C}} - 1\right)L^3\right]^{\frac{1}{3}} \tag{12.23}$$

$$G_{IIC} = \frac{9P^2 a_e^2}{16b^2 E_f h^3} \tag{12.24}$$

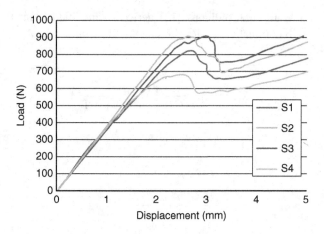

Figure 12.5   Experimental results of load–displacement ENF tests.

Four 3ENF tests were performed on a Zwick tensile testing machine as shown in Figure 12.4 with the specimen dimensions. The load–displacement results of ENF tests are shown in Figure 12.5. Table 12.3 summarises the test results in mode II loading. The results from CCM and DBT are very close and CBTE gave the highest estimate for mode II delamination toughness.

12.5   Mixed Mode I/II Delamination Testing: Mixed-Mode Bending Test Analysis and Results

The MMB test is a combination of mode I (DCB) and mode II (ENF) tests. The mixed mode bending (MMB) experimental set-up and the apparatus dimensions are shown in Figure 12.6.

Five different methods were used to analyse MMB test results: simple beam theory (SBT), corrected beam theory (CBT), compliance based beam method (CBBM), simple beam theory with elastic foundation (SBTEF) and ASTM 6671/D6671M-06. The MMB test is a combination of two tests, the DCB and ENF; typically the two tests are used to characterise mode I and mode II.

$$P_I = \left(\frac{3c - L}{4L}\right) P \tag{12.25}$$

$$P_{II} = \left(\frac{c + L}{L}\right) P \tag{12.26}$$

$$G_I = \frac{12a^2 P_I^2}{b^2 h^3 E_{11}} \tag{12.27}$$

Table 12.3   Results of interlaminar fracture toughness in mode II testing (ENF).

Method of analysis	CCM	DBT	MBT	CBBM	CBTE
G_{IIC} (J/m^2)	1620 ± 16	1680 ± 27	1460 ± 19	1855 ± 26	2044 ± 27

$$G_{II} = \frac{9a^2 P_{II}^2}{16b^2 h^3 E_{11}} \tag{12.28}$$

$$\frac{G_I}{G_{II}} = \frac{4}{3} \left(\frac{3c - L}{c + L} \right)^2, c \geq \frac{L}{3} \tag{12.29}$$

$$G_I = \frac{12 \left(a + h \, |\Delta_I| \right)^2 P_I^2}{b^2 h^3 E_1} \tag{12.30}$$

$$G_{II} = \frac{9 \left(a + 0.42h \, |\Delta_I| \right)^2 P_{II}^2}{16b^2 h^3 E_{11}} \tag{12.31}$$

$$\frac{G_I}{G_{II}} = \frac{4}{3} \left(\frac{3c - L}{c + L} \right)^2 \left(\frac{a + h\Delta_I}{a + 0.42\Delta_I h} \right)^2 \tag{12.32}$$

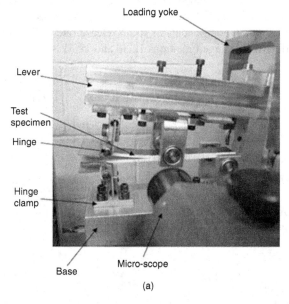

(a)

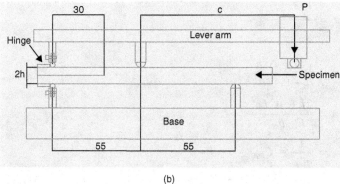

(b)

Figure 12.6 (a) Mixed mode bending experimental set-up. (b) Apparatus dimensions (all dimensions in mm), C = 42 mm and 97 mm.

The SBTEF is based on Kanninen's [24] assumption that each arm is a beam supported by an elastic foundation.

$$G_I = \frac{12P_I^2}{b^2h^3E_{11}}\left[a^2 + \frac{2a}{\lambda} + \frac{1}{\lambda^2}\right]$$

(12.33)

where

$$k = \frac{2bE_{22}}{h}$$

(12.34)

$$\lambda = \left(\frac{3k}{bh^3E_{11}}\right)^{\frac{1}{4}}$$

(12.35)

Reeder and Crews Jr [18] modified the beam theory equations further more for strain energy release rate by taking into account the shear deformation energy associated with bending by adding the shear deformation components of strain release rate to Equation 12.33. The contribution of shear effect will change Equations 12.30 and 12.31 to Equations 12.36 and 12.37.

$$G_I = \frac{3P^2(3c-L)^2}{4b^2h^3L^2E_{11}}\left[a^2 + \frac{2a}{\lambda} + \frac{1}{\lambda^2} + \frac{h^2E_{11}}{10G_{13}}\right]$$

(12.36)

$$G_{II} = \frac{9P^2(c+L)^2}{16b^2h^3L^2E_{11}}\left[a^2 + \frac{0.2h^2E_{11}}{G_{13}}\right]$$

(12.37)

ASTM D6671/D6671M-06 [23] explains the procedure for measuring the mixed mode delamination fracture toughness where the mode mixtity G_{II}/G_T depends on a correction for delamination length and for rotation at the crack front. The equations presented in the ASTM are based on research by Williams [12], Wang and Williams [13] and Kinloch and Wang [25].

$$c = \frac{12\beta^2 + 3\alpha + 8\beta\sqrt{3\alpha}}{36\beta^2 - 3\alpha}L$$

(12.38)

$$\alpha = \frac{1 - \dfrac{G_{II}}{G}}{\dfrac{G_{II}}{G}}$$

(12.39)

$$\beta = \frac{a + \Delta_I h}{a + 0.42\Delta_I h}$$

(12.40)

$$C_{cal} = \frac{2L(c+L)^2}{E_{cal}b_{cal}t^3}$$

(12.41)

$$C_{sys} = \frac{1}{m_{cal}} - C_{cal}$$

(12.42)

$$E_{1f} = \frac{8(a_0 + \Delta_I h)^3 (3c-L)^2 + \left[6(a_0 + 0.42\Delta_I h)^3 + 4L^3\right](c+L)^2}{16L^2bh^3\left(\dfrac{1}{m} - C_{sys}\right)}$$

(12.43)

$$G_I = \frac{12P^2 (3c - L)^2}{16b^2 h^3 L^2 E_{1f}} (a + \Delta_I h)^2 \tag{12.44}$$

$$G_{II} = \frac{9P^2 (c + L)^2}{16b^2 h^3 L^2 E_{1f}} (a + 0.42\Delta_I h)^2 \tag{12.45}$$

$$G_T = G_I + G_{II} \tag{12.46}$$

$$\frac{G_{II}}{G_T} = \frac{G_{II}}{G_I + G_{II}} \tag{12.47}$$

The compliance based beam method follows the Timoshenko beam theory (TBT). C_{sys} is calculated using the slope of calibration curve and compliance calibration specimen. There are a few issues within this method such as stress concentration and root rotation effects, and at the crack tip it influences compliance that is not accounted for in the beam theory. Oliveira et al. [21] states that when using this method on wood the flexural modulus is significantly specimen dependant due to the wood heterogeneity and this requires measurements of the modulus for each test specimen. Therefore a corrected flexural modulus is estimated using the initial compliance C_{0I} and the crack length a_0 which also considers the root rotation effects, shown in Equation 12.49, where Δ_I is given in Equations 12.12 and 12.13 using E_{fl} instead of E_{11}. A repetitive process should be used to obtain a converged value of E_{fl} using Equations 12.12 and 12.13. During the crack growth, Fracture Process Zone (FPZ) develops at the crack tip. To account for the dissipated energy in this region, a corresponding crack in mode I, a_{eqI}, is estimated from the existing specimen compliance by using Equation 12.54. a_{eqI} is the equivalent crack length in mode I.

The FPZ influences the specimen compliance and the effect is accounted for via the equivalent crack length. The solution of cubic Equation 12.52 can be obtained by standard math tools such as MATLAB software [21].

$$C_I = \frac{8a^3}{E_{11}bh^3} + \frac{12a}{5bhG_{13}} \tag{12.48}$$

$$E_{fl} = \left(C_{0I} - \frac{12(a_0 + h|\Delta_I|)}{5bhG_{13}} \right)^{-1} \frac{8(a_0 + h|\Delta_I|)^3}{bh^3} \tag{12.49}$$

$$G = \frac{P^2}{2b} \frac{dC}{da} \tag{12.50}$$

Using Equation 12.44 leads to

$$G_I = \frac{6P_I^2}{b^2 h} \left(\frac{2a_{eqI}^2}{h^2 E_{fl}} + \frac{1}{5G_{13}} \right) \tag{12.51}$$

where

$$\alpha a_{eqI}^3 + \beta a_{eqI} + \gamma = 0 \tag{12.52}$$

$$\alpha = \frac{8}{bh^3 E_{fl}}; \beta = \frac{12}{5bhG_{13}}; \gamma = -C_I \tag{12.53}$$

$$a_{eqI} = \frac{1}{6\alpha} A - \frac{2\beta}{A} \tag{12.54}$$

where A is given by

$$A = \left(\left[-108\gamma + 12\sqrt{3\left(\frac{4\beta^3 + 27\gamma^2\alpha}{\alpha}\right)} \right] \alpha^2 \right)^{1/3} \tag{12.55}$$

A similar procedure can be used for mode II. Equation 12.56 is used for compliance where $C_{II} = \delta_{II}/P_{II}$. The displacement δ_{II} is obtained from $\delta_{II} = \delta C + \delta_I/4$ [21], displacement δ_C is measured at the specimen mid-span by the loading lever; displacement δ_I is measured at the edge of the specimen. The corrected flexural modulus is obtained using the crack length and initial compliance. a_{eqII} is the equivalent crack length in mode II.

$$C_{II} = \frac{3a^3 + 2L^3}{8E_{11}bh^3} + \frac{3L}{10bhG_{13}} \tag{12.56}$$

$$E_{fII} = \frac{3a_0^3 + 2L^3}{8bh^3}\left(C_{0II} - \frac{3L}{10G_{13}bh}\right)^{-1} \tag{12.57}$$

Equivalent crack accounting for the fracture process zone (FPZ) effects is estimated from Equations 12.58 and 12.59.

$$a_{eqII} = \left[\frac{C_{IIcorr}}{C_{0IIcorr}}a_0^3 + \frac{2}{3}\left(\frac{C_{IIcorr}}{C_{0IIcorr}} - 1\right)L^3 \right]^{1/3} \tag{12.58}$$

where

$$C_{IIcorr} = C_{II} - \frac{3L}{10G_{13}bh} \quad \text{and} \quad C_{0IIcorr} = C_{0II} - \frac{3L}{10G_{13}bh} \tag{12.59}$$

$$G_{II} = \frac{9P_{II}^2 a_{eqII}^2}{16E_{fII}b^2h^3} \tag{12.60}$$

The load–displacement results of MMB tests are shown in Figure 12.7. Table 12.4 summarises the test results in mode I/II loading. The results of CBT and SBTEF are close and ASTM is the lowest estimate of mixed-mode delamination toughness.

12.6 Fracture Failure Envelope

The failure envelope for the GFRP is shown in Figure 12.8 where G_{IC} is the vertical axis (from DCB-CBT) and G_{IIC} is the horizontal axis (from ENF-CBBM) together with the results of

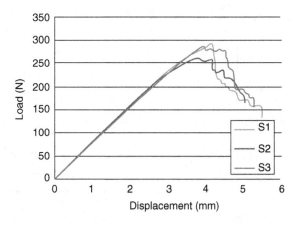

Figure 12.7   Experimental results of load-displacement for MMB tests at $c = 42$ mm.

Table 12.4   Results of MMB experiment $C = 42$ and 97 mm.

Method of analysis	$C = 42$ mm				$C = 97$ mm			
	G_I (J/m²)	G_{II} (J/m²)	G_T (J/m²)	G_{II}/G_T (%)	G_I (J/m²)	G_{II} (J/m²)	G_T (J/m²)	G_{II}/G_T (%)
CBT	500 ± 9	645 ± 9	1145 ± 15	56 ± 2	358 ± 21	103 ± 21	461 ± 13	22.3
ASTM	332 ± 11	429 ± 11	761 ± 11	56 ± 2	279 ± 11	80 ± 11	359 ± 11	22.3
SBT	434 ± 15	607 ± 15	1041 ± 15	58 ± 2	312 ± 21	97 ± 13	409 ± 21	23.7
SBTEF	484 ± 14	612 ± 15	1096 ± 15	56 ± 2	347 ± 21	98 ± 21	445 ± 21	22

MMB C97 and MMB C42. Inside the envelope, the material is safe from delamination under full range of mode mixity and outside of the envelope, it will delaminate.

Figure 12.9 shows variation of G_T versus G_{II}/G_T. The DCB results are shown at $G_{II}/G_T = 0$ and the ENF results are at $G_{II}/G_T = 1$. Between these two extreme is mix mode I/II loading. Two sets of MMB at $G_{II}/G_T = 0.22$ (C97) and $G_{II}/G_T = 0.56$ (C42) have been shown. As

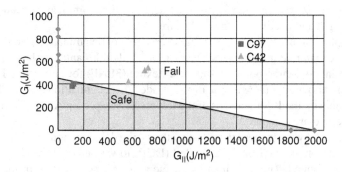

Figure 12.8   Failure envelope for the GFRP composite.

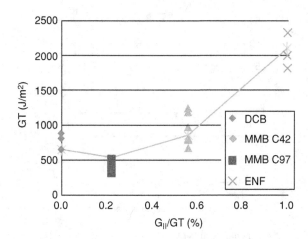

Figure 12.9   Variation of total delamination fracture toughness, G_T, with respect to mode mixity ratio G_{II}/G_T.

from pure mode I interlaminar delamination to pure mode II, the resistance to delamination increases.

12.7   Conclusion

Laminated FRP composite materials are used in many parts of automotive structures. These materials are susceptible to delamination, separation of the plies in the low resistance thin resin-rich interface between adjacent layers particularly under compressive loading, impacts or free-edge stresses. The existence of high stress gradients near geometric discontinuities in composite structures such as holes, cut-outs, flanges, ply drop-offs, stiffener terminations, bonded and bolted joints promote delamination initiation. Other causes of delamination are the existence of contaminated fibres during the manufacturing process, insufficient wetting of fibres, curing shrinkage of the resin and out of plane impact.

As there is always possibility of crash and impact of cars, the delamination characterisation of laminated materials is quite important. The car designer must be aware of the behaviour of delaminated structure and take into account the delamination properties during design stage of the vehicle as the presence and growth of delamination significantly reduces the overall buckling strength and bending stiffness of a structure.

In this chapter, three test methods for characterisation of delamination fracture toughness of FRP composite materials are discussed. These tests are double cantilever beam (DCB) for pure mode I delamination, three-point end-notched flexure (3ENF) for pure mode II delamination and mixed-mode bending (MMB) for mixed mode I/II delamination. For each test method, various available methods for analysis of the test results are presented and the results obtained from each method are compared.

It is shown that, by using these three test methods, the delamination fracture envelope under various loading conditions, from pure mode I (DCB) to various mode mixity to pure mode II (ENF) can be obtained. The delamination envelope is an essential tool for the application of the FRP material in different products.

Nomenclature

A_1	slope of a/b versus $C^{1/3}$
a	crack length
a_0	initial crack length
a_e	corrected crack length
b	specimen width
C_1	compliance
C	experimental compliance
C_{cal}	compliance of calibration specimen
C_{sys}	system compliance
C_C	corrected compliance
C_0	initial compliance of ENF graph
C_{0C}	initial corrected compliance
E_{cal}	modulus of calibration bar
E_f	flexural stiffness
E_{11}	modulus of elasticity in the fibre direction
E_{22}	transverse modulus
F	correction fracture for shortening of arm in DCB test
G_T	total strain energy release rate
G_I	mode I strain energy release rate
G_{II}	mode II strain energy release rate
G_{II}/G_T	mixed mode ratio
G_{12}	lamina longitudinal shear modulus
G_{13}	lamina transverse shear modulus
h	half of the specimen thickness
k	stiffness of elastic foundation
L	span length in MMB test
m	slope of C versus a^3
n	slope of log C versus log a
P	load
P_I	mode I load
P_{II}	mode II load
Δ	function of delamination length from $C^{1/3}$ versus a
Δ_I	crack length correction in relation for shear deformation in DCB test
Δ_{II}	crack length correction in ENF test
δ	deflection

References

[1] Williams, J.G. (1989) End corrections for orthotropic DCB specimens. *Composites Sciences and Technology*, **35**, 367–376.
[2] Adams, D.F., Carlsson, L.A. and Pipes, R.B. (2002) *Experimental Characterization of Advanced Composite Materials*, 3rd edn, CRC Press, London, ISBN-10: 1587161001.
[3] Davies, P., Blackman, B.R.K. and Brunner, A.J. (1998) Standard test methods for delamination resistance of composite materials: current status. *Applied Composite Materials*, **5**, 345–364.

[4] Ozdil, F. and Carlsson, L.A. (1999) Beam analysis of angle ply laminate DCB specimens. *Composites Science and Technology*, **59**, 305–315.

[5] Blackman, B.R.K., Kinloch, A.J., Paraschi, M. and Teo, W.S. (2003) Measuring the mode I adhesive fracture energy, GIC, of structural adhesive joints: the results of an International Round-Robin. *International Journal of Adhesion and Adhesives*, **23**, 293–305.

[6] Gordnian, K., Hadavinia, H., Mason, P.J. and Madenci, E. (2008) Determination of fracture energy and cohesive strength in mode I delamination of angle-ply laminated composites. *Composite Structures*, **82**(4), 577–586.

[7] ASTM (2007) D5528-01. Standard Test Method for Mode I Interlaminar Fracture Toughness of Unidirectional Fibre Reinforced Polymer Matrix Composites, pp. 261–270. ASTM, Philadelphia.

[8] Russell, A.J. (1982) On the Measurement of Mode II Interlaminar Fracture Energies. Material Report 82. Defence Research Establishment Pacific, Victoria, BC, Canada.

[9] Russell, A.J. and Street, K.N. (1982) Factor Affecting the Interlaminar Fracture Energy Of Graphic/Epoxy Laminates. Progress in Science and Engineering of Composite, ICCM-IV, Tokyo, Japan, pp. 279–296.

[10] Murri, G.B. and O'brien, T.K. (1985) Interlaminar GIIC Evaluation of Toughened Resin Matrix Composites using The End-Notched Flexure Test. 26th AIAA/ASME/ASCE/AHS Structures, Structural Dynamics, and Materials Conference, Orlando, Florida.

[11] Carlsson, L.A., Gillespie, J.W. and Trethewey, B. (1986) Mode II interlaminar fracture toughness in graphite/epoxy and graphite/PEEK composites. *Journal of Reinforced Plastics and Composites*, **5**, 170–87.

[12] Williams, J.G. (1989) The fracture mechanics of delamination test. *Journal of Strain Analysis*, **24**, 207–214.

[13] Wang, Y. and Williams, J.G. (1992) Corrections for mode II fracture toughness specimens of composite materials. *Composites Science and Technology*, **43**, 251–256.

[14] Ozdil, F., Carlsson, L.A. and Davies, P. (1998) Beam analysis of angle ply laminate end notched flexure specimens. *Composites Science and Technology*, **58**, 1929–1938.

[15] Russell, A.J. and Street, K.N. (1985) Moisture and Temperature Effects Can the Mixed-Mode Delamination Fraction of Unidirectional Graphic/Epoxy, Delamination and Rebinding of Material, ASTM STP 876, Philadelphia, pp. 349–356.

[16] Russell, A.J. and Street, K.N. (1983) Moisture and Temperature Effects on the Mixed-Mode Delamination Fracture of Unidirectional Graphic/Epoxy, DREP Technical Memo 83–22, Defence Research Establishment Pacific, Victoria, BC, Canada.

[17] Russell, A.J. and Street, K.N. (1985) Moisture and Temperature Effects on the Mixed-Mode Delamination Fracture of Unidirectional Graphite/Epoxy. Delamination and debonding of materials, ASTM STP 876 (ed. W.S. Johnson), Philadelphia, pp. 349–370.

[18] Reeder, J.R. and Crews, J.R. Jr (1990) Mixed-mode bending method for delamination testing. *AIAA Journal*, **28**(7), 1270–1276.

[19] Ozdil, F. and Carlsson, L.A. (1999) Beam analysis of angle ply laminate mixed mode bending specimens. *Composites Science and Technology*, **59**, 937–945.

[20] Tenchev, R.T. and Falzon, B.G. (2007) A correction to the analytical solution of the mixed-mode bending (MMB) problem. *Composites Science and Technology*, **67**, 662–668.

[21] Oliveira, J.M.Q., de Moura, M.F.S.F., Silva, M.A.L. and Morais, J.J.L. (2007) Numerical analysis of the MMB test for mixed-mode I/ II wood fracture. *Composites Science and Technology*, **67**, 1764–1771.

[22] de Morais, A.B. and Pereira, A.B. (2006) Mixed mode I+II interlaminar fracture of glass/epoxy multidirectional laminates – Part 1: analysis. *Composites Science and Technology*, **66**, 1889–1895.

[23] ASTM (2006) D6671/D6671M-06 Test Method for Mixed Mode I-Mode II Interlaminar Fracture Toughness of Unidirectional Fiber Reinforced Polymer Matrix Composites. ASTM, Philadelphia.

[24] Kanninen, M.F. (1973) An augmented double cantilever beam model for studying crack propagation and arrest. *International Journal of Fracture*, **9**(1), 83–92.

[25] Kinloch, A.J., Wang, Y., Williams, J.G. and Yayla, P. (1993) The mixed-mode delamination of fibre-composite materials. *Composites Science and Technology*, **47**, 225–237.

Part Four

Case Studies and Designs

13

Metal Matrix Composites for Automotive Applications

Anthony Macke[1], Benjamin F. Schultz[1], Pradeep K. Rohatgi[1] and Nikhil Gupta[2]

[1] *Center for Composite Materials, Materials Engineering Department, University of Wisconsin-Milwaukee, Milwaukee, WI 53201, USA*
[2] *Composite Materials and Mechanics Laboratory, Mechanical and Aerospace Engineering Department, Polytechnic Institute of New York University, Brooklyn, NY 11201, USA*

13.1   Automotive Technologies

13.1.1   Current Landscape

Automobiles are an integral part of the modern society. In 2012, there were 239 million cars in active use in the United States, which translates into nearly one car per person over the age of 18. The development of cities and the interstate system in the 1950s resulted in the present suburban culture, which has a heavy reliance on automobiles. It is expected that the use of personal automobiles will increase dramatically around the world over the next decade as the purchasing power of developing nations such as China, India and Brazil increases. In these countries currently less than 10% of the population owns an automobile. Other nations of the world will follow suit as industrialisation spreads and brings prosperity to more societies.

Conventionally, automobiles use an internal combustion (IC) engine and are powered by gasoline fuel. In the twentyfirst century, several challenges are faced by gasoline powered automobiles for the reasons outlined below:

- There is an increasing awareness about pollution and global warming around the world. Automobiles are identified as a major source of emissions released in the atmosphere. Although there has been progress in developing cleaner burning gasoline, the demand for immediate and substantial reduction in emissions has been rapidly mounting.

Advanced Composite Materials for Automotive Applications: Structural Integrity and Crashworthiness,
First Edition. Edited by Ahmed Elmarakbi.
© 2014 John Wiley & Sons, Ltd. Published 2014 by John Wiley & Sons, Ltd.

- Keeping pace with the demand for reducing emissions, nations around the world are adopting stricter standards for fuel economy (km/l or miles/gallon) of automobiles. In the proposed standards in the United States, the fuel economy of popular compact and mid-sized cars is required to improve by about 40% by the year 2025. Achieving this steep goal is a significant challenge.
- The price of oil has increased significantly over the past decade. Oil has frequently been trading at over US$100/barrel for the past five years. Consequently, the gasoline prices have increased from about US$1/gallon (1 gallon = 3.78 l) in the early 2000s to over US$4/gallon in 2011 and beyond. A fourfold increase in fuel prices has resulted in an increased demand for more fuel efficient automobiles.

These challenges are expected to be addressed by progress in three different areas, all of which are developing in parallel:

- Alternative technologies to power automobiles
- Increasing the efficiency of current gasoline engines
- Reduction in vehicle weight.

A brief discussion of the present state in all these areas is presented below to identify the potential for success and the present state of the art.

13.1.2   Alternative Technologies

Several alternative technologies are being developed that seem promising. These technologies are broadly divided into three categories, as briefly discussed below.

13.1.2.1   Hybrid Vehicles

This is presently the most widely used technology, where an IC engine is paired with an electric motor. The vehicle is primarily powered by the electric motor whenever possible and the IC engine is used as the backup. The Toyota Prius was the first mass marketed hybrid car, available since 1997. Now almost every major automobile company has this type of vehicle in its lineup. The major limitation of this technology is that the presence of two engines simultaneously increases the vehicle weight and the actual potential of emission reduction is not fully exploited because of increased fuel consumption due to increased weight of engine, battery and other components. For example, the curb weights of the 2012 Ford Escape gasoline and hybrid engine are 3231 and 3652 lb, respectively (3000 lb = 1361 kg). The city/highway fuel ratings of these models are, respectively, 23/28 and 43/31 miles/gallon (20 miles/gallon = 8.5 km/l). At a comparable vehicle weight, the fuel economy of the hybrid model could be even better. In these cars the lightweight structural materials and weight reduction of the engine block can help in fully realising the advantage of hybrid technology.

13.1.2.2   Electric Vehicle

Electric vehicle technologies have been available for the past few years. Tesla Motors has developed cars that are powered by a stack of rechargeable batteries. However, the weight of the battery pack and heat dissipation issues are important for such an approach. These cars have been available in production models but their cost close to or upwards of US$100 000 has limited their mainstream adoption. In 2010 Nissan and Chevy started selling their electric models, Leaf and Volt, respectively. The initial concerns of electric vehicles were their short driving range on one charge and long charging times. However, the new models with upwards of 60 miles range in one charge and a charging times of ≤6 h have made their urban usage a reality. Challenges related to an absence of charging stations in public or work places still hamper their usage.

An important challenge for these vehicles is to reduce the size and weight of the battery pack. Technologies are required to develop batteries with very high current densities and small size to reduce the vehicle weight and increase the usable space.

13.1.2.3   Fuel Cell or Hydrogen Vehicles

In fuel cell technology, the chemical energy of a fuel, such as hydrogen, is converted into electricity to power the motor. Although the development of fuel cell cars goes back to the Chevrolet Electrovan in 1966, the technology is still not available in mass produced automobiles. The Honda FCX-Clarity and Mercedes-Benz F-cell are examples of the vehicles currently being road tested. Lack of infrastructure for refuelling is a limitation for large scale testing of this technology. Currently, the driving range of these vehicles between refuelling is reaching 200 miles (438 km), which makes their adoption possible for normal urban and suburban driving. General Motors has tested fuel cells that weigh about 220 lb less than a standard four cylinder engine, which is useful in reducing the vehicle weight and fuel consumption. The driving range of these vehicles can be further increased by reducing the structural weight of the vehicle by using lightweight composites.

13.1.3   Promise for Lightweight Materials

Irrespective of the technology used in powering the vehicle, lightweight materials can provide a self-accelerating spiral of benefits: (i) lighter vehicles will require less fuel for the same driving distance, which will reduce the amount of fuel carried by the vehicle, (ii) reduction in fuel tank capacity will further reduce the weight and require a smaller engine, (iii) the engine block being amongst the heavier components of the car, a smaller engine will cause further reduction in the weight of the vehicle, which brings us back to the starting point of the cycle for the next iteration. Advancements in the technologies to improve the automotive engine capacity will be augmented by the benefits obtained from lightweight materials.

Several approaches can be taken to use lighter materials and reduce the vehicle weight:

- Replacing heavier materials such as steel with lighter materials such as aluminium or magnesium.
- Replacing conventional materials such as steel with new high strength varieties of the same material to using lesser quantity.
- Replacing monolithic materials with high performance composite materials, especially those made of lightweight metals such as aluminium and magnesium.

The possibility of obtaining a combination of several desired characteristics through materials engineering provides an edge to composite materials over other possible solutions to weight reduction. The following discussion will provide an introduction to advanced lightweight composites that are either currently used or have promise for future usage in automobiles.

13.1.4   Metal Matrix Composites

A general classification spectrum for composite materials is presented in Figure 13.1. As seen in Figure 13.1, the term metal matrix composites (MMCs) can cover a broad range of materials. As the basic definition, MMCs are defined as a metallic matrix reinforced with a second phase. The second phase may be in the form of continuous or discontinuous fibers, whiskers or particles. The reinforcement may have one or more dimensions in the micro or nano size range. The interface between the matrix and the reinforcement plays an important role in defining several properties of the composite, for example tensile strength, Young's modulus, and thermal conductivity. Engineering the reinforcement–matrix interface is a part of the composite synthesis process.

In some earlier classifications, alloys were not included in the MMC definition. Two examples of alloys are shown in Figure 13.2, where the microstructures of Al-Si alloy A356 and Mg-Al-Zn alloy AZ91D were taken by optical microscope. Both these micrographs show two-phase structures. Initially, it was argued that microstructures that are governed by phase diagrams should not be classified as composites. However, amongst a wide variety of material

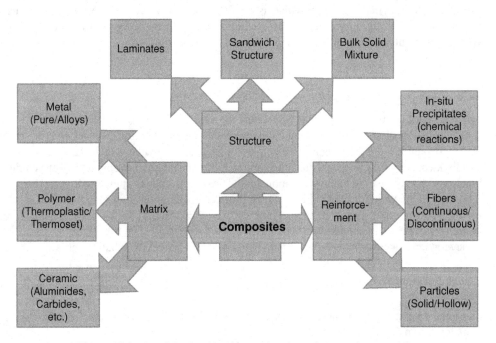

Figure 13.1 Possible classifications and variety of composite materials.

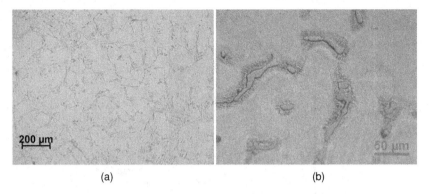

(a)   (b)

Figure 13.2 Optical microstructures of (a) A356 Al-Si alloy and (b) AZ91D Mg-Al-Zn alloy. Both micro structures are two phase, where hard particles are present in a softer matrix.

systems, narrowly excluding a few compositions is difficult. Now these two-phase alloys are commonly lumped into the category of in situ composites, where the second phase precipitates from the single phase material/melt during the composite synthesis.

One of the most appealing aspects of MMCs is that they are highly customisable for a given set of properties. MMCs can be tailored to exhibit high wear resistance, thermal conductivity, electrical conductivity, strength, low weight and some even show promise as self-lubricating or self-healing materials. Several of these properties can be customised simultaneously. By adjusting the volume percentage of reinforcement, different cost and performance profiles can be realised. Reinforcements can also be selectively applied to create MMCs where only a portion of the component is reinforced, while the remaining material in the component is simply the matrix. This technique is used to reduce cost by only using the reinforcement where it is useful. Reinforced cylinder liners are an example of this practice. Such composites with composition gradient can be synthesised by gravity assisted settling methods or centrifugal casting. Two or more types of reinforcements can be mixed to achieve a hybrid MMC that gains its properties from the combination of constituent materials. For example, adding graphite (Gr) and SiC to aluminium results in a composite which is high strength, wear resistant (due to SiC) and self-lubricating (due to Gr).

Although MMCs are extremely versatile and have appeared in high profile applications such as the fuselage structure of the space shuttle orbiter [1] and disk brakes of high end performance cars such as the Lotus Elise [2], use of this category of materials is not yet widespread. Additional costs are incurred in both developing the MMC and developing the process to make the products using MMCs because MMCs are more complex and relatively new compared to alloys. Since there are so many different combinations to consider for MMCs, modelling and simulation studies can help in narrowing down the choices for material types and their volume fractions for a desired set of properties. The availability of material selection charts, for example Figure 13.3, can also help in visualising the application domains for MMCs. This kind of chart can be plotted between any two desired properties as per the requirement of applications. In addition, charts where strength, modulus, thermal conductivity and other properties are plotted with respect to density are also useful in the material selection process.

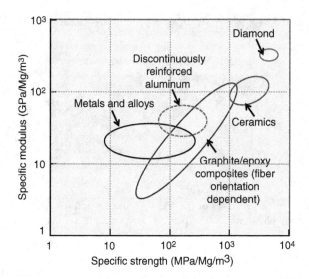

Figure 13.3 Trends in specific modulus with respect to specific strength for a variety of engineering materials. The properties domains include: yield strength for metals and alloys, tensile strength for composites and compressive strength for ceramics.

One of the leading areas of development is utilising MMCs for thermal management applications. Aluminium and copper are the two most likely candidates for heat sink applications, considering their good thermal properties. Although aluminium has only one-half the thermal conductivity of copper, it has one-third the density and, therefore, is a good candidate for heat sinks where mobility or portability is of concern. Adding high conductivity particles to aluminium increases its conductivity and makes them viable for commercial products.

13.1.5   Cost–Benefit Analysis

Cost–benefit analysis and the development of stable pricing models are important for businesses. Significant fluctuations in the prices of primary metals in recent years have made the forecasting of trends difficult. However, the use of low cost ceramic fillers and reinforcements can help to lower dependence on the metal and make trends more stable. For example, in several types of composites, industrial waste such as fly ash has been used. Fly ash is freely available and the cost associated with this material is primarily the transportation and beneficiation to remove impurities and undesired materials. Fly ash can replace up to 50 vol% aluminium in MMCs, depending on the required properties, and can provide a more stable cost model for MMCs. Further development in such lightweight materials can help in reducing the cost associated with their transportation, which can be a significant part of their overall cost.

Some examples can be useful in understanding the cost–benefit analysis for composites. These examples are based on model prices in 2012. In recent years fluctuations of over 50% in the price of primary metals have been observed and such changes can affect the calculations. Any increase in the price of metals will only add to the benefits of using composites as per the examples presented below.

In the first case, suppose a cast iron brake rotor is to be replaced with an Al/30 vol% SiC_p composite rotor. Here, for replacement, the volume of the rotor will be the same, but the composite rotor will be lighter due to the density difference. The densities of cast iron, aluminium and SiC are 7.8, 2.7 and 3.2 g/cm^3, respectively. The resulting density of the composite would be 2.85 g/cm^3. The prices of cast iron, aluminium and SiC_p are considered, respectively, to be 0.65, 2.00 and US\$1.50/kg. The price of SiC_p is found to vary from 1 to US\$15/kg depending on the particle size and purity. In the case of composites, the particle size is not a limitation and particles from nanometre to millimetre size can be used. If the cast iron rotor weighs 10 kg, then the total weight of the same sized composite rotor would be 3.65 kg. The cost of raw materials for the cast iron rotor is US\$6.5, whereas the cost for the composite rotor is US\$6.7. The calculation shows that the raw material cost is only 3% higher for the composite material. If the price of SiC_p is taken substantially higher, at US\$12/kg, the price of the rotor will double. Still, the weight saving and improvement in the working life of the rotor may ultimately prove to be cost effective. Fluctuations in the price of raw materials in the short or long term can cause greater perturbations in the cost model. There may be additional cost associated with composite processing and machining. The increase in the processing cost in commercial foundry based processes such as stir mixing and casting is only small. However, 63.5% weight reduction per rotor can provide a significant saving in terms of decreased fuel consumption and reduced emissions over the life of the rotor. Composite rotors are also known to last longer than cast iron rotors and have better corrosion resistance, thermal conductivity and high-temperature dimensional stability, which improve vehicle handling.

In the second case, suppose an engine cover or any other component of aluminium is to be replaced with an Al/50 vol% fly ash composite. Fly ash has mainly the transportation cost, which is estimated to be about US\$0.05/kg. Any local beneficiation process will add to this cost, so the actual cost of fly ash is taken as US\$0.1/kg. The density of commonly used fly ash in composites is in the range 0.4–0.8 g/cm^3. Here it can be assumed close to the upper limit, at 0.75 g/cm^3. The weight fractions for aluminium and fly ash are calculated as 78.26 and 21.74%, respectively. Compared to a 1 kg part that is made of aluminium and costs US\$2.00 in raw materials, the raw material cost for the same part made of composite will be only US\$1.74. The 14% saving is significant in this regard. In many cases 14% may be enough to cover any overhead and processing cost related to procuring and handling an additional raw material as well as the cost of additional steps related to composite synthesis. The savings in the transportation cost of aluminium and lighter finished parts will add to the cost benefits.

From the point of view of cost–benefit analysis, the initial or short term cost of composite materials may be high, but in analysis based on life cycle their light weight and higher performance may payoff more than the initial cost. The life cycle cost analysis was not the main focus during the days of US\$1/gallon gasoline. However, now the operating cost of a vehicle has become a significant part of the ownership cost of an automobile over its entire service life.

There are several indirect benefits of composite materials. It is estimated that the production of 1 kg of aluminium in ingot form requires 37.26 kWh of energy. Aluminium production is also known to generate significant amounts of SO_x, NO_x, CO_2 and CO. Reducing the requirement of primary aluminium will result in further savings in energy and emissions that are tied to the production of aluminium. These savings should also be counted towards the benefits of using fillers in primary metals. It also seems reasonable that, if composites offer the same level of performance as the matrix metal, their use can still be justified based on the reduced demand and consequently emissions of the primary metals.

(a)   (b)

Figure 13.4   (a) A stir-mixer mounted on a furnace that can be added to a foundry for low cost processing and (b) four different impeller designs for effectively breaking the particle clusters and uniformly dispersing particles in the matrix [29].

Although the discussion provided above is specific to aluminium, similar calculations can be conducted for other matrix metals and reinforcement materials to find the benefits in terms of costs. Including realistic cost estimates of composite synthesis methods and the machining of parts can further improve the accuracy of the calculation. The cost of establishing new infrastructure may be significant for composite manufacturing processes. New processes such as laser methods require expensive equipment that may be prohibitive for switching to these technologies for the production of automotive components. This is where foundry based solidification methods are promising. If only a stir mixer or melt infiltration equipment is added to standard foundries used in the automotive industry, the processing of composites can be conducted using the same infrastructure and the synthesis cost increase can be minimised.

A stir mixer is shown in Figure 13.4a. This is a small equipment and can be integrated within the foundry process flow to disperse particles in liquid melt between the furnace and moulding sections. Specially designed impellers (Figure 13.4b) are now available to work effectively with various particle sizes, volume fractions and melt viscosities. Figure 13.5 shows melt infiltration equipment. This equipment can be used to produce net or near-net shaped automobile parts. Several variations of squeeze cast technology are now available and industries can select the most suitable option for their requirements. Net or near-net shaped processing reduces machining and finishing costs for fabricated parts and is versatile for manufacturing a variety of composite castings.

In most cases, a significant part of the cost, which is mostly overlooked in initial analysis, in switching manufacturing technologies is trained professionals. The foundry based composite synthesis methods can be handled by the same traditional worker with only a little experience of the extra steps. Adaptation by the existing workforce can help in keeping the product cost comparable.

Figure 13.5 A lab scale gravity and squeeze caster.

13.2 Reinforcements

Common reinforcements for MMCs include graphite (Gr), alumina (Al_2O_3), silicon carbide (SiC), cubic boron nitride (cBN), hexagonal boron nitride (hBN), titanium diboride (TiB_2), molybdenum disulfide (MoS_2), carbon nanotubes (CNT), carbon fibers, fly ash cenospheres, graphene and numerous nano-particles, some of which are listed in Table 13.1. The most common matrix used is aluminium, but magnesium, copper, iron, titanium and lead alloys are also used as matrix materials.

A brief discussion of some of the materials that are commonly used as reinforcements is presented below.

13.2.1 Solid Ceramic Reinforcements

Alumina (Al_2O_3) or aluminium oxide is a hard ceramic material that is the naturally occurring oxide on the surface of all aluminium. Alumina exhibits low to moderate density (3.95 g/cm^3), is white in appearance and comes in the form of nano- and microparticles, hollow balloons, whiskers and fibers. Nextel (3M trade name) fibers are a popular reinforcement. The melting point of alumina is over 2000°C, which makes it stable for liquid state processing of composites for any type of matrix material.

Table 13.1   Common MMC reinforcement materials.

Reinforcement	Description	Type	Density (g/cm^3)	MMC application
Alumina (Al$_2$O$_3$)	White ceramic aluminium oxide	Particle, balloon, whisker, fiber, nano	3.95	High strength, low weight syntactic foams, wear resistance
Silicon carbide (SiC)	Hard crystal structure of silicon and carbon	Particle, balloon, whisker, fiber, nano	3.20	High strength, wear resistance
Diamond	Super-hard cubic structured carbon crystal	Particle	3.52	Wear resistance, thermal conductivity, high strength, low CTE
Graphite/carbon	Hexagonal carbon formation	Flake, balloon	2.20	Self-lubrication
Graphene	Single atom thick carbon sheet	Sheet	1.80	High strength
Carbon nanotubes	Single atom thick carbon tube	Micro or nano length tube	1.30	High strength
Cubic boron nitride	Super-hard cubic structured boron–nitrogen crystal	Particle	3.45	Wear resistance, thermal conductivity, high strength, low CTE
Hexagonal boron nitride	Hexagonal boron–nitrogen formation	Flake	2.10	Wear resistance
Fly ash	Silicon dioxide (SiO$_2$) and calcium oxide (CaO)	Hollow, porous, or solid particle		Low weight syntactic foams, dampening
Titanium boride (TiB)	Super-hard hexagonal lattice ceramic	Particle, whisker	4.52	Wear resistance, high strength
Titanium carbide (TiC)	Super-hard cubic structured ceramic	Particle	4.93	Wear resistance
Molybdenum disulfide (MoS$_2$)	Trigonal prism molybdenum sulfur structure	Flake	5.06	Wear resistance

Silicon carbide (SiC) is a ceramic made of a strong crystal structure of alternating Si and C atoms. SiC exhibits low density (3.21 g/cm^3), high strength (220 GPa) and good thermal conductivity (370 W/mK). SiC ranges from colourless to black depending on impurities and is available in the form of particle, whiskers and fibers. SiC has excellent specific toughness and is used for making ceramic plates for bullet-proof ballistic vests, because of its weight savings due to its lower density compared to steel or even alumina plates. The thermal conductivity of SiC is leading to applications of this reinforcement in aluminium matrix composites for thermal management applications.

Cubic boron nitride (cBN) shares nearly all its materials properties with diamond. This reinforcement may be beneficial when interactions between materials should be avoided, such as aluminium and carbon which form aluminium carbide.

Hexagonal boron nitride (hBN) has similar properties to graphite and is a dry lubricant, but with boron and nitrogen atoms. It has a low density (2.1 g/cm^3). Like graphite, the in-plane material properties (due to primary bonds) of hBN are better than through the thickness properties (due to van der Waals bonds). Like cBN, hBN can be a good alternative when interactions between carbon and the matrix should be avoided.

13.2.2   Hollow Reinforcements

Fly ash is a waste byproduct of coal burning power plants. Environmental regulations require the industry to collect lightweight ash particles to prevent their release into the environment. Typically, these waste products are disposed of in landfills, which is not an environmentally friendly solution. Development of the application of fly ash as fillers can be a cost-effective solution. Although the chemical composition of fly ash depends on the different impurities in the coal that is being burnt, the major constituents are SiO_2, Al_2O_3, CaO and oxides of iron [3–5]. A large number of other elements are also present in trace quantities. Fly ash cenospheres are hollow and therefore exhibit extremely low true particle density (0.2–0.8 g/cm^3). A sample of fly ash cenospheres is shown in Figure 13.6a. Particles of the desired size range can be selected from the bulk using sieving methods. Fly ash can be coated with metals to provide several benefits: (i) to cover the pores open to the surface (Figure 13.6b), (ii) to avoid undesired reactions between the elements present in cenospheres and matrix metals, for example the formation of brittle carbides and (iii) to promote strong bonding at the particle–matrix interface. Such inexpensive materials (US$40–50/ton in bulk industrial quantities), which are also a burden on the environment, can provide the benefits of low density and low cost to composites.

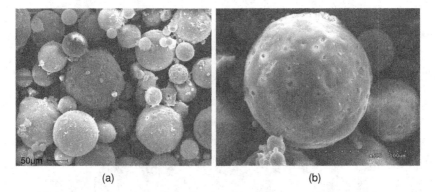

(a)   (b)

Figure 13.6   Fly ash cenospheres commonly used as fillers in metals to create lightweight composites. (a) A wide size distribution can be seen in cenospheres, which are usually hollow. (b) Some of the cenospheres may have porous walls. Materials for imaging courtesy of Dr. Gary Gladysz and William Ricci, Trelleborg Offshore Boston.

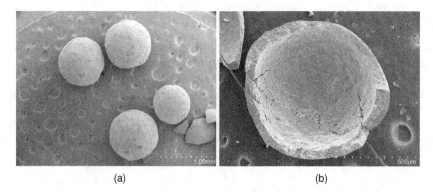

(a)   (b)

Figure 13.7   (a) Hollow SiC balloons on a substrate and (b) a broken balloon showing wall thickness. These balloons were produced by Deep Springs Technologies, OH, USA. Courtesy of Oliver M. Strbik III.

Silicon carbide (SiC) is now available as hollow balloons. These lightweight particles, shown in Figure 13.7a, are useful in developing composites for weight-sensitive applications. A broken balloon is shown in Figure 13.7b, which shows a uniform wall thickness of the particle. Such balloons are available in a large range of diameters and wall thicknesses and can be customised as per the requirement of the applications. SiC balloons can be very useful where high thermal conductivity of SiC is required but the composite needs to have high strength and low density.

Alumina microballoons are available in a wide variety of sizes, with consistent wall thickness. They are used in weight-sensitive applications and can be used in high volume percentages to create high strength syntactic foams [6]. Alumina is highly compatible with aluminium and is commonly used in aluminium metal matrix composites. Figure 13.8 shows several 600 μm diameter alumina microballoons.

Figure 13.8   Alumina hollow balloons.

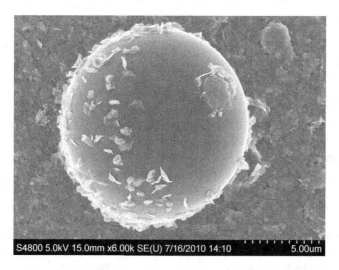

Figure 13.9   A carbon hollow microballoon.

Carbon microballoons are typically graphitic carbon, available in a wide variety of sizes and frequently have exceptional surface smoothness and shape (Figure 13.9). Figure 13.10 shows that diameter to wall thickness ratios can be quite high. These spheres can be used in a number of applications, including lightweight and self-lubricating composites, in a wide variety of materials [7]. Carbon spheres can be coated with a ceramic such as silicon carbide or a metallic coating in order to prevent undesired reactions, like the formation of aluminium carbide, and to help improve wetting between the matrix and the reinforcement.

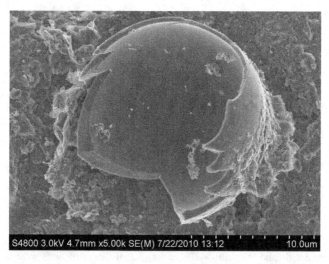

Figure 13.10   A carbon hollow microballoon coated with silicon.

13.2.3   Carbon Based Materials

Carbon nanotubes (CNT) are effectively super-strong tubes made of one single atom thick carbon layers, or graphene, which are rolled or structured into tube shapes [8, 9]. CNTs exhibit exceptional material properties, low density (1.3 g/cm^3) and in certain configurations can achieve tensile strengths 100 times that of stainless steel. A schematic representation of single- and multi-walled CNTs is presented in Figure 13.11.

Graphite is composed of carbon and its basic unit is stacked sheets of graphene. Graphite is a dry lubricant material due to its ability to shear thin layers of material off to coat surfaces. A sample of graphite flakes used in composites is shown in Figure 13.12. Graphite has a low density (2.2 g/cm^3) and is dark grey in colour. Its use in self-lubricating structures is promising. However, the formation of carbides may be undesirable in some MMCs and process control is required for such systems.

Diamond is composed entirely of carbon, constructed in cubic atomic arrangements. Although natural diamonds are available, synthetic diamonds are typically used for industrial applications. Diamond is one of the hardest known materials, has low to moderate density (3.54 g/cm^3), excellent isotropic thermal conductivity (2300 W/mK) and is an electrical insulator. The high cost of diamond is a limiting factor in developing large-scale applications of this reinforcement.

Graphene is sheet of carbon a single atom thick in a hexagonal structure [10]. The structure of graphene sheets is represented in Figure 13.13. Graphene sheets have a single atomic layer

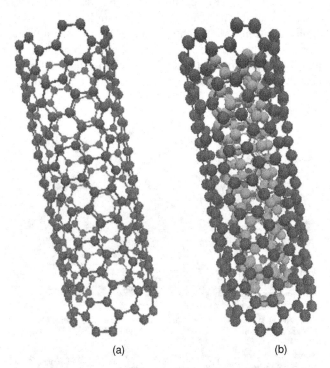

(a)   (b)

Figure 13.11   (a) Single and (b) multi-walled carbon nanotubes (structures were created using VMD code). Image courtesy Dr. Dinesh Pinisetty, Cal Maritime Academy.

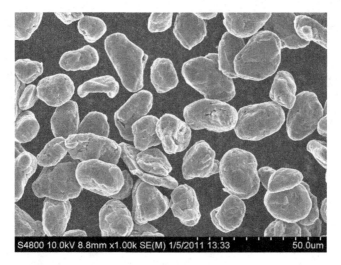

Figure 13.12 Graphite flakes (produced by Superior Graphite Co.; FormulaBT series). Courtesy of Superior Graphite Co., IL., USA.

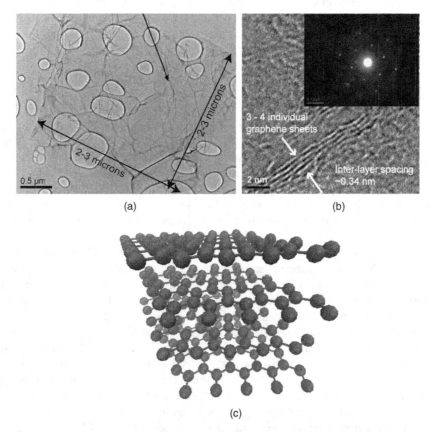

Figure 13.13 (a) Transmission electron micrograph of graphene platelets on a TEM grid. (b) High resolution TEM indicates graphene layer thickness of ~2 nm. The graphene appears to be comprised of ~3–4 graphene sheets with an interlayer spacing of ~0.34 nm. Inset shows the measured electron diffraction pattern. (c) Representation of graphene sheets comprised of carbon atoms, image courtesy Dr. Dinesh Pinisetty, Cal Maritime Academy. TEM images courtesy of Prof. Nikhil Koratkar, RPI, NY, USA.

thickness but their length and width can be several microns, as shown in Figure 13.13a. The spacing between layers is about 3.4 nm, as shown in Figure 13.13b. This layered structure is illustrated in Figure 13.13c. This is one of the strongest known materials comprising only primary carbon bonds in the structure and a very high surface area to bond with the matrix. Exfoliation of layers increases the surface area available for bonding with the matrix. Graphene exhibits extremely high thermal and electrical conductivity values for the in-plane direction. Producing large quantities of graphene for industrial purposes is still in early stages and the cost is relatively high compared to most other reinforcements.

13.3 Automotive Applications

13.3.1 Powertrain

The powertrain in modern automobiles operates under severe conditions. Bearings experiencing hundreds of thousands of rotations during a daily commute, piston assemblies travelling at a mean piston speed of 20 m/s and combustion products exceeding 2500 °C are commonplace in the internal combustion engine [11]. There are a number of components that can take advantage of MMCs superior thermal, wear and strength properties.

13.3.2 Cylinder Liner

Engine blocks and heads that were previously made with cast iron are now being designed and manufactured in aluminium, resulting in significant weight savings. For example, in 2011 Ford changed its engine block on the Shelby GT500 from a cast iron block to an aluminium block, which resulted in a 102 lb weight savings [12]. Due to the poor wear characteristics of aluminium, the cylinders must be lined with a better wear-resistant material. Common solutions include using pressed-in or cast-in-place cast iron cylinder liners. The cast iron is durable, but is over twice as heavy as the aluminium it replaces and has poor thermal characteristics. Poor heat transfer properties in the cylinder require more robust cooling solutions, which also adds size and mass to the engine.

One solution is to use an in situ MMC, composed of hypereutectic aluminium-silicon alloy that relies on the silicon in the hypereutectic range solidifying in a hard, crystal structure to operate as a wear surface instead of the aluminium. Since the silicon is present in higher quantities than the eutectic point, the silicon will precipitate out of the solution in blocky cubes of silicon. The liner then goes through what is called an exposure step where the aluminium is mechanically or chemically treated so that the aluminium is removed to create a recess such that the silicon is approximately 1–10 µm higher than the aluminium [13]. The exposure step prevents the aluminium from smearing over the silicon, provides silicon islands for the piston rings to glide over and provides a surface for oil to be in reserve on the cylinder walls for proper lubrication. Hypereutectic cylinder liners provide superior heat flow out of the cylinder while decreasing the overall weight of the engine compared to cast iron liners and they have been used by several different automotive manufactures, including General Motors, Porsche, Audi, Mercedes Benz, Volkswagen and BWM under various trade names. Honda has used a process for generating a MMC cylinder liner called fiber reinforced metal (FRM) which uses carbon and alumina fibers to reinforce the aluminium at the cylinder wall liner. This technique relies

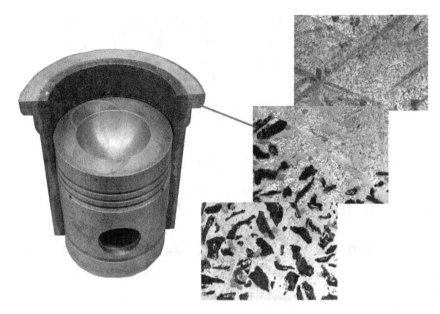

Figure 13.14   Graphite reinforced aluminium piston and cylinder liner developed at UWM. The inset microstructure shows graphite particles concentrate in the inner periphery of the liner [29].

on the strength of semi-continuous fibers to reinforce the walls, allowing for very thin, high strength liners and was used in several cars, including the Honda Prelude and Acura NSX [14].

For lower stress applications, aluminium alloy cylinder liners containing dispersed graphite particles have been used to provide solid lubrication (Figure 13.14). The graphite containing aluminium has a lower friction coefficient and wear rate and does not seize under boundary lubrication. The cylinder liner can be cast in a single step by a centrifugal casting process to concentrate the graphite particles only in the inner periphery where they are needed to provide solid lubrication. The pistons and liners of aluminium–graphite have been tested in gas and diesel engines and exhibit reduced friction coefficients and wear rates. As graphite shears under wear conditions, it creats a surface of graphite, reducing the wear rate of the liner. The measured friction coefficient of aluminium–graphite composites is as low as 0.2. The application of this material for cylinder liners in lightweight aluminium engine blocks will enable engines to reach operating temperatures more quickly while providing superior wear resistance, improved cold start emissions and reduced weight. The aluminium based composite liners can be cast in place using conventional casting techniques.

Research is underway to create hybrid MMCs that would incorporate super-hard, wear resistant particles like diamond or cubic boron nitride and self-lubricating particles like graphite or hexagonal boron nitride platelets. This would allow for similar wear resistant islands in the matrix after an exposure process to provide a surface for the piston rings to glide over, similar to hypereutectic Al-Si alloys, but have the added protection in instances if the aluminium was ever contacted, as the self-lubricating particle would also be contacted and coat the surface, preventing any further wear. This material would have higher thermal conductivity than any hypereutectic alloy, since both diamond and cBN have exceptionally high isotropic

S4800 20.0kV 12.6mm x300 SE(L) 7/26/2011 15:12   100um

Figure 13.15 Cubic boron nitride (dark grey) reinforced aluminium light grey) MMC, developed at UWM.

thermal conductivity (~2000 W/mK), allowing for better heat removal capabilities through the cylinder walls and therefore a more compact cooling system. This material (Figure 13.15) would create a cylinder liner with half the mass of a cast iron liner while achieving three times the thermal capabilities, allowing for not just the initial weight savings, but also the weight savings associated with less complex cooling systems.

13.3.3   Piston

Currently, pistons are made with Al-Si alloys in both hypo- and hypereutectic ranges using both casting and forging methods. Silicon is an important component in the piston material composition, as it reduces the coefficient of thermal expansion (CTE), or the amount the piston would increase in size as temperature increases. A piston with a high CTE would require a larger gap between the piston and the cylinder so that adequate sealing would occur at operating temperatures. This would introduce excessive wear and poor emissions, as there would be more volume for unburned gasses to be trapped until the engine was up to temperature. It would be ideal to have a CTE match between the cylinder and the piston, which would provide exactly the same expansion between the parts over any given heat range. An example of aluminium/graphite piston is shown in Figure 13.16. Carbon fiber, which is typically used in polymer matrix composites, also has applications in metal matrix composites. Woven tubes of carbon fiber can be infiltrated with aluminium to create cast pistons with very low CTE, while improving strength; and since the carbon fiber is 40% lighter than the aluminium it displaces, it creates a lightweight piston which reduces reciprocating mass in the engine, improving both performance and efficiency, while reducing unburned gasses due to poor tolerances below optimal operating temperatures. Since the reciprocating mass in an automotive engine needs to be properly counterbalanced by reducing the reciprocating mass, weight savings is also realised in the counterbalance.

Figure 13.16 Aluminium/graphite piston, permanent mould cast.

13.3.4 Connecting Rod

Connecting rods experience significant loading in the automotive engine in both compression and tension, but need to be lightweight to reduce reciprocating mass. Connecting rods currently are made from steel, aluminium, or titanium, although each has its own advantages and disadvantages. An ideal connecting rod would be one with the low fatigue failure rate and strength of steel combined with the low weight profile of aluminium and near net shape of powdered metal products. Aluminium/graphite connecting rods have been cast and tested for small engines (Figure 13.17).

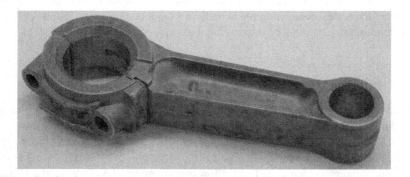

Figure 13.17 Aluminium/graphite connecting rod for a small gas engine.

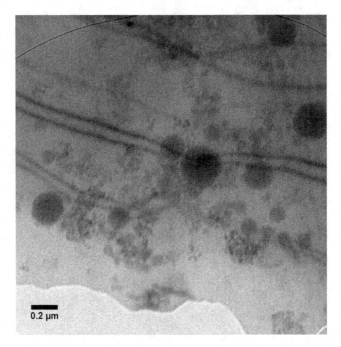

Figure 13.18   TEM of stir-cast A206 reinforced with 2 vol% 47 nm Al_2O_3 produced at UW-Milwaukee [15].

Nano-particle reinforced aluminium products are being developed that look to be an excellent alternative to existing materials for connecting rods. These materials are being developed utilising both powder metallurgy processes and casting processes incorporating stir-mixing and/or ultrasonic particle incorporation (Figure 13.18) [15]. The results are materials that have a higher modulus due to reduced grain size and improved resistance to fatigue failure because the nano-particles pin the dislocations.

3D weaves of carbon fiber or Nextel fabrics infiltrated with aluminium or magnesium are also being explored (Figure 13.19). By using high-tech fabrics that are known for their superior strength to weight ratio, connecting rods made of a net shape preform of the 3D fabric could be infiltrated to create a connecting rod at half the weight of aluminium, but twice the strength of steel, resulting in performance improvement at reduced weight.

13.3.5   Main and Other Bearings

Automotive main bearings are typically made from a lead-rich copper alloy that provides enough durability and a smooth rotating surface. For environmental reasons, there are efforts to remove lead from copper alloys and graphite is being considered as a replacement for lead due to its ability to contribute a self-lubricating surface in MMCs. This would benefit the environment and replace dense lead with a low density material, saving weight. Graphite-rich copper bearing materials have been developed through centrifugal casting techniques that can

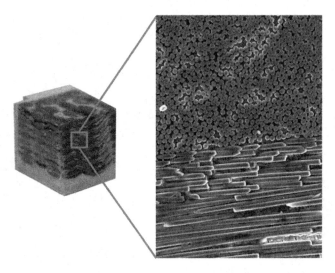

Figure 13.19   3D carbon fiber weave infiltrated with magnesium.

be used for bearing applications (Figure 13.20). Copper–graphite MMCs show significantly lower wear rates and at a lower density compared to copper–lead counterparts (Figure 13.21).

For further weight savings, a hybrid silicon carbide and graphite reinforced aluminium is being developed (Figure 13.22). This material would provide a smooth, tough surface for the crank to rotate, with excellent wear resistance and possessing self-lubricating properties at significantly less weight compared to copper–lead or copper–graphite components. This practice could be extended to other applications in the powertrain, including cam shaft journals, rocker arms, oil and water pump bearings and wrist pins.

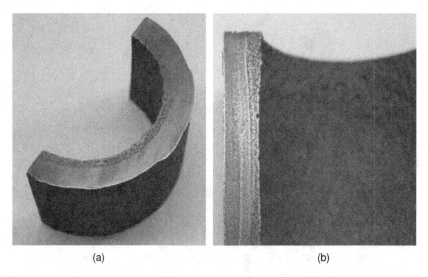

(a)   (b)

Figure 13.20   Centrifugally cast copper/graphite gradient material. The lower density of graphite particles makes them move to the inner surface, where lubrication is required during service conditions.

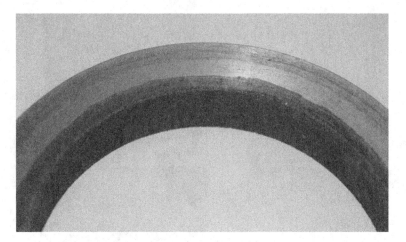

Figure 13.21 Centrifugally cast aluminium/graphite bearing.

13.3.6   Crankshaft

The crankshaft is essentially the backbone of the internal combustion engine. It is responsible for harnessing the power being released by the controlled explosions occurring in the combustion chamber and converting that linear motion to a rotational motion useful in propelling a vehicle. The crankshaft experiences high levels of cyclical loading which makes aluminium a poor material in this application, despite the fact that a lightweight component would improve performance. Aluminium alloys will inevitably fail when exposed to cyclical

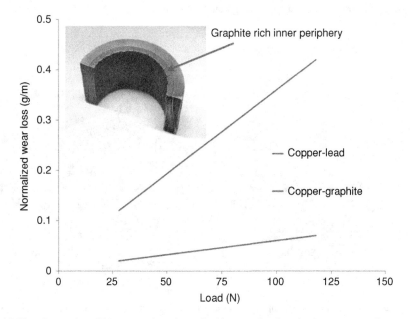

Figure 13.22 Copper/graphite composites show significant reductions in wear compared to copper/lead counterparts [29].

Figure 13.23   Aluminium/SiC-graphite crankshaft.

loading, so ductile iron and forged steel and, in some special cases, titanium can be used like in the Porsche GT3 RS. An example of an aluminium crankshaft reinforced with SiC and graphite particles is shown in Figure 13.23. It shows promise in laboratory scale testing but has not yet been field tested.

13.3.7   Valvetrain

Engine valves experience intense loading as they rapidly open and close every two revolutions of the engine. They must maintain a metal on metal seal, while several MPa of pressure tries to escape, withstand several thousand degrees C as combustion occurs across the valve face and exhaust valves must survive as hot exhaust gases flow across nearly the entire valve. Valve springs must be able to quickly close the valve to prevent valve float, while minimising resistance when opening the valve to prevent excessive valve train losses.

Stainless steel valves are typical, titanium valves are used for high performance applications. Hollowing the valve stem can save weight and even filling the hollowed stem with sodium is practiced to help with heat transfer through the valve, out through the valve seat and into the cylinder head. Although sodium is a suitable solution, there are safety concerns with sodium as it reacts violently when exposed to oxygen. In instances of engine failure, such as a timing belt failure, the valve can come in contact with the piston violently, causing the valve to break. A failure could cause the sodium to combust, potentially igniting the engine oil and causing a catastrophic engine fire.

Toyota has developed a special high temperature titanium MMC for exhaust valves that relies on the generation of in situ TiB particles to reinforce the matrix (Figure 13.24). This material increases Young's modulus and reduces grain size [16]. The reduced weight compared to stainless valves has the added benefit of reducing the spring tension required to operate the valve, which further reduces valvetrain losses.

13.3.8   Engine Accessories

Many components on the powertrain require stronger materials than plastics, but do not experience high levels of stress or where plastic may not be an ideal solution due to heat or chemical interaction concerns. These include items like accessory brackets, alternator housings, valve covers, oil pans, intake manifolds, front engine covers, engine pulleys and water pump housings (Figures 13.25–13.27). Fly ash reinforced aluminium is an excellent

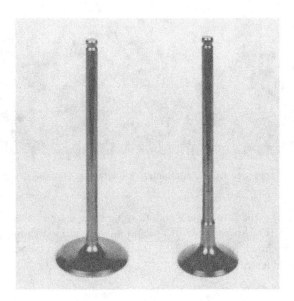

Figure 13.24   Ti/TiB intake and exhaust valve made by Toyota. (courtesy of Toyota Central Research and Development Laboratories, Reprinted with permission of ASM International. All rights reserved. www.asminternational.org) [14].

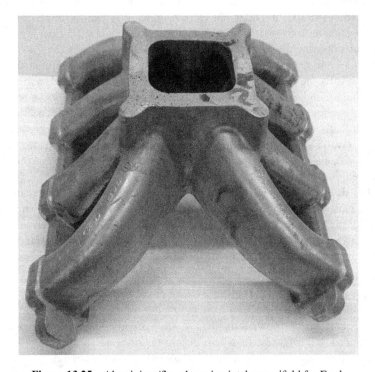

Figure 13.25   Aluminium/fly ash engine intake manifold for Ford.

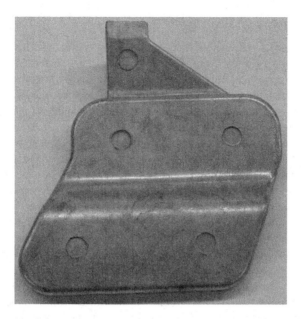

Figure 13.26   Aluminium/fly ash engine mount.

candidate to be used for these components because of the light weight [17]. These composites maintain a significant amount of strength from the matrix [18, 19]. A component made of a 40 vol% fly ash reinforced aluminium yields a density of 1.69 g/cm^3, therefore a part made from 1 kg of aluminium would only weight 625 g, a 37.5% weight savings, while benefiting from improved heat transfer compared to plastics.

Figure 13.27   Aluminium/fly ash motor cover similar to those found on alternators.

13.3.9   Drivetrain and Suspension

While the powertrain is responsible for generating the power, the drivetrain is responsible for getting the power to the wheels. This consists of the transmission, driveshafts, axles, differentials or transaxles and numerous mounts. The suspension is responsible for keeping the tyres in contact with the road while reducing energy transfer to the passenger compartment. Nearly every component of the suspension and drivetrain is made of metal, therefore MMCs can be used to decrease mass of the vehicle.

Suspension systems control the car's tyres' ability to stay in contact with the road and limit the amount of disturbances that are transferred to the passenger cabin. Since the majority of the suspension components are unsprung weight, the reduction of mass in these components will improve performance and have a positive ripple effect as they require less robust spring and dampening systems to control their motion.

Control arms require materials that have good torsional properties as they are flexed continuously both when the car is in motion and when the car is stationary. Some basic systems rely on a single control arm and a torsion beam or leaf-spring connected to the axle to control motion. More complex systems use multiple links between the chassis and wheel hub to keep the wheels square to the road as the wheel travels up and down. More complex systems tend to weigh more, although most systems components can use improvements through weight reduction. High strength aluminium alloys reinforced with SiC can be incorporated to replace steel components, and localised self-lubricating bushing journals can be used to eliminate binding in the system.

13.3.10   Transmission Housing

Many transmission housings are currently produced using high-pressure die castings in aluminium to achieve a lightweight, durable housing. It has been shown that low volume percent fly ash (up to 12 vol%) can be incorporated into die casting operations with improved tensile strength versus the base alloy. This can be used to create lighter transmission housings, converter or bell housings, tail housings, transaxle housings and differential housings. This change would result in a 10% mass reduction in these components while improving the damping effect gained from using metal foams.

13.3.11   Differential Housing

In a car where power is delivered to the rear wheels, a differential is required. A differential is a mechanical device that can change the axis of rotational motion through a gear set, in this case converting the torque from the engine driveshaft to power the two axles that connect to the wheels. The gear set in the differential is selected to the application and that gear ratio is referred to as the final drive.

Differentials are typically made from cast iron or aluminium, depending on the application. Cast iron units are more robust but have significant mass, while aluminium units are lighter but often require more preload to account for the material's thermal expansion, which can hurt fuel economy.

The Cadillac ATS is a focused lightweight sports sedan that uses weight saving techniques to trim the car's preproduction weight from 3700 to 3400 lb. Despite this dramatic cut,

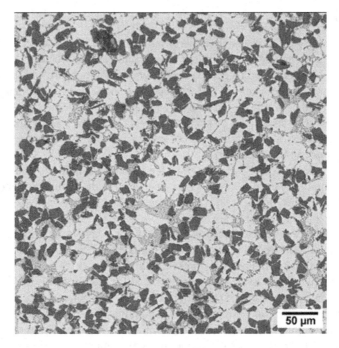

Figure 13.28   Al-SiC MMC (Duralcan F3S.20S, Courtesy of Rio Tinto Alcan).

the engineers decided to opt for the cast iron unit. This was because they were able to show, despite the weight, the unit performed better than the aluminium housing, due to better preload tolerances.

Improvements in this product could be made by using a high strength aluminium alloy with a wear-resistant MMC localised at the bearing journal areas. SiC reinforced aluminium MMCs have a low CTE to make it suitable for bearing journals [20]. Duralcan has developed a possible material suitable for this application which uses 20% SiC particles in A359 (Figure 13.28), which shows a twofold reduction in abrasion resistance and a 10-fold reduction in wear resistance when compared to A356 and a twofold reduction compared to cast iron.

13.3.12   Driveshaft

Driveshafts present a particularly interesting challenge when selecting materials. They need to be lightweight, have good torsional rigidity and span fairly long distances. If a driveshaft is required to span a distance longer than the material can support, it must be made in multiple sections with additional supports and bearing structures, which add weight and cost.

GM was faced with this dilemma in 1997 when the driveshaft for their C/K truck line required an expensive and heavy two-piece driveshaft design if they used conventional steel. They instead designed a one-piece driveshaft using an alumina reinforced aluminium MMC incorporating 20 vol% reinforcement, which reduced weight by 50% and cost by 20% compared to the steel unit [21]. This technology has also been used in the Chevrolet Corvette and the Ford Crown Victoria Police Interceptor models [22].

13.3.13 Brake

Brake rotors and drums are typically made from heavy cast iron, which are suitable for the task but are poor at moving heat from the system and can contribute to a condition called fade, where the braking materials exceed their operating temperature and no longer function properly, or in worst case scenarios fail to function at all. Using aluminium disks would be beneficial because they would be lightweight and have much better heat removal capabilities compared to cast iron drums and disks, but due to a very poor wear characteristic they are not suitable. Similar to cylinder liners, MMCs can greatly improve the wear characteristics of aluminium in brake applications [23].

MMCs have already found their way into the automotive industry with alumina or SiC reinforced aluminium brake disks which have been used in the Lotus Elise and the Plymouth Prowler. Similarly, rear drum brakes have been used on hybrid cars such as the Toyota RAV4-EV.

These materials are effective because they allow for a hard wear surface, excellent heat transfer capabilities and are lightweight. This benefits the car by reducing brake maintenance and reduces or eliminates the need for brake cooling ductwork. Since the mass is unsprung and rotating, a lighter brake system reduces the need for more robust suspension design and reduces drivetrain losses compared to much heavier cast iron components.

Further improvements can be made by incorporating ultra-hard particles with increased thermal capabilities into aluminium MMCs. Aluminium-cBN composites would provide one on the hardest wear surfaces possible, while the combination of aluminium and cBN would create a material that can exceed the thermal conductivity of copper or silver. This would allow for brake systems with a much smaller disk diameter since less material is needed to extract heat while generating a significant friction surface while significantly reducing weight. Aluminium/SiC brake drums are also attracting interest due to their light weight (Figure 13.29).

Figure 13.29   Aluminium/SiC brake drum for large vehicles.

13.3.14   Mount

Engines, transmissions and differentials or transaxles all need to be firmly mounted to the chassis of the car. Current practices include rubber bonded to steel carriers, rigid steel or aluminium mounts and complex hydraulic mounts. Bonded rubber mounts are effective at absorbing much of the vibration before it is transferred to the passenger cabin, but are prone to failure over time. Rigid mounts are much less likely to fail, however they can deliver a jolting ride. Hydraulic mounts are an excellent compromise, as they can adjust to a firm or soft setting depending on driving conditions, but they add complexity, cost and weight. Fly ash reinforced aluminium shows promise as an alternative to commonly used mount materials, as the syntactic metal foam has superior dampening characteristics, is lighter than solid aluminium and is less expensive to produce because less aluminium is used [24].

13.3.15   Impact Zone

Frontal impact members for cars are typically closed channel steel members that absorb energy and divert it away from the passenger compartment. Foam filled channels are now used for several of these applications [25–27]. To increase those component's capabilities, the channels can be filled with aluminium syntactic foam with high vol% hollow microspheres. When the channel begins to yield and crumple due to the imposed forces, the syntactic foam begins to absorb energy. The foam wants to rapidly expand outward, but is contained within the channel to prevent this from occurring, allowing large amounts of energy to be quickly absorbed.

13.3.16   Electronics

Electronics are becoming a highly integrated component of the automobile. Advanced electronics carefully control the entire engine management system, including fuel injection, ignition timing, cam phasing and a number of emission valves, solenoids and controls. Advanced traction, stability and even launch control systems use a number of sensors that feed inputs into a control unit. Climate controls that used to consist of simple slide levers are now elaborate, computer-controlled networks; and navigation and infotainment systems are becoming increasingly standard. As these systems become both more advanced and prevalent and increases in computational power occur, those systems need heat management systems that can quickly remove heat without adding weight. Copper is an excellent heat sink material but is quite heavy. Aluminium is lightweight and frequently used, but has high CTE and as alloying elements are introduced, thermal conductivity steadily drops.

SiC is very promising in thermal management applications [28]. SiC reinforced aluminium MMCs have been developed for the electronics industry and have similar thermal conductivity to aluminium, but have a much lower CTE that matches well with that of electronic components.

Hybrid vehicles in particular can benefit significantly from the application of MMCs. The battery systems are controlled by advanced computer systems, to regulate the battery and power deliver system, in addition to cooling systems to maintain a particular battery temperature. Diamond reinforced aluminium MMCs have been created that exhibit excellent heat sink capabilities, displaying excellent results in the three major criteria: CTE, thermal conductivity and density. Results show that, by using a bimodal mixture of small and large diamond micropowders (52 and 360 μm mean particle size, respectively), a material can be made with

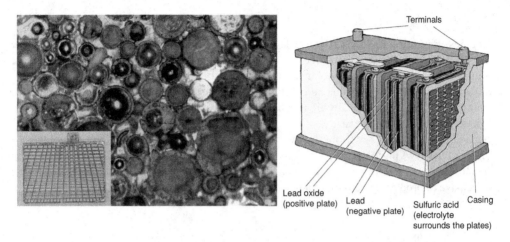

Figure 13.30 Pb-MMC incorporating fly ash developed at UWM, typical lead acid battery grid plate (inset) and battery cutaway diagram. Car battery illustration by Max-Karl Winkler, NSRC, courtesy Smithsonian Science Education Center [29].

one-third the density and double the thermal conductivity of copper, while maintaining a CTE of approximately 3.0 at room temperature.

13.3.17   Battery

The lead acid battery is the battery of choice when it comes to automotive applications. It can provide large amounts of energy in short bursts, which is required to start an engine, can provide power to electronics in the vehicle and can be easily recharged through an alternator. The main disadvantage is that the battery is constructed of a series of lead alloy plates, which are very heavy. To reduce the weight of batteries, research is being conducted to incorporated metal coated fly ash particles into a lead alloy matrix (Figure 13.30). This has resulted in a lead MMC that has electrical conductivity values similar to those of lead alloys found in batteries, but at half the density by replacing lead with a very low density, hollow cenosphere [29, 30].

13.4   Conclusion

MMCs have come a long way in the past two decades in automotive applications. Aerospace applications had demonstrated that MMCs would perform well in automotive applications but cost was a limiting factor for a long time. The development of MMCs with low cost reinforcements, net shaped MMC component processing and the development of new synthesis methods have significantly reduced the cost of MMC components and have made them competitive in automotive applications. Increasing fuel prices in the past decades have also helped in offsetting the high cost of MMC automotive components by reducing the vehicle weight. Close control over the microstructure and development of multifunctionality in MMC components have also helped in enabling their new applications. Aluminium matrix composites are now widely used in brake rotors, brake drums and body panels in cars. Some high-end cars have started using magnesium matrix composites and magnesium alloys for further weight reduction. New opportunities need to be identified in the automotive sector where heavier components can be replaced with lighter components. For example, electrical vehicles will benefit by reducing the weight of the battery pack. The use of hollow particles or fly ash cenospheres may have the potential to reduce the weight of batteries. Significant research focus on nanocomposites has resulted in a large number of studies on processing, microstructure and processing these composites. Applications of metal matrix nanocomposites have not yet materialised, which can be the next area of opportunity.

Acknowledgments

This chapter is based on work supported by the United States Army Research Laboratory under Cooperative Agreement No. W911NF-08-2-0014 and by the United States Army TARDEC through grant TACOM-W56HZV-04-C-0784 of P.K.R. The material is also in part based on work supported by the National Science Foundation under Grant No. OISE–0710981. N.G. acknowledges support by the Office of Naval Research grant N00014-10-1-0988 and Army Research Laboratory cooperative agreement W911NF-11-2-0096. The views, opinions and conclusions made in this document are those of the authors and should not be interpreted as representing the official policies, either expressed or implied, of the Army Research Laboratory or the United States Government. The United States Government is authorised to reproduce and distribute reprints for government purposes, notwithstanding any copyright notation herein. Dr. Dung D. Luong, Dr. Dinesh Pinisetty and Ronald L. Poveda are acknowledged for generating several figures and micrographs.

References

[1] Rawal, S. (2001) Metal-matrix composites for space applications. *Journal of the Minerals, Metals and Materials Society*, **53**(4), 14–17.

[2] Chawla, N. and Chawla, K. (2006) Metal-matrix composites in ground transportation. *Journal of the Minerals, Metals and Materials Society*, **58**(11), 67–70.

[3] Rohatgi, P.K., Gupta, N. and Alaraj, S. (2006) Thermal expansion of aluminium-fly ash cenosphere composites synthesized by pressure infiltration technique. *Journal of Composite Materials*, **40**(13), 1163–1174.

[4] Thipse, S.S., Schoenitz, M. and Dreizin, E.L. (2002) Morphology and composition of the fly ash particles produced in incineration of municipal solid waste. *Fuel Processing Technology*, **75**(3), 173–184.

[5] Vassilev, S.V. and Vassileva, C.G. (2007) A new approach for the classification of coal fly ashes based on their origin, composition, properties, and behaviour. *Fuel*, **86**(10–11), 1490–1512.

[6] Balch, D.K., O'Dwyer, J.G., Davis, G.R. *et al.* (2005) Plasticity and damage in aluminium syntactic foams deformed under dynamic and quasi-static conditions. *Materials Science and Engineering: A*, **391**(1–2), 408–417.

[7] Gladysz, G., Perry, B., McEachen, G. and Lula, J. (2006) Three-phase syntactic foams: structure-property relationships. *Journal of Materials Science*, **41**(13), 4085–4092.

[8] Coleman, J.N., Khan, U., Blau, W.J. and Gun'ko, Y.K. (2006) Small but strong: a review of the mechanical properties of carbon nanotube–polymer composites. *Carbon*, **44**(9), 1624–1652.

[9] Spitalsky, Z., Tasis, D., Papagelis, K. and Galiotis, C. (2010) Carbon nanotube–polymer composites: chemistry, processing, mechanical and electrical properties. *Progress in Polymer Science*, **35**(3), 357–401.

[10] Zhu, Y., Murali, S., Cai, W. *et al.* (2010) Graphene and Graphene Oxide: Synthesis, Properties, and Applications. *Advanced Materials*, **22**(35), 3906–3924.

[11] Pulkrabek, W.W. (2003) *Engineering Fundamentals of the Internal Combustion Engine*, Prentice Hall, Upper Saddle River, NJ.

[12] Ford (2010) Ford Shelby GT500 Supercharged 5.4-Liter Aluminium-Block V-8. Ford and Media, New York.

[13] Meara, T. (2008) *New Honing Options For Hypereutectic Aluminium Cylinder Bores*, Modern Machine Shop, New York.

[14] Hunt, W.H. and Miracle, D.B. (2001) *ASM Hand Book: Automotive Applications of Metal-Matrix Composites*, ASM, New York, pp. 1029–1032.

[15] Schultz, B.F., Ferguson, J.B. and Rohatgi, P.K. (2011) Microstructure and hardness of Al_2O_3 nanoparticle reinforced Al–Mg composites fabricated by reactive wetting and stir mixing. *Materials Science and Engineering: A*, **530**, 87–97.

[16] Leyens, C. and Peters, M. (eds) (2003) *Titanium and Titanium Alloys: Fundamentals and Applications*, Wiley-VCH, Weinheim.

[17] Rohatgi, P.K., Kim, J.K., Gupta, N. *et al.* (2006) Compressive characteristics of A356/fly ash cenosphere composites synthesized by pressure infiltration technique. *Composites Part A*, **37**(3), 430–437.

[18] Rohatgi, P.K. and Weiss, D. (2003) Casting of aluminium-fly ash composites for automotive applications. SAE Technical Paper 2003-01-0825.

[19] Rohatgi, P., Gupta, N., Schultz, B. and Luong, D. (2011) The synthesis, compressive properties, and applications of metal matrix syntactic foams. *JOM: Journal of the Minerals, Metals and Materials Society*, **63**(2), 36–42.

[20] Mizumoto, M., Tajima, Y. and Kagawa, A. (2004) Thermal expansion behavior of Sicp/aluminium alloy composites fabricated by a low-pressure infiltration process. *Materials Transactions*, **45**(5), 1769–1773.

[21] Evans, A., San Marchi, C. and Mortensen, A. (2003) *Metal Matrix Composites in Industry – An Introduction and a Survey*, Springer/Kluwer Academic, New York.

[22] Miracle, D.B. (2005) Metal matrix composites – from science to technological significance. *Composites Science and Technology*, **65**(15–16), 2526–2540.

[23] Rehman, A., Das, S. and Dixit, G. (2012) Analysis of stir die cast Al-SiC composite brake drums based on coefficient of friction. *Tribology International*, **51**(9), 36–41.

[24] Wu, G.H., Dou, Z.Y., Jiang, L.T. and Cao, J.H. (2006) Damping properties of aluminium matrix–fly ash composites. *Materials Letters*, **60**(24), 2945–2948.

[25] Santosa, S., Banhart, J. and Wierzbicki, T. (2001) Experimental and numerical analyses of bending of foam-filled sections. *Acta Mechanica*, **148**(1), 199–213.

[26] Chen, W. and Nardhi, D. (2000) Experimental study of crush behaviour of sheet aluminium foam-filled sections. *International Journal of Crashworthiness*, **5**(4), 447–468.

[27] Kirkpatrick, S.W., Schroeder, M. and Simons, J.W. (2001) Evaluation of passenger rail vehicle crashworthiness. *International Journal of Crashworthiness*, **6**(1), 95–106.

[28] Molina, J.M. (2011) SiC as base of composite materials for thermal management, in *Silicon Carbide– Materials, Processing and Applications in Electronic Devices* (ed. M. Mukherjee), InTech, New York.

[29] Macke, A., Schultz, B. and Rohatgi, P. (2012) Metal matrix composites offer the automotive industry an opportunity to reduce vehicle weight, improve performance. *Advanced Materials and Processes*, **170**(3), 19–23.

[30] Deqing, W., Ziyan, S., Hong, G. and Lopez, H.F. (2001) Synthesis of lead–fly-ash composites by squeeze infiltration. *Journal of Materials Synthesis and Processing*, **9**(5), 247–251.

14

Development of a Composite Wheel with Integrated Hub Motor and Requirements on Safety Components in Composite

Nicole Schweizer and Andreas Büter

Fraunhofer Institute for Structural Durability and System Reliability (LBF), Bartningstrasse 47, 64295 Darmstadt, Germany

14.1 Introduction

Lightweight design is becoming increasingly important for automotive engineering. Fibre reinforced plastics (FRPs) have a wide range of advantages compared to metal-based materials, in addition to the weight saving aspect. In this study, a wheel of carbon fibre reinforced plastics with an integrated electric motor was developed. In addition to the development of a safety component in composites with all its challenges, the development of adapted testing methods for composite wheels is essential. The main focus of the development of the wheel was to achieve the optimum in lightweight potential considering structural durability. During the project, the technical challenges of multifunctional design were considered in the whole product life cycle. Further, a general outlook regarding the validation of structural durability and system reliability of composite safety components is given.

14.1.1 Lightweight as a Key Technology for Automotive Engineering

Over the last decade, lightweight design has become an important criterion for developments in automotive engineering. fibre reinforced plastics (FRP) are being increasingly utilised in high-volume automotive production. Safety relevant components in particular, such as

Advanced Composite Materials for Automotive Applications: Structural Integrity and Crashworthiness,
First Edition. Edited by Ahmed Elmarakbi.
© 2014 John Wiley & Sons, Ltd. Published 2014 by John Wiley & Sons, Ltd.

chassis parts and wheels, are potential fields of application for the various FRPs due to the possibilities that they offer. The wide range of variants of these materials is already immense and continues to grow. Material systems exist for almost any application, spanning the range from those that are very easy to process to those that are particularly effective. It is therefore not only the mere weight saving potential that makes these materials so attractive. In contrast to metallic lightweight materials, FRPs also offer an additional degree of freedom, through which the material and component properties can be influenced. The types of matrices and fibres influence the properties as do their volume ratio and the orientation of the fibres in the component. Weight specific factors, such as durability and stiffness, and also production costs can be ideally balanced, depending on the profile of requirements.

In general, lightweight design means the reduction of weight while maintaining sufficient stiffness, stability and strength. In the context of a developed lightweight structure, the structural durability and reliability are important aspects for the evaluation over the structure's life time.

The stability of a structural lightweight system is influenced by stiffness, damping and mass distribution. Hence, parameters in the design process such as material, geometry, construction and, indirectly, costs have strong influences on the structural dynamic behaviour, as well as on the buckling behaviour (see Figure 14.1).

Further, strength against static and cyclic mechanical loads also makes a fatigue design necessary. The chosen material, the geometry, the manufacturing process, the environmental conditions such as temperature, humidity, fluids and so on and, indirectly, the costs influence the fatigue strength of the lightweight system and have to be taken into account.

The costs associated with complaints, inspections and/or maintenance are generally related to the costs for detailed investigations during the design process. One goal of the life cycle cost analysis here is to find an optimum. Therefore, the aim for a classic durable lightweight design is always to find an economic compromise between weight reduction on the one hand and a safe and reliable design on the other hand.

Factors that influence structural weight, in addition to the component's service loading, are the material, the geometry or shape and the manufacturing process including the method of construction. The objective of lightweight design "based on the material" is to find the best material in respect of the application. Important parameters in evaluating different materials are the density rated material properties, which can be derived for different tasks. For example, the design of tension loaded beams commonly uses density rated stiffness or density rated tensile strength as the material selection values. The better the material fits to the application the lighter will be the component.

Based on this and on account of their high specific strength and stiffness, the use of fibre reinforced plastics – especially carbon fibre reinforced plastics (CFRP) – offers the advantages of lightweight design, increased structural damping and improved damage tolerance. The higher specific tensile strength and the larger specific elasticity modulus enable a significant reduction of the vibrating mass of moving components. Through a lower vehicle mass, a reduced driving power is needed for the achievement of comparable driving performance. Hence, lower pollutant emissions are achieved. Furthermore, the heavy battery weight for hybrid and electric vehicles makes the lightweight design concept essential. The large weight saving is important in order to be able to achieve acceptable driving ranges in traffic. A reduction of the weight consequently results in a reduction of the weight dependent driving

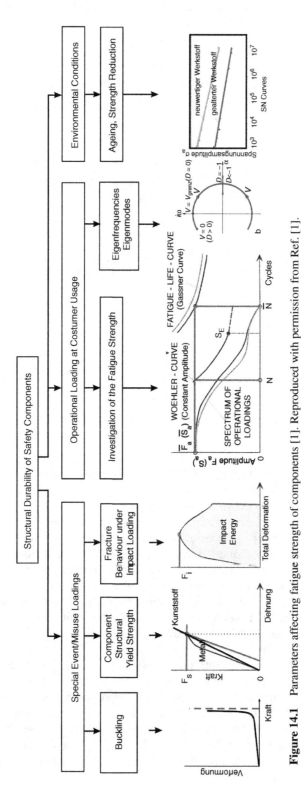

Figure 14.1 Parameters affecting fatigue strength of components [1]. Reproduced with permission from Ref. [1].

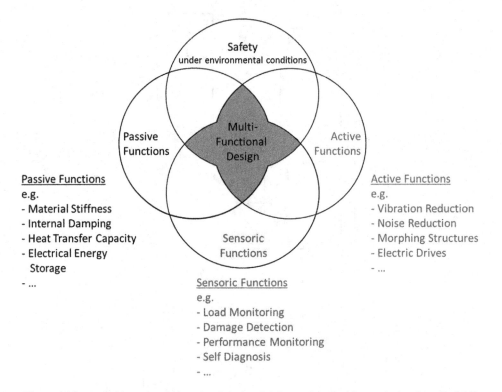

Figure 14.2   Definition of multifunctional design [1]. Reproduced with permission from Ref. [1].

resistance of the vehicle, making the lightweight design a key technology in automotive development. In addition to the excellent design freedom, other advantages of lightweight design are driving comfort and functional integration.

Considering the use of new material systems, "multi-functional design", a new special category of lightweight design, becomes possible. The idea is to achieve weight reduction due to the structural combination of different multiple passive, active and/or sensory functions in one component (see Figure 14.2). The number of parts in a component can be reduced because of the integration of functions. However, the manufacturing process itself can become more complex but the associated costs usually are mitigated by the reduced number of joints and so on. Further advantages are lower total costs, reduced manufacturing effort, smaller space and lower weight. Therefore, multifunctional design is a system-related lightweight design, which is more complex in evaluation.

Based on the concept of multifunctional design, the process for development of a FRP wheel with integrated electrical motor will be shown. Section 14.2 describes the challenges and requirements of vehicle wheels made from composites. Section 14.3 demonstrates the development of a composite wheel with an integrated electrical motor, based on the outcomes from Section 14.2. Multifunctional design is a system-related lightweight design, which is more complex in evaluation. In Section 14.4, the requirements and difficulties of reliability analysis of multifunctional systems will be shown through a number of examples.

14.2   Wheels Made from FRPs

The idea of manufacturing car wheels from FRPs is not novel. As shown in the images on the left and centre of Figure 14.3, the entire development of FRP vehicle wheels is ultimately based on the development of fibrous wooden wheels.

The use of polymeric materials for wheels was tested as long ago as the 1970s and 1980s. For example, in the 1970s wheels made of sheet moulding compound (SMC) by Firestone (see Figure 14.4) were tested at the Fraunhofer Institute for Structural Durability and System Reliability (LBF) in Darmstadt, Germany, and an improved testing procedure for rotating bending moment tests (RMBT), described in Section 14.2.4.1, was derived from the test series.

In 2001, BTE Hybrid-Tech GmbH introduced car wheels made from SMC (see Figure 14.5). Up to the present day, these wheels have successfully covered a distance of 250 000 km. RBMT were carried out on SMC wheels in 2006 at BTE Hybrid-Tech GmbH to demonstrate their fatigue strength (static wheel load 620 kg). The duration of the RBMT was extended by about a factor of ten compared to the standard tests, in order to demonstrate the potential of the material. As with the LBF tests on the Firestone SMC wheels, there was a slow, gradual decrease in stiffness for all three tests. Surface cracks appeared within the course of the full-load tests, but these did not lead to failure. Apart from the loss of stiffness, the retaining screws were the weakest point. Some of them failed and needed to be replaced several times. Under partial loading, the stiffness decreased by about 11% within the measured period.

14.2.1   Structural Durability of Lightweight Wheels Made from FRP

Vehicle wheels are heavily loaded safety components, whose manufacturing quality has to be inspected and whose structural durability must be proven by experimental testing. Testing procedures for metal wheels have been adapted and verified for an adequate testing of the material used. The fact that wheels made from plastics are a new step of "lightweight design

| Wooden spokes and rim, tire consisting of rubber | Büssing steel wheel with pipe-spokes, wooden rim and iron tire | Composite truck wheel (Dynawheel, Prins Dokkum) |

Figure 14.3   Truck wheels from Dynawheel at the beginning of the twentieth (on the left and in the centre) [1] and at the beginning of the twentyfirst century (on the right). Reproduced with permission from Ref. [2].

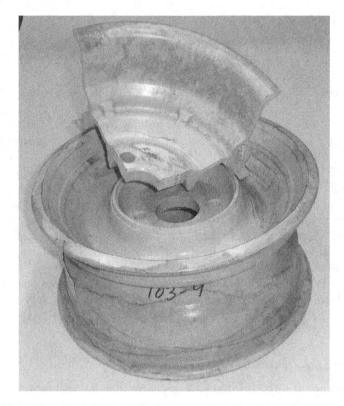

Figure 14.4 Glass fibre reinforced SMC wheel produced by Firestone, 1970s and 1980s.

of wheels" is a result of the uncertainty relating to the transfer of the testing methods to the behaviour of plastic materials. This again originates from the diversity of plastic materials, whose properties vary greatly, combined with the fact that plastics fail in very different ways compared to metals. From the point of view of structural durability, the following items require further investigation, in order for FRP wheels to receive approval, for example, for validation:

- Design philosophy ("safe life" with "damage tolerance")
  - How can the damage tolerance be demonstrated?
  - How can reliable proof tests be performed?
- Design concept allowing for the fibre reinforcement
  - How would the components be dimensioned?
- Quality assurance/damage evaluation (effects of defects)
  - How could manufacturing errors or damages be identified non-destructively?
  - What influences would such errors or damages have on service life?
- Assessment of components in use
  - The "how-to" of real-world component inspection – non-destructive testing (NDT)

In the following, based on experience relating to metallic wheels, an analysis and a first approach for a customised design rule for FRP is proposed.

Figure 14.5   Glass fibre reinforced SMC Wheel produced by BTE Hybrid-Tech GmbH, 2001.

14.2.1.1   Requirements on Composite Wheels with Respect to Fatigue

For the validation of the structural durability and system reliability of safety components made from FRPs, the material characteristics with their very different failure mechanisms compared to metals, manufacturing properties and component performance must be considered. "Ageing", named after the physical and chemical behaviour of plastics (DIN 50 035 part 1), is a very important process, which also has to be investigated. *Physical ageing of plastics* refers to the changes in the material (e.g. crystallisation or residual stresses), non-effective on the chemical structure. *Chemical ageing* occurs, for example, due to radiant exposure with high energy radiation or reaction with external metal ions. As a result, the mechanical properties can change – this may generally feature a more brittle behaviour in altered situations.

For a significant verification of the structural durability of safety components, all influences reducing the stiffness must be investigated. Table 14.1 shows the significance of operational loads on structural durability. Using simplified testing methods, all these types of loading, with all their respective damage-relevant consequences, must be taken into account.

Hence, for a reliable structural durability analysis of wheels, *deformations as they occur in operation* must be determined on the whole assembly of the rolling wheel, including the influences of wear and long service time. A *realistic sequence* (mix of loads), relating to different drive conditions (cornering, straight line driving, rolling under static wheel load

Table 14.1 Significance of operational loads for structural durability.

Type of loading	Damage relevant consequences
Intermittent loading, special events, peak loads (e.g. driving over curb stone)	Local damage, stiffness decrease, changed load path, changed fracture behaviour, changed functionality
Alternating amplitudes and high number of cycles	Fatigue, mechanical ageing, changed load path, friction corrosion (slip joint, screwed/threaded joint)
Variable amplitude loading	Fatigue behaviour can be influenced by load time history
Static loading, high centrifugal forces, humidity or changes in temperature	Mean stresses, changed internal stresses, creeping/relaxation, varied condition of fit, additional loads
Multi-axial loading	Multi-axial stress conditions with varied durability behaviour
Type of loading (bending, tension, pressure, torsion)	Changed fatigue behaviour and life expectancy
Environmental conditions	Chemical ageing

on smooth surface etc.) must also be considered. Furthermore, the *accurate correlation* of different load components and also the *damage equivalence* of test program and operational use are pre-conditions for fatigue strength examination.

14.2.2   Operational Strength Verification of Wheels

As part of the wheel suspension, vehicle wheels in use are subjected to cyclic loads caused by driving on poor condition roads, on straight roads, around curves, during braking and acceleration.

The resulting loading at a reference point of a wheel is additionally caused by internal stresses that result from manufacture, mounting, temperature and tyre pressure.

The different ways in which forces can be transferred is the reason why tyre design is a key to wheel rim loading. A further important factor is the stiffness and tension of the attached parts, consisting of brakes, connections and bearings. The stiffness of the wheel also has a non-negligible effect on the wheel hub. Therefore, the most important criterion for a reliable verification of structural durability is that the wheel deformation must be simulated correctly, something that cannot be achieved with simplified methods and sweeping assumptions.

In general, the wheel stresses are multiaxial. Therefore, considering the external loading, these multiaxial cyclic stresses are fundamentally composed of two parts, that is: (i) a stress under static wheel loading that varies periodically with the revolution of the wheel as it moves along an ideal smooth road surface and (ii) superimposed additional stresses created by the alternation of radial and lateral forces on the wheel, generated by operational use.

Based on the determined stress–time history, the frequency distributions (or stress spectrum) for each system point of the wheel can be deduced [3], (see Figure 14.6).

The stress–time history and the spectrum of the operational loading are caused by dynamic wheel loads in straight-line driving and road undulation, which are superimposed on the static wheel load, and by quasi-static wheel loads in cornering. Depending on the system point of the wheel and the superimposition of part collectives, different shapes of spectra are the result. These spectra are the basis for stiffness evaluation and for optimising the verification of

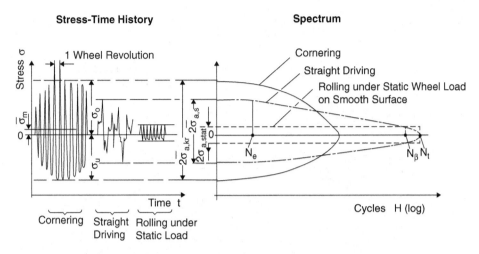

Figure 14.6   Stress–time history and design spectrum of wheels [2].

structural durability. From the superimposition of the part spectra (here: cornering, straight-line driving and rolling under static wheel load on a smooth surface), the so-called total design spectrum (Figure 14.7) is the result. In special cases, part collectives for braking and acceleration or special use must be considered, for example off road/open country or manoeuvring/ranking [3].

Figure 14.8 shows design load spectra (approx. 300 000 km) and the damage equivalent test spectra (approx. 10 000 km) for a car wheel. Based on such load data, measured on the vehicle, an experimental simulation of operational loading can be carried out for the rotating wheel. Further important influences are the effective wheel forces, force transfer from tyre to wheel

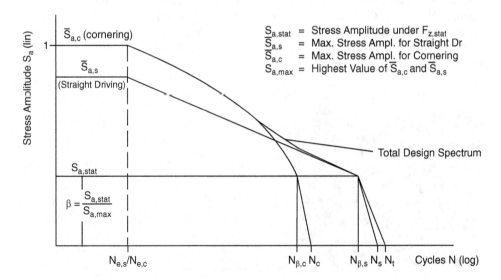

Figure 14.7   Generalised description of wheel and hub design spectra [3].

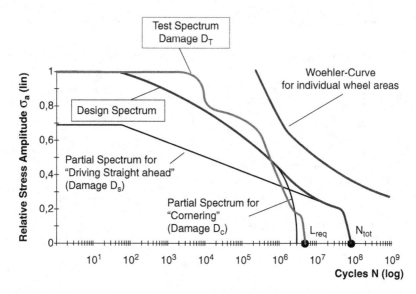

Figure 14.8   Basic design spectrum for 300 000 km of service and test spectrum [4].

rim (principally determined by tyre design) and the stiffness of the attached parts. With the estimated Woehler curve, also called the S/N curve, which is generated for individual wheel areas (e.g. with the "rotating bending moment test") and additionally with the test spectrum, a cumulative frequency distribution for total life evaluation and fatigue life evaluation can be determined.

Hence, for simulation of loads which correspond to the wheel operational loads, the biaxial wheel testing facility BIAX (Zweiaxiale Radprüfeinrichtung, or ZWARP) is used at Fraunhofer LBF to make a quick and economically justified inspection of structural durability (see Figure 14.9). This facility is also used by numerous wheel and vehicle manufacturers in Europe and the United States [4]. The evaluation of metallic wheels by means of this specific testing facility is already state of the art, while the evaluation of FRP wheels is still to be developed, because of the different material behaviour.

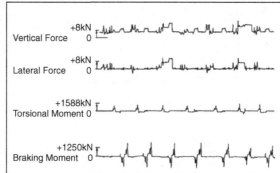

Figure 14.9   Result of a structural durability examination [4].

14.2.3 Evidence of Operational Stability of Car Wheels Made from Plastic

Wheels made from FRPs have great potential for lightweight design and they are highly damage tolerant. Therefore, they can ideally be used for car wheels. In considering safety for FRP wheels and conventional metal wheels, the same level must be proven. In addition to special event and misuse loads, environmental effects such as temperature, humidity and ageing have to be taken into account for FRP wheels. Environmental effects are dependent on the particular operational location.

FRPs exhibit very different failure mechanisms compared to metallic materials. This result, in very individual failure behaviour, is dependent on the laminate structure (number of layers, fibre orientation, fibre volume content) and the fibre/matrix material, which defines the anisotropic behaviour in stiffness and strength.

Therefore, in general, the fatigue strength distribution of components made from FRPs is determined by the manufacturing quality and design (fibre lay-up). Further, the resulting damage mechanisms can lead to quite different failure mechanisms. However, the damage-tolerant behaviour of FRPs, which is already proven for tested composite wheels, can be applied to a variety of applications.

For heavily stressed composite structures, which are almost without exception subject to high fatigue loads, the damage mechanisms mentioned and their properties are dependent on the type of loading [4]. This means that any damage will not necessarily lead to the total failure (break) of the structure, as shown in Figure 14.10 using the results of three rotating bending moment tests on SMC wheels from the Firestone company. The rupture is generally preceded by a loss of stiffness, depending on the structure and the material composition.

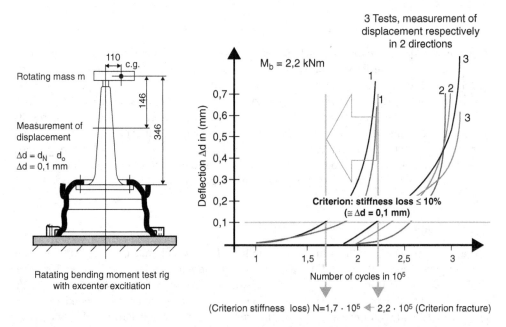

Figure 14.10 Rotating bending moment tests of Firestone SMC wheels [4]. Reproduced with permission from Ref. [1].

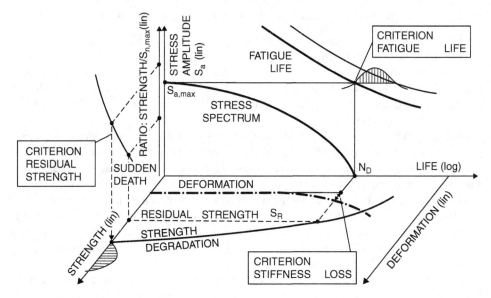

Figure 14.11   Failure criteria for design evaluation of composite wheels [5].

This presupposes a more or less slowly progressing damage to the heterogeneous fibre/matrix material system.

14.2.4   Results of Fatigue Tests on Composite Wheels

As explained above, factors such as fibre orientation and fibre volume content have a decisive influence on the material specific strength. Moreover, the damage mechanisms, which are different to those of metals, can lead to fundamentally different kinds of failure. In the first approximation, the results of the tests on the existing high-performance wheel rims substantiate the robustness and damage tolerance of fibre composite components. Based on these results, and taking particular consideration of the damage tolerance, an initial experimental structural durability verification concept ("worst case") could be derived for plastic wheels (proof test), though a differentiation must be made between three different kinds of failure criteria: loss of stiffness, absence of residual strength and rupture (Figure 14.11).

14.2.4.1   Fatigue tests on CFRP wheels

For a first assessment of the structural durability of car wheels made from FRPs, two aluminium/magnesium FRP hybrid designs (each with the wheel rim made from FRP and the spoke wheel centre from Al or Mg) made by two different manufacturers were tested in the biaxial wheel testing facility (Figure 14.9). To verify the damage tolerance, pre-existing damage was induced by an impact loading of the rim flange with a standard punch device. With this impact loading and its load level according to the static wheel load, local damage after hitting a curbstone could be simulated (see Figure 14.12).

Figure 14.12   Evaluation of a pre-damaged FRP hybrid wheel [5].

After pre-damage at the rim flange, generated by an impact load by a standard punch device, the investigations with the biaxial wheel test rig (ZWARP) were carried out. The test results can be summarised as follows:

- The FRP is very damage tolerant, if designed properly.
- Despite an increased static wheel load chosen for the test program (dimensioning load of 850 kg) and the pre-existence of visible damage at the wheel rim flange (based on the increased static wheel load), the wheel design was capable of withstanding 4600 BIAX kilometres. This equates to approximately 140 000 km on the road.
- After pre-damage (simulation of hitting a curbstone) and application of the regular dimensioning load of 650 kg, the FRP wheel was able to withstand 10 000 BIAX kilometres with no damage. Following this, the metal wheel centre was replaced, the wheel was pre-damaged (curbstone crossing) again and was exposed to a further BIAX test. Neither of the pre-damages led to any visible damage.
- Cracks first appeared in the aluminium wheel centre in both wheel designs.
- Failure mode of the wheel: failure of the seal in the area around the screw connections, air loss due to a crack in the rim well.

Furthermore, SMC wheels, which had been applied on different vehicles for approximately 250 000 service kilometres on the road, were made available to Fraunhofer LBF and were analysed for damage and compared with three identical, unused wheels. NDT tests, simple flat base wheel rolling tests with the BIAX test facility, rotating moment bending tests and actual BIAX tests of fibre reinforced plastic wheels were and will still be carried out.

The objective is to be able to identify damages that occur in real operation and to compare these with damage that occurs in experimental simulations. This should provide an answer to the question as to what extent the real damage mechanisms are obtained by the experimental loading simulation. Reasonable test procedures must activate all damage mechanisms, or they should precisely reproduce the relationships between the different loading types and the damage effects they create.

In this respect, the rotating bending moment test is a greatly simplified test method for wheels. The RMBT should principally only be used for characterising wheels at certain points. These points or areas are dependent on the design of the wheel and must therefore be examined and defined before testing.

Hence, to demonstrate the damage tolerance and to establish a testing method, the following BIAX tests were carried out. The testing procedure, derived from the tests performed on the two aluminium/magnesium FRP hybrid wheel designs, serves a basis for these tests:

- Conduction of an impact pre-loading (hitting a curbstone) at three points on the rim flange, at least one of the pre-loadings must be so high that visible damage occurs.
  - Determination of the ratio of this impact load to the dimensioning load to quantify the reserve.
  - Verification of the damage tolerance versus damage from special event load.
- First BIAX test with dimensioning load (static wheel load).
- Further impact loading (curbstone crossing) at three points on the rim flange.
  - Verification of the damage tolerance versus damage from operational loads.
- Second BIAX test on the same wheel with dimensioning load.
  - Verification of the damage tolerance versus damage from special event and operational loads.

14.3   Development of a Composite Wheel with Integrated Electric Motor

Lightweight design is becoming increasingly important for automotive engineering. FRPs have a wide range of advantages compared to metal-based materials, in addition to the weight saving aspect. In this study, a wheel of CFRPs with an integrated electrical motor was developed. The main focus of the development was to achieve the optimum in lightweight potential, considering structural durability. During the development, the technical challenges of multifunctional design were considered in the whole product life cycle. Finally, a general outlook regarding the validation of structural durability and system reliability of composite safety components is given.

14.3.1   CFRP Lightweight Wheel with Integrated Electrical Motor – Characteristic Data

The CFRP lightweight wheel with a size of 6.5 × 15.0 in (16.5 × 38.1 cm, without the CFRP housing to integrate the hub motor, without metal parts, such as sleeves for bearing and screws and without motor components) has a weight of approximately 3.5 kg. The motor housing is not directly connected to the rim, but to the inner area of the wheel axle. This prevents radial or lateral loads, especially shocks caused by a rough road or by curbstone crossing, from being transferred directly to the hub motor. Another advantage of the separation of the

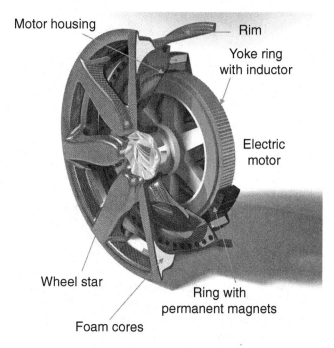

Motor housing

Rim

Yoke ring
with inductor

Electric
motor

Wheel star

Ring with
permanent magnets

Foam cores

Figure 14.13   Section of the CFRP wheel with motor housing and integrated electrical motor [6].

load paths is that the rim can be more flexible than if it were directly connected to the hub motor. In order to increase the flexural rigidity at a constant weight, foam cores are inserted into the spokes. A smaller, commercially available hub motor is used as the electrical motor. The hub motor, consisting of a ring with permanent magnets (external rotor) and a yoke ring with electromagnets (stator), has a motor capacity of 4 kW and a drive voltage of 2.0×24.5 V (Figure 14.13).

14.3.2   Development Process

In this section, the process of the development of a CFRP wheel with integrated hub motor is described, with all its technical challenges of development and multifunctional design. The manufacturing was prototypal – the main focus, achieving the optimum in lightweight design, was reached successfully.

14.3.2.1   Technical Challenges for Multifunctional Design

Multifunctional design is the integration of multiple passive, active or sensory functions into a component. Due to this functional integration, the component has a reduced number of separate parts. Further advantages of multifunctional design are lower costs, lower manufacturing effort and lower weight. On the one hand, through the integration of functions, the manufacturing process can be more complex. On the other hand, manufactured multifunctional components

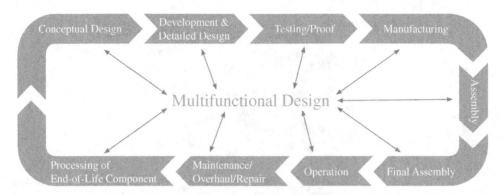

Figure 14.14 Product life cycle and its interaction with multifunctional design [1]. Reproduced with permission from Ref. [1].

have, for instance, a lower number of joints. The integration of functions into lightweight components brings with it some challenges in manufacturing, but it also has many advantages and it has a very great lightweight potential. For electromobility, multifunctional design means the synergetic integration of electrical functions into lightweight components. The philosophy and other examples of multifunctional design are described in Refs. [7, 8]. The application of multifunctional design has a great influence on a component and on every step of its product life cycle (see Figure 14.14).

This includes project planning from the concept phase and manufacturing, operation and the means of processing the end of life component. Thus, the final decision for the systematic application of (active) multifunctional design can only be made if the whole component life cycle is considered. The main focus of the presented project was to achieve a prototype composite wheel with integrated hub motor with the optimum in lightweight potential, considering structural durability and system reliability. Therefore, in the case of the presented CFRP wheel, the product life cycle was finished with the step "Assembly" (Figure 14.14.).

14.3.2.2 Design of the Wheel

As a first step in the development process, a design study based on current existing plastic wheels was carried out. Subsequently, solution variants were determined for the possible design methods to mount and attach the hub motor and for the optimum utilisation of the offered material properties. From these solutions, the rough concept for a CFRP wheel with integrated electrical motor was developed. For the definition and the optimisation of the laminate, a surface model (with no thickness information at this stage) was designed. Additionally, the virtual model served to generate the negative and positive mould for the manufacturing process. To obtain an optimised design, which considers the material properties, soft radii and flowing transitions were used. Hence, stress peaks or stiffness changes in the part were avoided by continuous fibre routing in line with the flow of forces.

The procedure for laminate definition, laminate optimisation and, finally, for the derivation of the moulds begins with the subdivision of the surface model into zones with different laminate

Figure 14.15   Surface model subdivided into zones for different laminate definitions [6].

definitions (Figure 14.15). This step facilitates the subsequent calculation and optimisation of the laminate by using the finite element method (FEM).

The laminate of the individual zones can very easily be predetermined and adjusted using a table for the specific design parameter of each zone. A FEM mesh of shell elements was produced on the basis of the surface model with zone geometries. The layer structure of the respective zone is transferred to the individual elements during the crosslinking. The laminate of the individual element here is based on a coordinate system related to each zone.

In order to reproduce the stresses in the FEM model according to reality, a numerical reproduction of the tyre as the load bearing element between road loads and local stresses is essential. Since the main focus with regard to evaluation is on the wheel, the FEM modelling of the tyre could take place in an abstract, simplified manner [9]. To simulate the interaction between tyre and wheel, a simplified but realistic load distribution was applied. Therefore, the tyre was reduced to its characteristic properties, not least to keep the size of the model as small as possible.

Simulations were carried out for the loads occurring when driving on rough roads and cornering. Based on the assumption that the components are thin-walled, the calculation used the shell theory, which only considered in-plane stress. The in-plane stresses and shear forces were evaluated for each layer in directions parallel and perpendicular to the fibres. The exploitation of the material could therefore be optimised by using several loops. The required wall thicknesses of the individual zones were determined with the laminate optimisation (Figure 14.16).

The calculation of this prototype CFRP wheel was made under simplified conditions because the software utilised – LBF.WheelStrength, developed specifically for wheel design – is based on the evaluation of isotropic materials and does not yet take into account the anisotropic material properties existing in the FRP wheel. There is still need for action here. Work is being done on this issue at present. In the next step, a volume model was created from the

Figure 14.16 FE model of the wheel thickness distribution for calculated load cases [6].

surface model using the wall thickness information. The mould cores for the tool could then be derived from this volume model. The final design step was to create the shape of the cuts of the individual layers with a draping simulation.

14.3.2.3 Manufacturing

A carbon fibre prepreg system was used in order to generate the required volume content of fibres, to guarantee a uniform resin–fibre distribution and to keep the equipment costs lower than for, for instance, infusion processes. A standard system from the company Hexcel was selected: M49 as the epoxy resin base with a "high-strength carbon fibre" as the technical fibre. A 2 × 2 twill weave with sufficient draping properties was selected as the weave of the fibre mat for the wheel rim and spokes. By using twill weave fabric with 50% of fibres in the 0° direction and 50% in the 90° direction, respectively, the two main directions are covered at the same time [10].

Unidirectional (UD) fibres were selected for the rim cylinder due to their geometric simplicity. The wheel that displays a complex three-dimensional geometry is produced here in one piece. For economic reasons, a two-part mould made of closed cell rigid foam with a polyurethane base was used for the prototype manufacturing (Figure 14.17). The mould was sealed and wetted with release agents before the fibre mats were applied, in order to prevent the epoxy resin matrix from adhering to the mould surface.

A corresponding ply book, listing the number of layers with their respective orientation and any other relevant manufacturing data, was drawn up for the structure of the woven fabric layers of the individual sections. The layer structure was a result of the simulation carried out in the design process. The layers were defined in accordance with the stresses and strains of the wheel rim, occurring in the 0°, 45°, 90° and –45° orientations. The resulting stacking sequence was then used for the manufacturing. Intermittently, a vacuum was built up over the

Figure 14.17 Moulds made of closed cell rigid foam with polyurethane base for manufacturing the prototype wheel [11].

mould during the application of a defined number of layers in order to press the layers together and therefore achieve a higher quality of the component. The final vacuum build-up with the corresponding sequence of layers was cured for 2 h in an autoclave at a temperature of 120 °C and a pressure of 3.5 bar (350 kPa) and annealed at a temperature of 50 °C for 16 h.

The finished component was extracted, separated from both halves of the mould and then finished. The wheel, housing and wheel hub were bonded to each other after mounting the valve and fitting the tyre. Finally, the electrical motor components were attached and the whole system was put into operation (Figure 14.18).

Figure 14.18 Completed composite wheel with integrated hub motor [11].

14.4 Multifunctional Design – Requirements regarding Structural Durability and System Reliability

The development and qualification of multifunctional safety systems for operational use have to be carried out with regard to the knowledge and methods of the structural durability and system reliability. The effects of the defects and failures of the entire system, as well as the failure type analysis, are very significant for the development phase of multifunctional design, since the defects and failures of any subsystem can also lead to a breakdown of the entire system.

14.4.1   Reliability Analysis of Multifunctional Systems

Reliability refers to the probability that failure of the entire system will not occur until after a defined period of time in operation. At this point, the "failure" is a function that can no longer be carried out. The uniqueness of the composite wheel with integrated hub motor above, however, lies in the associated *multifunctionality*. For such systems, the "failure" in the system reliability analysis must be newly defined. It has to be determined when there will be a failure or, rather, which functions of a subsystem are of importance for the usefulness of the entire system (see Figure 14.19).

An important part of the design of multifunctional components is the *reliability analysis* of the system. While there must be a concrete application for the quantitative examination, the qualitative examinations of the multifunctional system can also be generally performed in the first approximation.

In the following section, the special features of the qualitative and quantitative reliability analysis are demonstrated using the above-defined example of a CFRP wheel with integrated hub motor under special consideration.

14.4.2   Qualitative Reliability Analysis of Multifunctional Systems Performed on CFRP Wheel with Integrated Hub Motor under Operation

Design, which can take on various functions equally, can also fail in more than one way (see Figure 14.20). In order to meet all demands and to be able to specify the term "failure" in association with a function, the different functions must first of all be examined during the reliability analysis. However, for the qualitative reliability analysis performed on the

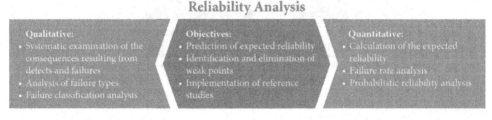

Figure 14.19   System reliability analysis of multifunctional design [1]. Reproduced with permission from Ref. [1].

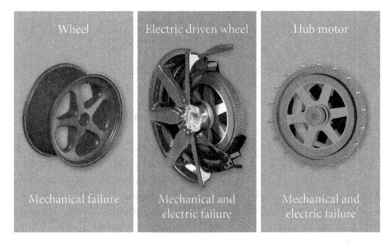

Figure 14.20 Multifunctional design: functional failure [1]. Reproduced with permission from Ref. [1].

multifunctional design of safety components and systems, all of the various functions have to be considered.

The term "load bearing function" (see Table 14.2) summarises the mechanical requirements of the entire system (here: the wheel as a safety component). The structure, comprised of various materials, must withstand all inner and outer loads and environmental conditions while remaining safe to operate. The requirements placed on the electrical components of the active system are summarised under the term "activating function" [here: motor driving the wheel (motor characteristic)]. This means that the use of the active structure, which depends upon the activation of the hub motor, must be ensured during the service life of the complete structure.

The term "interacting function" summarises the requirements on the active system through a special consideration of the interaction [12] between the mechanical and electrical components of the structure (here: operating point of the motor). These interactions include the fixed air gap of the hub motor when the electric wheel is moved and the influences of the mechanical structure on the requirements of the system's electrical components.

14.4.2.1 Quantitative System Reliability Analysis of Multifunctional Systems Performed on CFRP Wheel with Integrated Hub Motor under Operation

For a quantitative system reliability analysis, the failure characteristics of individual system components or subsystems are used. The loads specific to the operation are derived from the

Table 14.2 Multifunctional total system.

Component	Function
Wheel	Load bearing
Hub motor	Activation
Electric driven wheel	Interaction

$$\text{Failure Rate } \lambda\ (t) = \frac{\text{number of failure}}{\text{number of intact units}}$$

Figure 14.21 Definition of the Failure Rate.

availability or time in operation. With the help of the failure rate (see Figure 14.21), the actual system reliability analysis can be carried out.

The failure characteristics of electrical and mechanical systems differ, as shown in Figure 14.22. Figure 14.23 displays the special features of the failure characteristics, known also as the "bathtub curve" [6]. The key figures of the failure characteristics of each system component include their availability at a specific failure rate.

Figure 14.22 shows the failure characteristic for electrical systems. This depends on the environmental and maintenance conditions. A noticeable plateau in area 2 corresponding to random failures means that so-called wear out defects in this area are excluded. Consequently, for a system that is intact, the probability for its "survival" over the next time interval is of equal value at any given time. This assumption excludes failures that are caused by ageing, since this would make them time-dependent. Ageing can only be excluded if the wearing parts of a system are replaced sufficiently early. This assumption continues to require constant operational and environmental conditions.

In mechanical systems, the probability for "survival" over the next time interval varies with time t in interaction with the respective loads. The failure characteristic for mechanical

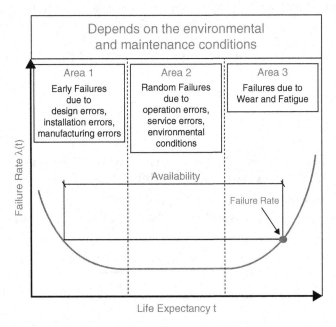

Figure 14.22 Failure characteristic for electrical systems, based on Ref. [13].

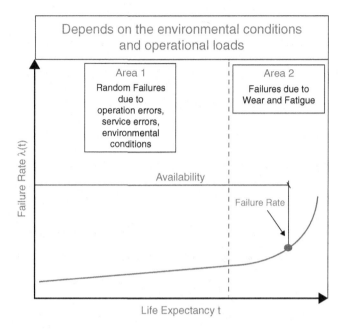

Figure 14.23   Failure characteristic for mechanical components [1]. Reproduced with permission from Ref. [1].

components depends on the environmental conditions and operational loads. Wear out defects (i.e. failures) which are caused by ageing must also be taken into account. Therefore, area 2, corresponding to random failures, shows a curve which is essentially dependent upon the individual operational loads (Figure 14.22).

The key figures of the failure characteristics necessary for a reliability analysis determine the availability corresponding to an acceptable failure rate [14]. The static and the dynamic behaviour of the component is ensured finally due to the safety distance between the loading and the load capability (see Figure 14.23). For a significant fatigue investigation of a component, a completely realistic simulation of all operating loads is needed. If the local stress is higher than the local load capability of material, design and condition of manufacturing, failure is the consequence. The loading and the load capability are subject to natural scatter. As shown in Figure 14.23, this changes throughout the lifetime. The height and the scatter of the local loading can vary due to changes of the system (e.g. changes caused by crack and delamination). The load capability of the entire component changes especially through influences that decrease the strength during operation, for instance reduction of fatigue life in terms of micro- or macro-structural damage.

The "area of failure" is a measure of the probability of failure. The safety distance has to be chosen so that the value of probability of failure is acceptable. In order to obtain a meaningful statement regarding the fatigue life, the associated probability of failure must be considered. The existing relationship between fatigue life and probability of failure is also illustrated in a

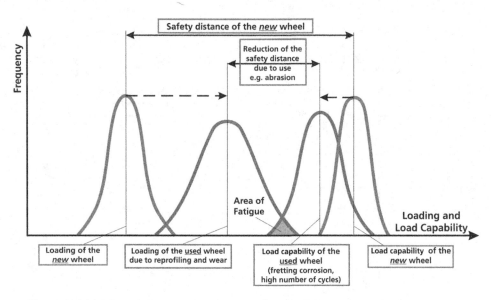

Figure 14.24 Safety of new and used wheels [1]. Reproduced with permission from Ref. [1].

comparison of the acceptable operational loading and the actual loading in the system on the fatigue life curve (Figure 14.24) [15].

The design spectrum and the special events describe all the loads that appear over the entire time in operation and thus are directly connected with the availability of the structure. If the assumption is made that, during the structural durability analysis, the amplitudes of the design data and the structural durability are statistically normally distributed, the probability of survival is also normally distributed. From these distributions, the life expectancy for a specific mathematical probability of failure can be determined. The mathematical probability of failure, which is only valid for a specific availability due to the design spectrum used, is directly connected with the failure rate and consequently all the necessary data for a reliability analysis is given with this information.

Figure 14.25 also includes the scatter band of acceptable operational loading; its scatter is influenced by material inhomogeneities, variation in material composition, variation in manufacturing procedure and variation in experimental procedure. The decrease of this scatter band with increasing fatigue life is indicated by the lines of the maximum stress amplitude of the spectrum, for the probability of survival at 99, 90, 50 and 10%. The scatter band of acceptable operational loading is compared with a scatter band, independent from the fatigue life. The scatter band of operational loading overlaps the scatter band of acceptable operational loading, which decreases with increasing cycles. The overlap visualises that, in a statistical selection process, the operational loading exceeds the individual fatigue life in a certain number of components, with fatigue cracking being the result (descriptively visualised through the overlap in Figure 14.24) [15]. In order to increase the usable lifetime of a component, the operational loading must be reduced and/or the loading capacity must be

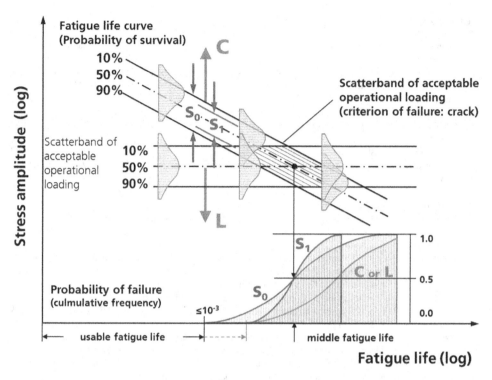

Figure 14.25 Result of a structural durability examination [1]. Reproduced with permission from Ref. [1].

increased. The usable lifetime can also be increased by influencing the scatterband. The lower the width of the scatterband, the longer is the usable lifetime.

14.5 Conclusion

In this chapter, a prototype composite wheel with integrated hub motor was developed. The objective of achieving an optimised lightweight design was reached successfully. Special material properties were used to develop a multifunctional component with reduced weight, increased structural damping and improved damage tolerance. For composite wheels, the same safety level as for metallic wheels has to be verified. In this context, the same operating conditions must be considered. At present, there are no definitive test guidelines for the approval of fibre reinforced wheels on German roads, but work on this issue is currently in progress.

Another challenge is the manufacturing in high production quantities. To cost-efficiently produce a high number of carbon fibre reinforced wheels, the use of other procedures (e.g. pressing method) is needed. However, with a shorter fibre length, the possible weight reduction is not fully exploited. Therefore, the aim is to find an acceptable compromise between weight reduction and manufacturing costs.

References

[1] Fraunhofer Institute (2012) Fraunhofer Institute for Structural Durability and System Reliability LBF Picture Library, www.lbf.fraunhofer.de (Accessed 8th July 2012).

[2] Fersen, O. (1987) *Ein Jahrhundert in der Automobiltechnik – Nutzfahrzeuge*, VDI Verlag, Düsseldorf, Germany.

[3] Grubisic, V. and Fischer, G. (1995) Criteria for light-weight design of wheels. *VDI Berichte*, **1224**, S.309–S.326.

[4] Büter, A., Jaschek, K., Türk, O. and Schmidt, M.R. (2008) Hochfeste kunststoffstrukturen – räder aus sheet moulding compound SMC. *MP Materials Testing*, **50** (1\2), S.28–S.36.

[5] Grubisic, V. (1996) *Experimental Techniques and Design in Composite Materials* **3**, 30–31 October, Caligari, Italy.

[6] Schweizer, N., Giessl, A., Schwarzhaupt, O. and Büter, A. (2012) Development of a Composite Wheel with Integrated Hub Motor. 15th European Conference on composite materials (ECCM15), 24–29 June, Venice, Italy.

[7] Büter, A., Gaisbauer, S. and Kraus, K. (2012) *Multi-Functional Lightweight Design. Recent Developments in Adaptronik*, Springer Verlag, Berlin.

[8] Kraus, K., Mayer, D., Herold, S. and Büter, A. (2012) Online processing of structural health assessment, in *Proceedings of the 13th Mechatronics Forum International Conference*, 1st edn (eds N. Jones and T. Wierzbicki). Trauner Verlag, Linz, Austria.

[9] Breitenberger, M. (2007) Numerische Analyse von Krafteinleitungsmechanismen in das System Rad-Reifen. Diploma Thesis, Fraunhofer Institute of Structural Durability and System Reliability LBF, Darmstadt, Germany.

[10] Büter, A. (2011) Multifunctional Design – Possibilities and challenges in the area of E-Mobility. Automotive Composites IQPC 2nd International Congress, 6–8 December, Munich, Germany.

[11] Schweizer, N., Giessl, A., and Schwarzhaupt, O. (2012) *Entwicklung eines CFK Leichtbaurads mit integriertem Elektromotor*, ATZ, Nr. **5** (114).

[12] Hanselka, H. and Melz, T. (2001) *Adaptronik als Schlüsseltechnologie für die Produktionstechnik*, ATZ.

[13] Bertsche, B. and Lechner, G. (1999) *Zuverlässigkeit im Maschinenbau*, Springer Verlag, Berlin.

[14] Büter, A., Melz, T. and Hanselka, H. (2002) Significance of Reliable Active Systems. International Conference on Adaptive Structures and Technologies ICAST, 7–9 October, Potsdam, Germany.

[15] Haibach, E. (2006) *Betriebsfestigkeit*, Springer Verlag, Berlin.

15

Composite Materials in Automotive Body Panels, Concerning Noise and Vibration

Peyman Honarmandi

Mechanical Engineering Department, The City College of the City University of New York, New York, USA

15.1 Introduction

In the last two decades, there has been a great increase in the use of composite materials in the automotive industry. They have replaced traditional metallic materials in the construction of structural components. The exceptional properties of composite materials, such as superior yield strength, damping and strength-to-weight ratios, make them very desirable. Despite these properties, one of the main concerns for designers is the behaviour and response of the composite component to noise, vibration and impact. Effective control of noise and vibration in composite structures requires the implementation of several methods such as absorption, barriers and enclosures, structural damping and vibration. In the automotive industry, all vehicles use damping materials, which are usually made of composites, to reduce structure-borne noise and prevent structural damage. The effectiveness of such damping treatments depends upon design factors such as the type of materials being used, environmental terrain and size of the treatment [1–18]. In this chapter, we first review typical composite materials used in the shells and panels of a vehicle. Second, we present two examples to show the modelling, modal analysis and damping treatment of composite structures used in body panels in the automotive industry.

15.2 Composite Materials in Automobile Bodies

Composite materials are now used extensively in the automotive industry in place of metallic materials. Sheet moulding compound (SMC), which is based on liquid unsaturated polyester

Advanced Composite Materials for Automotive Applications: Structural Integrity and Crashworthiness,
First Edition. Edited by Ahmed Elmarakbi.

(UP), has been increasingly used for automotive components including body panels, chassis parts and under the hood parts. The SMC composition generally includes UP resin (25–30%), fibrous reinforcing material (25–30%), mineral filler (40–45%) and other additives such as a release agent, pigment, thickening agent and so on. A low-pressure mouldable SMC, called LPMC, was introduced by the Scott Bader Company. Low pressure moulding compound (LPMC) is a new type of composite material that can be used for automotive body panels. LPMC has similar mechanical properties compared to conventional SMC but excellent mould ability due to a different thickening system.

The Scott Bader Company uses a unique thickening system in which crystalline polyester, called Crystic Impreg, is used as a thickening agent instead of the metal oxide in a conventional SMC formulation. The crystalline resin is dissolved in the liquid resin at 82 °C, processed into a paste, then converted into LPMC while maintaining a temperature over 50 °C. Other conditions like glass fibre loading and resin feeding are very similar to traditional SMC. Once converted to LPMC, the compound is allowed to cool down to room temperature, the crystalline resin becomes insoluble and the viscosity increases almost immediately. Therefore, an additional maturation period is not needed and the handling characteristics remain for a much longer time than metal oxide thickened SMC. The advantages of LPMC mainly come from the different thickening mechanism, which gives it flow characteristics that are superior to conventional SMC. Moreover, LPMC materials can be compression moulded at a much lower pressure (1–3 MPa), reducing tooling and maintenance costs.

15.3 Multilayer Composite Materials in Noise and Vibration Treatment

Automobiles encounter many scenarios where noise and vibration inside a cabin can affect the passenger comfort. Low noise and vibration inside the cabins of automobiles are often used as indicators of product acceptance quality. Consequently, engineers strive to achieve a comfortable environment for customer satisfaction. An acoustic package that is incorporated in vehicle design to control vibration usually consists of various components such as absorbers, barriers, dampers and isolators. Surface damping materials, which are effectively used in reducing structure-borne noise at frequencies beyond 100 Hz, are modelled and examined in this section. These surface damping treatments play a significant role in reducing structural panel vibrations. Typically, damping materials are applied on the door, roof, dash, floor and cab back panels of vehicles. For body panels, surface damping treatments are categorised into two types. The first type is an unconstrained or free layer damping treatment that consists of only the base layer and the damping layer, as shown in Figure 15.1.

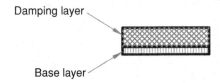

Figure 15.1   Unconstrained layer damping treatment consisting of two layers.

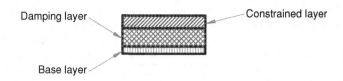

Figure 15.2   Constrained layer damping treatment consisting of three layers.

The second type of surface damping treatment is a constrained or multi-layer damping treatment; which contains more than two layers, as shown in Figure 15.2. In this figure, there is a base layer in which a viscoelastic-damping layer is applied and constrained by a third layer. Such arrangements are commonly used in partially constrained layer (PCL) damping treatments to provide extensional and shear damping.

In this section, these two different forms of composite material damping treatments are modelled as two degree of freedom (DOF) and three DOF systems, respectively. Results are shown in separate graphs to compare the damping treatment between two-layer and three-layer composite materials. Traditionally, experimental techniques or laser vibrometry are used to optimise design parameters such as material type, size and location of damping treatment. Experimental approaches can become cumbersome, time-consuming and prohibitively expensive since the structure needs to be excited for a large frequency range at all noise transfer paths to identify flexible regions on structural panels. To overcome the drawbacks of experimental approaches, engineers are challenged to find cost-effective analytical approaches that can be employed in early stages of the vehicle design process for optimal damping treatments. The modal analysis of a sectional panel provides an initial estimate of system parameters, including damping and stiffness coefficients for designers.

15.4   Case Studies

15.4.1   Case Study I: Modal Analysis of Vehicle Hood

The construction of the hood panel (engine cover, or bonnet) for passenger cars is an important example of the incorporation of the SMC procedure. A hood panel is typically made of two layers, the outer and inner panels. SMC hood panels can be seen in the Ford Lincoln Continental, Chrysler Sebring JX, Dodge Viper, GM Corvette, Camaro, EV1 and so on. The two fabrication formulations for the outer and inner panel of a typical car hood that are modelled in this case study were developed and introduced by the Hyundai Motor Company.

The outer panel is made of general-density low profile LPMC and the inner panel is made of low-density hollow glass micro-sphere filled LPMC. The outer and inner panels are moulded by using zinc alloy moulds and then bonded together with an adhesive at room temperature. Hyundai's reformulations of the LPMC compound for hood outer and inner panels are summarised in Table 15.1. It is noted that the resin consists of ortho-type UP, iso-type UP, crystalline saturated polyester and styrene monomer. The detailed explanations of materials, compounding and moulding, can be found in the technical published documents [8,10].

The tensile, flexural and impact properties of LPMC materials can be determined using a universal testing machine (UTM), following the standard procedures described in ASTM

Table 15.1 LPMC composite formulation for vehicle body panels [10].

Material	Standard LPMC panel		Hyundai prototype outer hood panel		Hyundai prototype inner hood panel	
	Volume fraction	Weight fraction	Volume fraction	Weight fraction	Volume fraction	Weight fraction
Resin	40.6	22.6	46.3	27.1	39.9	32.1
Glass fibre	18.4	25.0	17.4	25.0	17.6	35.0
Mineral filler	34.5	48.6	29.1	43.4	9.4	19.3
Micro-sphere	—	—	—	—	26.7	7.6
Additives	6.5	3.8	7.2	4.5	6.4	6.0
Total	100	100	100	100	100	100

D638, ASTM D790 and ASTM D256, respectively. The material properties of Hyundai hood panels are summarised in Table 15.2.

During the motion of a vehicle, the vibration induced between the road and the body frame or chassis is transmitted to the hood panel, causing noise and discomfort. The response of the hood is important for designers in estimating the stiffness and damping coefficients of latch/hook assembly, hinges, rubber-stop, as well as the hood panel itself. In this section, a simple model is proposed for the hood of a passenger car made of LPMC composite material. After estimations of system parameters, vibration analysis is conducted and the differential equations of motion are developed – this is the primary step for designers to obtain the total response of the hood and perform further analysis. Based on the response, one can design or improve the system parameters as needed.

As shown in Figure 15.3a, the hood panel is locked by latch/hook assembly from its front and connected by hinges from its back to the body. The rubber-stop parts are also located on the side of contact between hood and body and are used to reduce noise or isolate vibration. In Figure 15.3b, the four wheels of a car are represented by one wheel in our lumped model. In this model, the car hood and the engine are connected directly to the car frame, or to the chassis. The frame is also connected to the wheels by shock absorbers. m, k and c are the values of mass, stiffness and damping coefficients, respectively, whose descriptions are detailed in Table 15.3.

It should be noted that the effective stiffness, k_1, for instance, is the parallel summation of the stiffness coefficients of hinges, latch and rubber-stops that are in series with the structural

Table 15.2 Material properties of LPMC composite body panels.

Material property	Standard LPMC body panel	Hyundai prototype outer hood panel	Hyundai prototype inner hood panel
Tensile strength (MPa)	67	70	61
Flexural strength (MPa)	170	172	160
Flexural modulus (GPa)	1.20	1.13	1.01
Impact strength (J/m)	750	820	630
Specific gravity	1.95	1.82	1.31

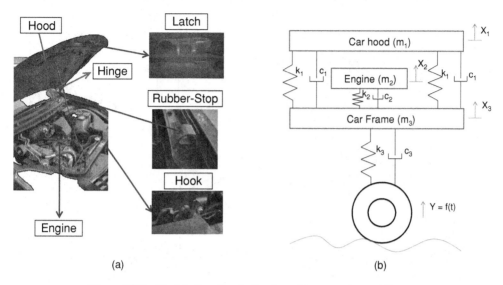

Figure 15.3 Model of car hood vibrating with car engine and frame.

stiffness of the composite hood. One may ignore the effect of stiffness of rubber-stops in comparison with the high stiffness coefficients of hinges and latch/hook assemblies. Likewise, the effective damping coefficient, c_1, is the parallel summation of the damping coefficients of hinges, latch and rubber-stops that are in series with the structural damping of the composite hood. One may also ignore the effect of damping of hinges and latch/hook assemblies in comparison with the higher damping coefficients of rubber-stops.

As shown in Figure 15.3b, the car is assumed to be subjected to an input displacement variation of $Y = Y_0 \sin(\omega t)$, which represents a sinusoidal model of road profile. One may assume that the maximum amplitude of this displacement is approximately 0.05 m with a wavelength of $L = 6$ m. It can also be assumed that the vehicle speed is $V = 40$ mph or 18 m/s.

Table 15.3 System design parameters of vehicle hood model.

Design parameter	Description
m_1	Effective mass of carbon fibre hood
m_2	Effective mass of engine block
m_3	Effective mass of body frame or chassis
k_1	Effective stiffness of hinges and latch/hook between hood and frame + structural stiffness of composite hood
k_2	Effective stiffness of mount rubber between engine block and frame
k_3	Effective stiffness constant of shock absorbers
c_1	Effective damping coefficient of rubber-stops + structural damping of composite hood
c_2	Effective damping coefficient of engine mount rubber
c_3	Effective damping constant of shock absorbers

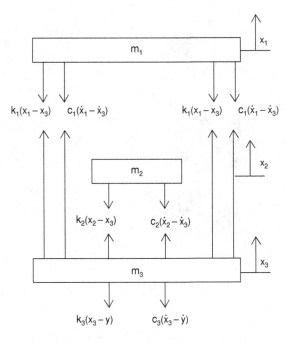

Figure 15.4   Free body diagram of the hood system.

The system is then analysed under the harmonic motion of the base wheel. The respective responses can be obtained and plotted for different velocities of vehicle.

To obtain the governing equations of motion for the model, a free body diagram should be considered along with Newton's second law, $\sum F = m_i \ddot{X}_i$, where $i = 1, 2, 3$. The free body diagram of forces is illustrated in Figure 15.4; applying Newton's law, the matrix form of equations of motion can be described as

$$[m]\ddot{\mathbf{X}} + [c]\dot{\mathbf{X}} + [k]\mathbf{X} = \mathbf{Y} \tag{15.1}$$

The equations of motion are obtained and expanded in matrix form in Equation 15.2

$$
\begin{bmatrix} m_1 & 0 & 0 \\ 0 & m_2 & 0 \\ 0 & 0 & m_3 \end{bmatrix}
\begin{Bmatrix} \ddot{x}_1 \\ \ddot{x}_2 \\ \ddot{x}_3 \end{Bmatrix}
+
\begin{bmatrix} 2c_1 & 0 & -2c_1 \\ 0 & c_2 & -c_2 \\ -2c_1 & -c_2 & 2c_1 + c_2 + c_3 \end{bmatrix}
\begin{Bmatrix} \dot{x}_1 \\ \dot{x}_2 \\ \dot{x}_3 \end{Bmatrix}
$$
$$
+
\begin{bmatrix} 2k_1 & 0 & -2k_1 \\ 0 & k_2 & -k_2 \\ -2k_1 & -k_2 & 2k_1 + k_2 + k_3 \end{bmatrix}
\begin{Bmatrix} x_1 \\ x_2 \\ x_3 \end{Bmatrix}
=
\begin{bmatrix} 0 & 0 & 0 \\ 0 & 0 & 0 \\ 0 & 0 & c_3 \end{bmatrix}
\begin{Bmatrix} 0 \\ 0 \\ \dot{y} \end{Bmatrix}
\tag{15.2}
$$
$$
+
\begin{bmatrix} 0 & 0 & 0 \\ 0 & 0 & 0 \\ 0 & 0 & k_3 \end{bmatrix}
\begin{Bmatrix} 0 \\ 0 \\ y \end{Bmatrix}
$$

Using $Y = Y_0 \sin(\omega t)$, Equation 15.2 can be written as

$$
\begin{bmatrix} m_1 & 0 & 0 \\ 0 & m_2 & 0 \\ 0 & 0 & m_3 \end{bmatrix} \begin{Bmatrix} \ddot{x}_1 \\ \ddot{x}_2 \\ \ddot{x}_3 \end{Bmatrix} + \begin{bmatrix} 2c_1 & 0 & -2c_1 \\ 0 & c_2 & -c_2 \\ -2c_1 & -c_2 & 2c_1 + c_2 + c_3 \end{bmatrix} \begin{Bmatrix} \dot{x}_1 \\ \dot{x}_2 \\ \dot{x}_3 \end{Bmatrix}
$$

$$
+ \begin{bmatrix} 2k_1 & 0 & -2k_1 \\ 0 & k_2 & -k_2 \\ -2k_1 & -k_2 & 2k_1 + k_2 + k_3 \end{bmatrix} \begin{Bmatrix} x_1 \\ x_2 \\ x_3 \end{Bmatrix} = \begin{Bmatrix} 0 \\ 0 \\ 1 \end{Bmatrix} A \sin(\omega t - \alpha)
$$

(15.3)

where A, α and ω are given by $A = Y_0 \sqrt{k_3^2 + (c_3\omega)^2}$, $\alpha = \tan^{-1}(-c_3\omega/k_3)$ and $\omega = 2\pi V/L$. Assuming initial conditions, the differential equations in Equation 15.3 can be solved numerically to obtain the responses. For example, the response of system that consists of the displacement responses of composite hood panel, engine block and chassis are obtained and plotted in Figure 15.3. For this analysis, the typical values of the system parameters are estimated as follows:

$m_1 = 9$ kg (20 lbs)	$m_2 = 272$ kg (600 lbs)	$m_3 = 136$ kg (300 lbs)
$k_1 = 22 \times 10^4$ N/m	$k_2 = 20 \times 10^4$ N/m	$k_3 = 3 \times 10^4$ N/m
$c_1 = 85$ N.s/m	$c_2 = 8851$ N.s/m	$c_3 = 1.15 \times 10^4$ N.s/m

It is noted that the responses in Figure 15.5 are directly dependent on the values of system parameters listed above; and the results can be engineered or optimised as required by the designer. From Figure 15.5, it is observed that the response of the mass of the hood and the mass of the frame follow the same trend and are similar in a steady-state period, while the response from the engine mass is different with generally a higher amplitude. This implies that, after a transient time, the car frame and the composite hood follow the same response in the steady-state region, which is very effective in overall noise reduction and control.

In Figure 15.6, the response of the system is plotted for the case in which the weight of the hood is doubled, that is $m_1 = 2 \times 9$ kg, while the rest of system parameters remain the same. As shown in Figure 15.4, if the weight of the hood panel is increased to the typical weight of steel hood ($m_1 = 18$ kg), it seems that there is no significant change in the response; however, the transient time will rise, causing an increase in noise. Moreover, the weight of the hood is a concern for the aerodynamics of the car because the composite hood reduces the weight of the car, which in turn improves fuel efficiency.

As shown in Figures 15.5 and 15.6, the hood closely follows the same response trend as the frame. If vibration is a concern for the car hood, the parameters of vibration isolators (i.e. rubber-stops) should be designed such to reduce the response.

It would be interesting to compare the results if, in the proposed model, the effective damping coefficient related to the composite hood panel increases. In Figure 15.7, the response is shown for the case of a higher effective damping coefficient, for instance $c_1 = 10 \times 85$ Ns/m. As implied from this figure, the response of the composite hood becomes over-damped with a

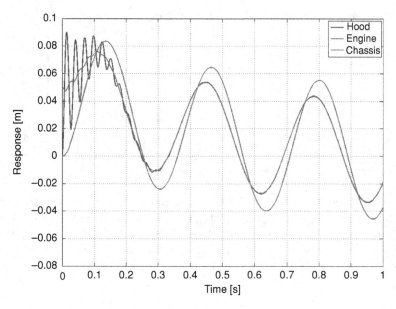

Figure 15.5   Displacement responses of composite hood panel, engine block and body frame.

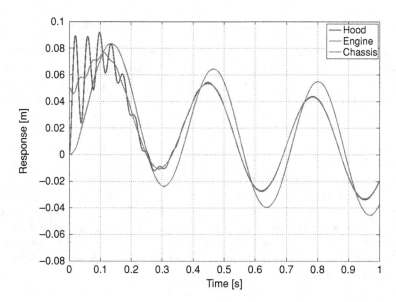

Figure 15.6   Displacement response of a system with twice the effective weight of the hood panel ($m_1 = 18$ kg).

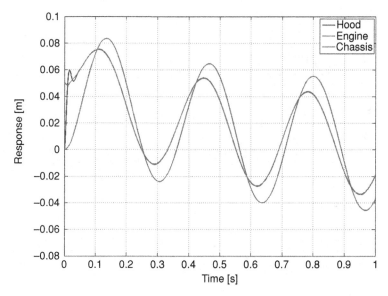

Figure 15.7 Displacement response of a system with ten times more effective damping coefficient of composite hood panel vicinity ($c_1 = 850$ N.s/m).

lower overshoot than in the former case. It is observed that the transient time is very short, and the hood panel quickly reaches the same response as the body frame. This advocates the minimum local vibration of the hood, effectively causing a reduction in noise.

However, if the effective damping coefficient reduces by an order of 10, $c_1 = 8.5$ N.s/m, both the amplitude and transient time increase, as shown in Figure 15.8. This may cause more noise and discomfort compared to the previous cases in Figures 15.5–15.7.

As demonstrated in Figure 15.9, if the effective stiffness coefficient of the composite hood panel, k_1, reduces by an order of 10, $k_1 = 22 \times 10^3$ N/m, the number of vibration cycles in the transient region will be less but there is a phase shift between input and hood response in the steady-state region.

However, by increasing this stiffness coefficient by an order of 10, $k_1 = 22 \times 10^5$ N/m, and keeping the rest of system parameters the same, the response of the hood panel improves in the steady-state region while there is noise in the transient state, as shown in Figure 15.10.

Although there is no significant change in the response of the engine block by changing the affiliated composite hood parameters, a designer can improve the engine block response by changing other system parameters. For instance, if one increases the effective damping coefficient of engine mount rubber threefold, $c_2 = 3 \times 8851$ N.s/m, and keeps the rest of the system parameters the same, the response of the engine block, as shown in Figure 15.11, will be improved while the response of the hood and frame remain almost the same as that in Figure 15.5.

Therefore, it should be noted that changing the values of k and c has a significant effect on the system response. These parameters may be incorporated into the composite material property of the hood panel and its affiliated structural components.

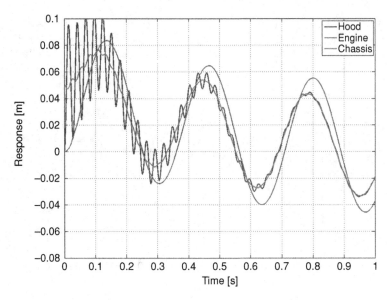

Figure 15.8 Displacement response of a system with ten times less effective damping coefficient of composite hood panel vicinity ($c_I = 8.5$ N.s/m).

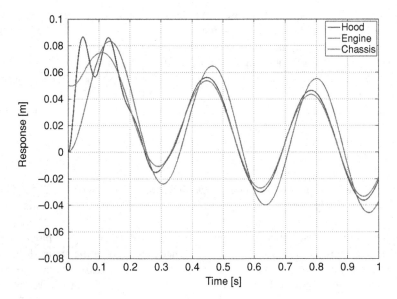

Figure 15.9 Displacement response of a system with ten times less effective stiffness coefficient of composite hood panel vicinity ($k_I = 22 \times 10^3$ N/m).

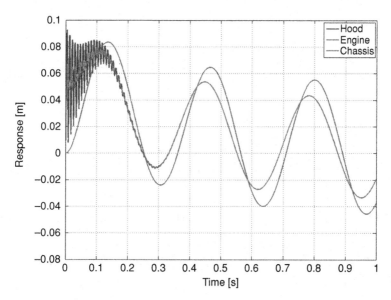

Figure 15.10   Displacement response of a system with ten times more effective stiffness coefficient of composite hood panel vicinity ($k_1 = 22 \times 10^5$ N/m).

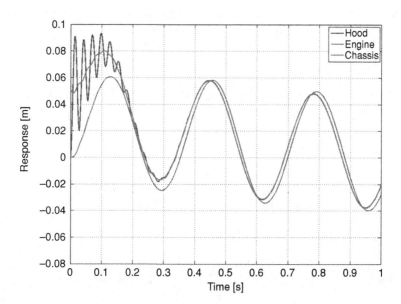

Figure 15.11   Displacement response of a system with three times more effective damping coefficient of engine block vicinity ($c_2 = 26\,553$ N.s/m).

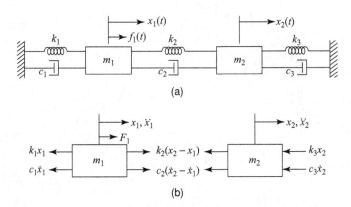

Figure 15.12   (a) Two DOF model of unconstrained layer damping treatment. (b) Free body diagram.

15.4.2   Case Study II: Modal Analysis of Two- or Three-Layer Damping Treatment

15.4.2.1   Unconstrained Layer Damping Treatment

The two-layer unconstrained damping treatment shown in Figure 15.1 can be modelled as a damped two degree of freedom (DOF) system, as schematically illustrated in Figure 15.12a.

The two masses m_1 and m_2 represent the base and the damping layers, respectively. This model uses a fixed-fixed beam approach to model the damping treatment. The free body diagrams of forces are shown in Figure 15.12b. Applying Newton's second law, that is $\sum F = m_i \ddot{X}_i$ where $i = 1, 2$, results in the equations of motion as

$$m_1\ddot{x}_1 + (c_1 + c_2)\dot{x}_1 - c_2\dot{x}_2 + (k_1 + k_2)x_1 - k_2x_2 = f_1(t) \tag{15.4}$$

$$m_2\ddot{x}_2 + (c_2 + c_3)\dot{x}_2 - c_2\dot{x}_1 + (k_2 + k_3)x_2 - k_2x_1 = 0 \tag{15.5}$$

which, in matrix form, are given by

$$[m]\begin{Bmatrix} \ddot{x}_1 \\ \ddot{x}_2 \end{Bmatrix} + [c]\begin{Bmatrix} \dot{x}_1 \\ \dot{x}_2 \end{Bmatrix} + [k]\begin{Bmatrix} x_1 \\ x_2 \end{Bmatrix} = \begin{Bmatrix} f_1(t) \\ 0 \end{Bmatrix} \tag{15.6}$$

where the mass, damping and stiffness matrices are specified as

$$[m] = \begin{bmatrix} m_1 & 0 \\ 0 & m_2 \end{bmatrix}, \quad [c] = \begin{bmatrix} c_1 + c_2 & -c_2 \\ -c_2 & c_2 + c_3 \end{bmatrix}, \quad [k] = \begin{bmatrix} k_1 + k_2 & -k_2 \\ -k_2 & k_2 + k_3 \end{bmatrix}$$

To solve the set of differential equations in Equation 15.6, numerical methods are desirable since it is usually tedious to proceed with analytical techniques. However, for low DOF, one may employ the impedance method to solve and obtain the responses analytically. In the impedance method, responses and force are assumed as complex variables, that is $x_j(t) = X_j e^{i\omega t}$ where $j = 1, 2$ and $f_1(t) = F_1 e^{i\omega t}$, which are then substituted into Equations 15.4 and

15.5. After simplifications, we have

$$\left[-\omega^2 m_1 + i\omega(c_1 + c_2) + k_1 + k_2\right] X_1 - (i\omega c_2 + k_2)X_2 = f_1(t) \tag{15.7}$$

$$\left[-\omega^2 m_2 + i\omega(c_2 + c_3) + k_2 + k_3\right] X_2 - (i\omega c_2 + k_2)X_1 = 0 \tag{15.8}$$

Equations 15.7 and 15.8 should be solved for X_1 and X_2 simultaneously. To obtain the actual responses, the real part of the complex variables should be obtained. Therefore, the actual responses could be defined as

$$x_j(t) = \text{Re}(X_j e^{i\omega t}) = \text{Re}(X_j \cos \omega t + i X_j \sin \omega t) \tag{15.9}$$

As aforementioned, analytical methods such as impedance, Laplace methods, etc, are appropriate for a low DOF system, but the numerical methods more promising, especially for high DOF systems.

15.4.2.2 Constrained Layer Damping Treatment

The three-layer constrained damping treatment shown in Figure 15.11 can be modelled as a three DOF system, as shown in Figure 15.13.

Using a free body diagram of system and applying Newton's second law, the differential equations of motion for this mode are obtained and given by

$$m_1\ddot{x}_1 = -k_1 x_1 + k_2(x_2 - x_1) - c_1\dot{x}_1 + c_2(\dot{x}_2 - \dot{x}_1) + f_1(t) \tag{15.10}$$

$$m_2\ddot{x}_2 = -k_2(x_2 - x_1) - k_3(x_2 - x_3) - c_2(\dot{x}_2 - \dot{x}_1) - c_3(\dot{x}_2 - \dot{x}_3) \tag{15.11}$$

$$m_3\ddot{x}_3 = +k_3(x_2 - x_3) - k_4 x_3 + c_3(\dot{x}_2 - \dot{x}_3) - c_4\dot{x}_3 \tag{15.12}$$

Arranging Equations 15.10–15.12 in matrix form, we obtain

$$[m]\begin{Bmatrix} \ddot{x}_1 \\ \ddot{x}_2 \\ \ddot{x}_3 \end{Bmatrix} + [c]\begin{Bmatrix} \dot{x}_1 \\ \dot{x}_2 \\ \dot{x}_3 \end{Bmatrix} + [k]\begin{Bmatrix} x_1 \\ x_2 \\ x_3 \end{Bmatrix} = \begin{Bmatrix} f_1(t) \\ 0 \\ 0 \end{Bmatrix} \tag{15.13}$$

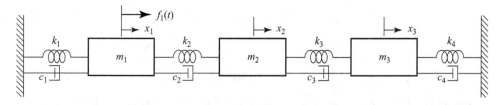

Figure 15.13 Three DOF model of constrained layer damping treatment.

where the mass, damping and stiffness matrices are

$$[m] = \begin{bmatrix} m_1 & 0 & 0 \\ 0 & m_2 & 0 \\ 0 & 0 & m_3 \end{bmatrix}, \quad [c] = \begin{bmatrix} c_1 + c_2 & -c_2 & 0 \\ -c_2 & c_2 + c_3 & -c_3 \\ 0 & -c_3 & c_3 + c_4 \end{bmatrix},$$

$$[k] = \begin{bmatrix} k_1 + k_2 & -k_2 & 0 \\ -k_2 & k_2 + k_3 & -k_3 \\ 0 & -k_3 & k_3 + k_4 \end{bmatrix}$$

Similarly, the impedance method may be attempted to solve the set of differential equations of Equation 15.13 analytically, but the numerical method based on Runge–Kutta techniques is employed in this chapter. The parameters used in this analysis are given as:

$m_1 = 2$ kg	$m_2 = 1$ kg	$m_3 = 2$ kg	$F_1 = 10$ N
$k_1 = 10 \times 10^3$ N/m	$k_2 = 500$ N/m	$k_3 = 500$ N/m	$k_4 = 10 \times 10^3$ N/m
$c_1 = 10$ N.s/m	$c_2 = 100$ N.s/m	$c_3 = 100$ N.s/m	$c_4 = 10$ N.s/m

For an unconstrained layer damping treatment, the free vibration and forced vibration response of two-layer composite material are obtained and plotted in Figures 15.14 and 15.15, respectively. Note that free vibration occurs when $f_1(t) = 0$. For the forced vibration, it is assumed that the vehicle moves with a speed of V = 40 mph or 18 m/s on a bumpy road. The excitation force is also assumed to be harmonic with the frequency of excitation of

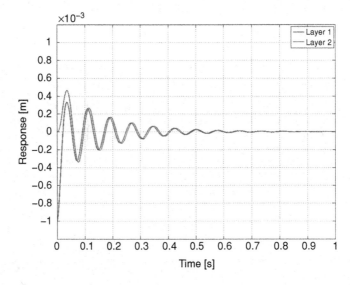

Figure 15.14 Free vibration response of two-layer composite panel in unconstrained damping treatment.

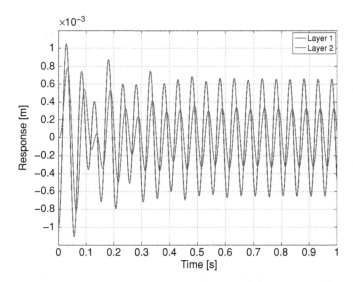

Figure 15.15 Forced vibration response of two-layer composite panel in unconstrained damping treatment.

20 Hz. As observed from Figure 15.14, the analysis shows that the response decays over time due to the damping treatment. Without the second layer, the effect of springs will be dominant, that is $c_2 \approx c_3 \approx 0$, and the response of panel will not be damped. By assuming a small amount of structural damping, the response will decay over a long period of time.

The benefit of damping treatment can be seen better through forced vibration analysis where there is external excitation on a two-layer panel. The forced vibration is generated by the motion of the vehicle on the road. As shown in Figure 15.15, the response of the second layer, which is transmitted to the other components and the passenger cabin, is much lower than the base layer in both transient- and steady-state responses. This helps reduce overall noise and improve fatigue resistance in the system. For instance, the amplitude of response in the second layer becomes 50% less than that of the first layer. This is seen in Figure 15.15.

For constrained layer damping treatment, the free vibration and forced vibration responses of a three-layer composite material are obtained and plotted in Figures 15.16 and 15.17, respectively. In free vibration analysis, as shown in Figure 15.16, the response of the third layer, which is transmitted to the other components and the passenger cabin, is improved. In comparison with the results shown in Figure 15.14, the overshoot and overall amplitude of response is less than that of a two-layer treatment although the decay time is slightly increased.

Further, from forced vibration analysis results shown in Figure 15.17, it is observed that the response of the third layer is improved substantially in both transient and steady-state modes. In steady-state mode, the amplitude of response is about 40% less than the base layer. In comparison with the forced vibration results of unconstrained damping treatment in Figure 15.15, although the transient time is increased slightly, the amplitudes of response are reduced effectively.

These analyses suggest that the damping treatment, in general, is effective in reducing and controlling the noise and vibration of body panels, which then results in better resistance

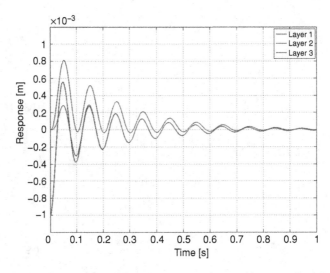

Figure 15.16   Free vibration response of three-layer composite panel in constrained damping treatment.

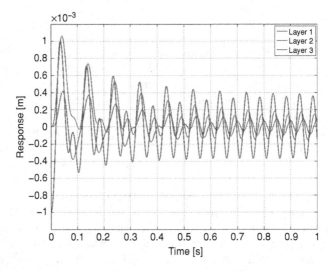

Figure 15.17   Forced vibration response of three-layer composite panel in constrained damping treatment.

against fatigue. The three-layer constrained damping treatment gives us better results than the unconstrained damping treatment; however, there is always a trade-off with manufacturing cost.

15.5   Conclusion

The use of composite materials has quickly become ubiquitous in the market for high-performance vehicles. In this chapter, the vibration analysis of the recent typical composite materials used in the body panels of automobiles is conducted for two separate case studies. The main goal of this chapter is to give readers a pathway for modelling, analysis and controlling vibration and noise in any segment of the body shell of a vehicle. A structural panel made of a conventional metal with a low amount of structural damping results in a system with a nearly undamped forced vibration – the total response will cause large oscillations that may lead to resonance for the given body panel. In contrast, it should be noted that if the structural panels are made of an appropriate composite material with adequate damping coefficient or damping treatment, the noise and vibration may become tolerable and consequently prevent passenger discomfort as well as structural fatigue.

References

[1] Subramanian, S., Surampudi, R., Thomson, K.R. and Vallurupalli, S. (2004) Optimization of damping treatments for structure borne noise reduction. *Journal of Sound and Vibration*, **38**, 14–17.

[2] Rao, M.D. and Gruenberg, S. (2002) Measurement of equivalent stiffness and damping of shock absorbers. *Experimental Techniques*, **26**(2), 39–42.

[3] Vinson, J.R. and Chou, T.W. (1975) *Composite Materials and Their Use in Structures*, John Wiley & Sons, Inc., New York, ISBN 978-0470908402.

[4] Reddy, J.N. and Miravete, A. (1995) *Practical Analysis of Composite Laminates*, 1st edn, CRC, New York, ISBN 978-0849394010.

[5] Abbas, W., Abouelatta, O.B., El-Azab, M.S. and Megahed, A.A. (2010) Application of Genetic Algorithms to the Optimal Design of Vehicle's Driver- Seat Suspension Model. Proceedings of the World Congress on Engineering 2010, June 30 – July 2, London, UK.

[6] Akanda, A. and Goetchius, G.M. (1999) Representation of Constrained/Unconstrained Layer Damping Treatments in FEA/SEA Vehicle System Models: A Simplified Approach. Noise and Vibration Conference and Exposition, 17–20 May, Traverse City, Michigan.

[7] Trindade, M.A. (2007) Optimization of passive constrained layer damping treatments applied to composite beams. *Latin American Journal of Solids and Structures*, **4**, 19–38.

[8] Shen, I.Y. and Reinhall, P.G. (2001) Surface Damping Treatments: Innovation, Design, and Analysis. Technical Report, 17 June, Seattle, University of Washington.

[9] Brockman, A., Clough, J., Mulji, K. and Hunter, N. (2001) Identifying the Effects of Stiffness Changes in a 5-DOF System. American Society for Engineering Education Annual Conference and Exposition, Los Alamos, New Mexico, USA.

[10] Choi, C.H., Park, S.S., Ahn, K.W. and Rhee, J.E. (2000) Low Pressure Molding Compound Hood Panel for a Passenger Car. FISITA World Automotive Congress, 12–15 June, Seoul, Korea.

[11] Park, C.K., Kan, C.D., Reagan, S.W. *et al.* (2011) Crashworthiness and Sensitivity Analysis of Structural Composite Inserts in Vehicle Structure. Eighth European LS-DYNA Users Conference, 23–24 May, Strasbourg, France.

[12] Singh, R., Kim, G. and Ravindra, P.V. (1992) Linear analysis of automotive hydro-mechanical mount with emphasis on decoupler characteristics. *Journal of Sound and Vibration*, **158**(2), 219–243.

[13] Nguyen, Q.S. (2000) *Stability and Nonlinear Solid Mechanics*, 1st edn, John Wiley & Sons, Inc., New York, ISBN 978-0471492887.

[14] Marshall, I.H. (1985) *Composite structures 3*, Elsevier Applied Science, New York, ISBN 978-0853343783.

[15] Donaldson, S.L. and Miracle, D.B. (2001) *ASM Handbook Composites Volume 21*, 10th edn, ASM International, New York, ISBN 978-0871707031.

[16] Kim, J.K. and Mai, Y.W. (1998) *Engineered Interfaces in Fiber Reinforced Composites*, 1st edn, Elsevier Applied Science, New York, ISBN 978-0080426952.

[17] Voyiadjis, G.Z. and Kattan, P.I. (2005) *Mechanics of Composite Materials with MATLAB*, Springer, Heidelberg, ISBN 978-3540243533.

[18] Jones, D.I.G. (2001) *Handbook of Viscoelastic Vibration Damping*, 1st edn, John Wiley & Sons, Inc., New York, ISBN 978-0471492481.

16

Composite Materials for Automotive Braking Systems

David C. Barton

School of Mechanical Engineering, University of Leeds, Leeds, LS2 9JT, UK

16.1 Introduction

This chapter addresses the use of composite materials in the design and manufacture of friction brakes for automotive applications. All road vehicles must have a safe and reliable braking system. Indeed it can be argued that the brakes are the single most important safety critical item on any vehicle. Although electrical and hydraulic brakes have found applications, especially for regenerative braking systems, the vast majority of road vehicles still rely on friction brakes where basically the kinetic energy of the vehicle is converted to heat via dry sliding friction between a rotating disc or drum and a stationary brake pad or shoe lined with a proprietary friction material. A typical vented brake disc is shown in Figure 16.1a and the corresponding brake pad is in Figure 16.1b. For a full description of the performance requirements of modern braking systems, the reader is referred to Ref. [1] from which some of the material contained in the present chapter is taken with the permission of the publishers.

In any conventional foundation brake, the relative rotation of the so-called "friction pair" under the action of the brake system actuation force is responsible for generating the frictional retarding torque required to decelerate the vehicle. Most friction pairs consist of a hard, usually metallic, rotating component and a relatively compliant "friction" material in the form of a lined brake pad or shoe, as indicated in Figure 16.1. The materials requirements for the rotating and stationary components of the friction pair are therefore quite different, as discussed below.

The front brake assembly is considered specifically in this chapter since this is generally the higher duty brake on any road vehicle. However, much of what follows applies equally to the rear brake, particularly if this is a disc rather than a drum. This chapter does not consider the brake calliper explicitly although there have been composite materials developments here, such as the use of metal matrix composites (MMCs) with selective reinforcement to

Advanced Composite Materials for Automotive Applications: Structural Integrity and Crashworthiness,
First Edition. Edited by Ahmed Elmarakbi.
© 2014 John Wiley & Sons, Ltd. Published 2014 by John Wiley & Sons, Ltd.

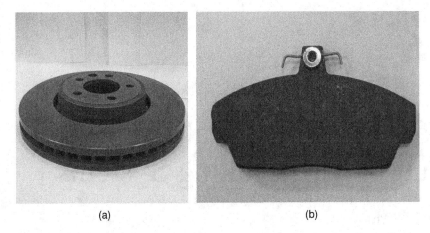

(a)   (b)

Figure 16.1   (a) Ventilated rotor. (b) Brake pad. Reproduced with permission from Ref. [1].

increase the specific stiffness of lightweight alloy racing callipers. However, here we concentrate on the rotor and stator materials since these represent the heart of any friction braking system. We start by considering the materials requirements for the rotor as this is where most of the recent composite materials developments have taken place.

16.2   Materials Requirements for Brake Rotors

Any rotor material must be sufficiently stiff and strong to be able to transmit the frictional torque to the hub without excessive deformation or risk of failure. However, the stresses arising from thermal effects are much higher than purely mechanical stresses and are more likely to give concerns over disc integrity. Thus the rotor material should have high volumetric heat capacity ($\rho.C_p$) and good thermal conductivity (k) in order to absorb and transmit the heat generated at the friction interface without excessive temperature rise. Further, the maximum operating temperature (MOT) of the material should be sufficiently greater than the maximum in-service surface temperature to ensure integrity of the rotor even under the most severe braking conditions. Ideally the rotor material should have a low coefficient of thermal expansion (α) to minimise thermal distortions such as "coning" of the rubbing surface. It should also have low density (ρ) to minimise the unsprung mass of the vehicle in order to improve ride and reduce road damage. It should be resistant to wear since generally it is far easier and cheaper to replace the friction pads or shoes rather than the rotor itself. Finally and most importantly, the rotor should be cheap and easy to manufacture.

The overwhelming majority of rotors for conventional automotive friction brakes are manufactured from grey cast iron (GCI). This material, also known as flake graphite iron, is cheap and easy to cast and machine in high volumes. It is a true composite in the sense that it is composed of two distinct materials (pearlitic iron and graphite flakes), the relative composition and exact form of which can be tailored to give distinct properties not found in either constituent material. Further discussion of these properties and how they can be optimised for a given application is given in Section 16.3 below.

Table 16.1 Typical physical properties of four candidate disc materials.

Disc material	ρ kg m^{-3}	Cp J kg^{-1} K^{-1}	$\rho.\,Cp$ kJ m^{-3} K^{-1}	K Wm^{-1} K^{-1}	$\alpha \times 10^{-6}$ K^{-1}	MOT (°C)
High carbon cast iron	7150	4316	3132	50	10	ca. 700
Generic 20% SiC-reinforced Al MMC	21 600	1600	2240	1160	17.5	ca. 450
Carbon–carbon composite	1750	1000	1750	40–150	0.7	> 1000
Carbon-reinforced SiC CMC	2300	1000	2300	10	4	> 900

Although GCI is a cheap material with good thermal properties and strength retention at high temperature, its density is high and, because section thickness must be maintained for both manufacturability and performance, cast iron rotors are heavy. Currently there are significant incentives to reduce rotor weights in order (a) to reduce emissions by improving the overall fuel consumption of the vehicle and (b) to aid refinement and limit damage to roads by reducing the unsprung mass. Thus much effort has been directed at investigating lightweight alternatives to cast iron. Three such alternatives which have received serious attention are aluminium metal matrix composites (MMCs), carbon–carbon (C–C) composites and carbon-reinforced ceramic matrix composites (CMCs), typical properties for each of which are displayed in Table 16.1 together with corresponding properties for a high-carbon cast iron [2].

When considering alternative materials or designs for disc brakes, reference can be made to the so-called "bucket and hole" analogy in which the rate of water flow into the bucket is taken to represent the heat flow into the disc and the height of the water level in the bucket represents the maximum temperature of the disc surface (Figure 16.2). A hole in the bucket represents the ability of the disc to lose heat to the surroundings. The volume of the bucket is therefore the heat capacity of the disc whilst the height is the MOT of the disc material. The question then is how close does the level of water in the bucket get to overflow?

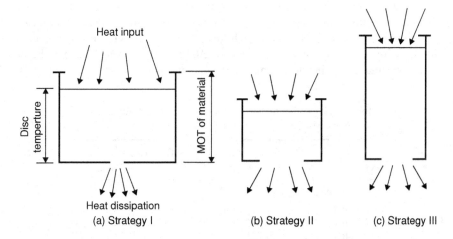

Figure 16.2 Different strategies for thermal management of friction brakes.

Consideration of this "bucket and hole" analogy with reference to the typical material properties in Table 16.1 allows three distinct strategies for brake rotor materials to be identified:

Strategy I: large diameter and relatively deep bucket with small hole (see Figure 16.2a). This implies a high volumetric heat capacity to store heat during braking and a relatively high MOT but only moderate conductivity of heat away from the rubbing surfaces. Current grey cast iron discs represent such a system but some steels may also meet these criteria.

Strategy II: smaller diameter and relatively shallow bucket but large hole (see Figure 16.2b). This implies smaller volumetric heat capacity and a relatively low MOT. Hence it is important to have high conductivity to transfer heat to other parts of the rotor and thence the surroundings in order to prevent temperature build-up at the rubbing surfaces. It can be seen from Table 16.1 that light alloy MMCs may fit this requirement.

Strategy III: even smaller diameter bucket but much deeper with moderately sized hole (see Figure 16.2c). This implies a material with high MOT, which can be allowed to run much hotter than current designs and so lose significant amounts of heat by radiation as well as more moderate amounts by conduction/convection. Carbon-reinforced composites are a possibility here either in the form of carbon–carbon or ceramic matrix composites.

Further details of each of these main classes of materials are given in the following sections, starting with the conventional grey cast iron.

16.3   Cast Iron Rotors

Grey cast iron (GCI) has a good volumetric heat capacity due mainly to its relatively high density and a reasonable conductivity due largely to the presence of the graphite (or carbon) flakes. The coefficient of thermal expansion is relatively low and the material has a MOT well in excess of 700 °C (but note that martensitic transformations at high temperatures can lead to hot judder problems). Although the compressive strength is good, the ultimate tensile strength is relatively low and the material is brittle and prone to micro-cracking in tension. As the proportion of flake graphite in GCI is increased, the tensile strength reduces but thermal conductivity increases, as shown in Table 16.2. Note that spheroidal graphite iron (SGCI) has

Table 16.2 Ultimate tensile strength and conductivity of some common cast irons.

Grade	Tensile strength (MPa)	Thermal conductivity at 300 °C (W/m K)
400/116 SGCI	400	36.2
250 GCI	250	45.4
200 GCI	200	416.1
150 GCI	150	50.5

a higher tensile strength than GCI but a much-reduced conductivity, which explains why it is rarely used for brake rotors.

Currently GCI grades used for disc brakes fall into two categories reflecting two different design philosophies:

> *Medium carbon GCI (e.g. grade 220):* These irons are used for small diameter discs such as on small and medium sized passenger cars. Such discs will run hot under extreme conditions, and good strength and thermal crack resistance at high temperatures are therefore required.

> *High carbon GCI (e.g. grade 150):* These grades tend to be used for larger vehicles where space constraints are not as limited. Discs are larger and, with the improved conductivity due to the high carbon content, will run cooler. Strength retention at high temperature is therefore not as critical and manufacturability improves with the higher carbon content.

Alloying elements can be applied to all grades of cast iron with the general effect of improving strength but at the expense of thermal properties and manufacturability. The most commonly used elements and their effects are as follows:

- Chromium – increases strength by stabilising pearlitic matrix at high temperatures (preventing martensitic transformations) but tends to promote formation of bainitic structures which cause casting/machining difficulties and can reduce pad life.
- Molybdenum – similar to chromium.
- Copper – increases strength without causing manufacturing difficulties.
- Nickel – as for copper but more expensive.
- Titanium – reported to influence friction performance but rarely used at significant levels.

However, as with most composites, it is the volume fraction and morphology of the second phase constituent (in this case, graphite) that allows the properties to be optimised for a given braking application.

16.4 Carbon Composite Rotors

16.4.1 Carbon–Carbon Composites

The first alternative rotor material to cast iron to receive serious attention was a carbon–carbon (C–C) composite, that is some form of carbon fibre embedded in a carbon matrix. Originally developed for both the rotor and stator of enclosed aircraft brakes where energy densities are very high and high temperatures are reached very quickly during aborted take-offs and landings, C–C composites were adopted as the standard braking material for F1 cars in the 1960s and were also introduced as a suitable clutch material to cope with the very demanding conditions at the start of a race.

It can be seen from Table 16.1 that carbon–carbon composites have the lowest density of any candidate material and can have a high thermal conductivity depending on the orientation of the carbon fibres. Their MOT is also very high, raising the possibility of using thin rotors, which run much hotter and lose heat by radiation as well as by conduction/convection. Also the very low coefficient of thermal expansion of carbon minimises thermal distortions. Thus there is the potential for very significant weight savings with carbon–carbon composite discs. However, currently available material has a poor low temperature friction performance and moreover is much more expensive to produce than metallic or ceramic matrix alternatives. Therefore the automotive applications of C–C composites are likely to remain confined to high performance race cars for the foreseeable future.

16.4.2   Ceramic Matrix Carbon Composites

As an alternative to pure carbon–carbon composite, carbon fibre reinforced silicon carbide ceramic composite (C-SiC) rotors have received much attention over the last few years. Their thermal capacity properties are similar to C–C but their thermal conductivity is relatively low. However, their maximum operating temperature is high (at least 900 °C) and they can be allowed to run much hotter and thereby lose heat from the rubbing surfaces by convection and radiation rather than by conduction to neighbouring parts. The problem has been in establishing an economically viable production route. The currently preferred process involves impregnating carbon fibre mats with polymer resin to form PMC "prepregs". These prepreg sheets can then be moulded into near net shape products (e.g. a brake disc) before pyrolysis (thermal degradation) of the resin to form a porous glassy carbon structure which can be machined closer to the final shape if necessary. Finally, the preform is impregnated with liquid silicon which reacts with the carbon to form SiC, the remaining silicon solidifying to form a Si/SiC matrix reinforced by carbon fibres and ready for final machining.

The carbon can be in the form of short randomly oriented fibres or continuous unidirectional fibres. Usually these are arranged in a layered structure in a brake disc with low volume fraction short fibre layers acting as a corrosion resistant friction surface and high volume fraction unidirectional layers at varying angles of orientation forming a stiff, strong substrate. The carbon fibres give the composite increased toughness and damage tolerance whilst the ceramic matrix provides the wear and oxidation resistance.

The tribological performance of C-SiC ceramic discs has been studied and it has been found that, with suitable pad compositions, high friction levels can be achieved with very little fade at elevated temperatures leading to reduced vehicle stopping distances from high speed. It is found that a stable tribo-layer is quickly established on the ceramic disc rubbing surface and this seems to consist crucially of iron oxide formed from ferrous ingredients in the brake pad, but hard silica nanoparticles and wear debris from the pad are also constituents of the tribo-layer [3]. The symmetric design of the ceramic rotor (with no "top hat" structure) and the low thermal expansion coefficient mean that there is little thermal deflection (minimal "coning" of the rubbing surface) and no reports of problems of disc thickness variations or judder.

Carbon-reinforced CMC discs have been introduced as options (at a considerable cost premium) on exotic road cars such as the Porsche 911 Turbo, the Mercedes CL55 AMG F1 and the Bentley Continental GT for which significant improvements in braking performance over cast iron discs are claimed with an overall brake system weight reduction of some 50%. The challenge now is for further reductions in the cost of manufacture by scaling up the process and the major European manufacture (Brembo SGL Ceramic Brake Systems) is aiming for an annual production capacity in excess of 100 000 rotors which should bring the costs down considerably. However, it still seems unlikely that such rotors will find their way onto mass-produced passenger cars in the near future.

16.5   Light Alloy Composite Rotors

Aluminium MMCs typically incorporate 10–30% by volume silicon carbide particle reinforcement within a silicon-containing alloy matrix. The resulting composite has a much lower density than cast iron and a significantly improved conductivity. Thus the thermal diffusivity ($k/\rho.c_p$) is much higher, which opens the possibility of lighter discs running cooler by being able to rapidly conduct heat away from the friction interface. However, aluminium MMCs have a low MOT (ca. 450 °C) and there are serious consequences if this MOT is exceeded, since complete surface disruption may then occur leading to extremely rapid pad wear. In theory, higher reinforcement contents or alternative reinforcing materials (e.g. alumina) could be used to increase the MOT, but the former cause severe casting difficulties whilst alumina reinforcement results in poorer thermal properties. Recently, good results were claimed for MMC rotors produced using spherical "fly ash" particles as the reinforcement and a pressure die casting technique [4].

Generally, the reinforcement of the alloy matrix can come in the form of particles, short fibres/whiskers or continuous fibres although the latter are less usual. Particles and short fibres can be mixed in with the molten alloy and cast using conventional techniques (for small reinforcement fractions) or by more advanced techniques such as squeeze casting. A continuous fibre preform can also be infiltrated with molten metal as for the SiC ceramic matrix discs. A powder metallurgy production route is also possible whereby the raw ingredients are milled into powders and consolidated through the usual powder compaction and sintering processes. Finally, a spray compaction route was recently developed whereby small SiC particles are spray injected together with the molten alloy into a mould. The latter is used to produce MMC discs from a high Si-containing alloy ($AlSi_{20}Fe_5Ni_2$) which has increased elastic modulus, hardness and wear resistance at elevated temperatures up to 500 °C compared with more conventional low or non Si-containing alloys. This process also gives a good distribution of the SiC particles and is claimed to be cost effective [5].

16.6   Evaluation of Composite Disc Materials

Ultimately, any new brake material or design must be validated by experimental trials on actual vehicles to allow accurately for model-specific parameters, such as the effect of body trim on rotor cooling. However, much can be learnt about potential new rotor materials or designs by numerical simulations of critical brake tests using finite element (FE) analysis. Such techniques require the rotor and/or stator geometry to be broken down into a number of small non-overlapping regions known as elements, which are assumed to be connected to one another at certain points known as nodes. A 2D axisymmetric finite element idealisation can be used as a first approximation, but for more accurate simulation of the heat flow and stresses a 3D model is desirable, such as the model of a back-vented brake disc, pads and pistons shown in Figure 16.3 in which 16-noded hexahedral elements were used throughout. Note

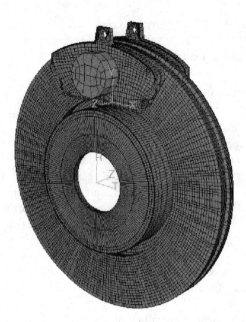

Figure 16.3   3D finite element model of back-vented brake disc.

that, in order to accurately simulate the heat loss from the rotor, it is sometimes necessary to include the wheel hub and other components in the model although this is not shown in Figure 16.3.

The heat input to the system is estimated from theoretical consideration and applied over the rubbing surface. The heat loss to the surroundings is specified by convective and sometimes radiative heat transfer conditions along relevant boundaries of the model. The temperatures predicted by a thermal analysis can be used as input conditions to a structural analysis in order to predict thermal deformations and stresses. If the pad is included in the model, the contact pressure distribution (and hence the distribution of heat input) can be estimated, leading to the possibility of a fully coupled thermal–structural analysis [6]. Note however that other work [7] has highlighted the importance of local contact conditions in determining the true surface temperatures at the interface and allowance should therefore be made for real surface topographies, including the effects of thermal distortion and wear.

In addition to details of geometry and material properties, accurate data on heat loss to other components and to the atmosphere are vital to allow accurate predictions of rotor temperatures using FE methods. So-called "cooling tests" conducted on actual vehicles fitted with representative brake rotors carrying rubbing or embedded thermocouples can generate such data. The rotor surface is first heated to a predetermined temperature by dragging the brakes and then allowed to cool whilst the vehicle is driven at constant velocity. By comparing the experimental rate of cooling with that predicted by the FE simulation for different boundary conditions, optimised heat transfer coefficients can be derived which are then assumed to apply for different rotor materials and are factored for the varying air stream velocity under different test conditions.

Two very different vehicle brake tests are often simulated to critically examine the maximum temperatures and integrity of new rotor materials or designs: (i) a long slow Alpine descent during which the brakes are dragged and the vehicle is subsequently left to stand at the end of the descent, (ii) a repeated high-speed autobahn stop with the rotor allowed to cool only moderately between stops. The former test determines the ability of the design to limit temperature build-up in the rotor by heat transfer to the atmosphere whilst the high-speed repeated stop examines the ability of the rotor material to withstand repeated thermal cycling and the ability of the friction pair to resist "fade" under these severe conditions.

Although FE analysis is a very powerful design tool in the initial stages of materials/design evaluation, friction performance cannot easily be predicted by this approach alone and there remains a requirement for dynamometer testing to determine the fade and wear characteristics of every new friction pair. The dynamometer can either be a full-scale device or a small sample rig in which the geometry and loading conditions are scaled to give an accurate representation of the actual brake. These tests not only give data on friction performance over a wide range of conditions but can also be used to determine the MOT of the pad and rotor materials by progressively increasing the temperature at the rubbing interface until some form of failure occurs.

As an example of the application of the above evaluation strategies, thermal FE analysis was combined with Taguchi techniques by Grieve *et al.* [2] to optimise the material and design parameters of an aluminium MMC passenger car disc. The heat transfer coefficients for the 3D FE model shown in Figure 16.3 were first derived from cooling test data on a test vehicle fitted with cast iron rotors and validated by comparing the results from simulation of a typical Alpine descent with those measured using rubbing thermocouples on the same test vehicle. The predicted maximum rotor temperature at 639 °C was very close to the measured temperature of 636 °C, giving confidence in the accuracy of the FE model results.

The two critical brake tests were then simulated: namely, a long slow Alpine descent and a sequence of high-speed, repeated autobahn stops. A generic MMC material model was derived based on the relevant properties of a range of available composites including both silicon carbide and alumina reinforced MMCs. The simulation results demonstrated the importance of thermal conductivity in the Alpine descent test and the dominance of volumetric thermal capacity in the repeated autobahn stop. The simulations also enabled key design parameters of the disc, notably the vent width and cheek thickness, to be optimised (Figure 16.4). Despite this, subsequent dynamometer and in-car tests of the optimised design indicated insufficient margin of the peak operating temperature over the maximum allowable surface temperature of even the high SiC content aluminium MMC selected for the design. The exercise clearly

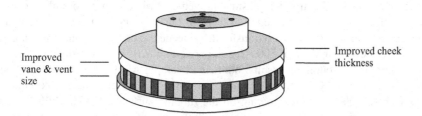

Figure 16.4   Modifications to disc design for MMC materials.

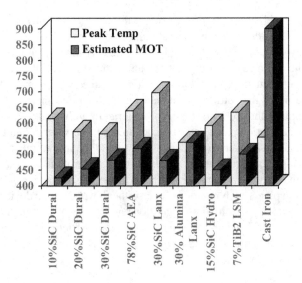

Figure 16.5   Predicted peak temperatures after a single snub from 60.0 to 2.716 m/s at 0.16 g.

demonstrated the advantage of cast iron over currently available MMCs (Figure 16.5), which suggests that the latter lightweight alternatives are not likely to prove acceptable for the majority of passenger car applications.

16.7   Surface Engineering of Light Alloy Brake Discs

It is apparent from the above that the main problem with light alloy and MMC brake rotors is the rapid deterioration of the rubbing surface due to high surface temperatures under arduous braking cycles. In other respects (strength, thermal conductivity, high specific heat capacity, processability, costs) light alloy rotors offer distinct advantages over more exotic alternatives. Therefore, it may be possible to utilise these good properties if the surface of the rotor can be protected by a temperature- and wear-resistant coating. Such a coating may also offer advantages for the friction performance of the brake (higher or more stable coefficient) and this must also be carefully considered in the design of any coating system.

One example from the recent past is a Brembo 30% SiC aluminium MMC motorcycle disc that was coated with 0.2 mm thick plasma-sprayed ceramic coating. Not only did this allow the SiC reinforcement to be reduced to 20% by volume, it enabled the use of standard pads with no running-in period (since no transfer layer needed to be formed) and gave much better wet performance with a MOT of 500 °C, compared with 420 °C for the standard MMC.

Based on these encouraging results, both small sample Al alloy discs and a full-size Al-MMC ventilated rotor were surface-treated using the plasma electrolytic oxidation (PEO) process to give a hard alumina coating on the rubbing surfaces [8]. The full-size disc was subjected to an arduous brake duty cycle on a brake dynamometer (see Figure 16.6) and the surface temperature of the disc was measured with a rubbing thermocouple. Plots such as in Figure 16.7 show the increased surface temperature of the coated disc compared to the

Figure 16.6   PEO coated aluminium MMC disc mounted on dynamometer.

monolithic aluminium MMC of identical design. The somewhat lower temperatures of a very similar geometry cast iron disc again illustrate the excellent thermal capacity properties of this latter material.

Under the most arduous test conditions, the surface temperature of the coated disc was raised to almost 500 °C, a temperature which would have destroyed the surface of the monolithic Al-MMC equivalent. On inspecting the coated disc after test, a definitive transfer layer was observed as was some evidence of surface pitting, probably as a result of the PEO processing, but there was no sign of any surface softening or disruption (Figure 16.8). This gave confidence

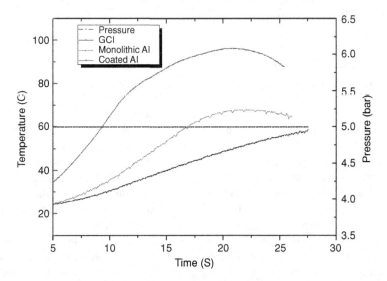

Figure 16.7   Rubbing surface temperatures for different discs tested on dynamometer.

Figure 16.8   PEO coated MMC disc after testing showing transfer layer.

that such a coated disc has potential applications in light passenger car brakes, especially in electric or hybrid vehicles [8].

16.8   Friction Material

16.8.1   Material Requirements

The brake pad or shoe represents the stationary part of the foundation brake assembly. Normally, a proprietary composite friction material is bonded to a steel backing plate or shoe platform (see Figure 16.1b). The primary function of the friction material is generally considered to be the production of a stable and predictable coefficient of friction to enable reliable and efficient braking of the vehicle over a wide range of conditions. In fact, it is the combined tribological characteristics of both rotor and stator materials (i.e. the "friction pair") which are responsible for the generation of the frictional torque. As for the rotor, the friction material must have sufficient structural integrity to resist mechanical and thermal stresses. This is particularly important for the bond between the friction material itself and the steel structure, which supports it, as a complete failure here could have disastrous consequences. The friction material should have a relatively high MOT to prevent thermal degradation of the surface although, due to the nature of its composition, the MOT of the pad material will always be lower than that of the disc. A low conductivity for the pad or shoe material is desirable to minimise the conduction of heat to other components of the system, in particular to the hydraulic fluid. The material should be reasonably wear resistant but not excessively so, since wear can be beneficial in promoting a uniform contact pressure distribution and preventing "hot spotting". Likewise, the elastic modulus of the material should be relatively low to give good conformity with a roughened or thermally distorted rotor surface. Finally, as for the rotor, the friction material should be cheap and easy to manufacture.

16.8.2   Overview of Friction Material Formulations

The friction material formulated to meet the above stringent requirements is invariably a complex composite consisting of a variety of fibres, particles and fillers bonded together in a polymeric matrix which is usually phenolic based. For many years, asbestos fibres were an important constituent of friction materials due to their excellent thermal and friction properties. For health and safety reasons, asbestos has now been replaced by other less harmful fibres, for example Kevlar, leading to the concept of non-asbestos organic (NAO) friction materials. Other important constituents of modern friction materials include abrasives (usually in the form of hard particles such as zirconium silicate) and lubricants designed to stabilise the level of friction under certain operating conditions. Metal particles and fibres are often added to give desirable mechanical and thermal properties and a particularly high metallic content gives rise to the concept of semi-metallic friction materials specified for certain demanding duty cycles. The exact composition of any friction material must be tailored to the application, and knowledge of the formulation is proprietary to the supplier. However, a typical composition of NAO friction material is shown in Table 16.3. For a more complete description of the use of composites in modern automotive friction materials, the reader is referred to Ref. [9].

16.8.3   Evaluation of Friction Material Performance

As stated above, a friction material for automotive braking applications must primarily maintain a stable level of friction over a wide range of vehicle operating conditions. It must also possess

Table 16.3   Typical composition of pad friction material.

Constituent	Percentage by weight
Whiting (chalk)	316
Bronze powder	15
Graphite	10
Vermiculite	16
Phenolic resin	16
Steel fibres	6
Rubber particles	5
"Friction dust"	5
Sand	3
Aramid fibres	2

adequate strength and wear resistance, suitable thermal properties and moreover should be cheap and easy to produce. The formulation of a suitable composite material to meet these requirements is a time-consuming specialised business and to date has been carried out on a mainly empirical basis using the results of numerous dynamometer tests on trial materials to arrive at a compromise solution. Indeed this formulation process has been considered by many to be more "black art" than science [10].

General composite design methodologies have some relevance to braking friction materials but, by and large, they do not address the fundamental tribological interactions that determine the behaviour of a friction pair. In fact, it is the surface properties of the friction material rather than its bulk properties that control these interactions. Elzey *et al.* [11] discussed this distinction between bulk and surface properties and argued that the latter are governed by area fractions of inclusions exposed at the surface rather than by the volume fractions responsible for bulk properties. However, when attempting to optimise the design of a friction material to meet the multi-faceted criteria required by the industry, both surface and bulk properties as well as non-physical parameters such as cost must be considered. Elzey *et al.* [11] proposed an intelligent selection of materials (ISM) approach to account systematically for the complex and often conflicting requirements of a modern friction material. They argued that friction and wear at the interface can be dealt with initially using the available theory of homogeneous materials subject to sliding contact, followed by the application of micro-mechanics to study multiphase (heterogeneous) materials. Their ISM methodology provides a framework to move from the micro-mechanical analysis to the overall characterisation and selection of the friction material.

Eriksson *et al.* [12] also recognised that friction at the rubbing interface is controlled by the areas of real contact between the rotor and friction material. In an earlier study, Bergman *et al.* [13] showed that the number of contact spots when extrapolated from measurements over a small area was in excess of 100 000 over a typical brake pad rubbing surface. These plateaux of contact are essentially formed around localised sites where harder particles are located near the surface. The rubbing interactions at the interface cause the softer constituents within the formulation to be worn away, leaving the harder inclusions exposed. The harder particles are therefore forced to carry a greater proportion of the load and it is the interactions of these particles both with the sliding counterface and with the underlying friction material matrix that control the dominant abrasive friction and wear mechanisms that operate at the interface.

16.9 Conclusion

The combination of a grey cast iron rotor and polymeric composite stator remains the friction pair of choice for most automotive braking applications. Although the technology is mature and costs are relatively low, the primary disadvantage of the cast iron rotor is its weight. Two alternative strategies for the design of braking materials have been identified and used to consider options for reducing rotor weights. One strategy is to use a relatively low conductivity but highly temperature-resistant material such as a carbon-reinforced CMC. Such materials give safe, predictable performance even under severe braking conditions but the costs currently prohibit their use except for the most exotic road and race cars. A much cheaper alternative relies on the high conductivity of light alloy MMCs to rapidly conduct heat away from the friction interface to prevent the build-up of high surface temperatures. However, simulation results coupled with dynamometer tests have demonstrated the difficulty such rotors may have in surviving certain critical brake tests. Light alloy discs coated with a wear- and temperature-resistant surface layer may offer an effective solution but a number of processing and performance issues remain to be solved. Simple theoretical models and more complex finite element analysis can be used to assess and rank alternative brake materials and to assess failure modes, but their final acceptance depends on thorough dynamometer and on-vehicle tests.

References

[1] Brooks, P.C. and Barton, D.C. (2001) Chapter 14: braking systems, in *An Introduction to Modern Vehicle Design* (ed. J. Hapian-Smith), Butterworth–Heinemann, New York, ISBN 07506 5044 3.

[2] Grieve, D., Barton, D.C., Crolla, D.A. *et al.* (1998) Design of a lightweight automotive disc brake using finite element and Taguchi techniques. *Proceedings of IMechE Part D*, **1998**, 245–254.

[3] Osterle, W., Deutsch, C., Rooch, H. and Doirfel, I. (2012) Friction Films on C-SiC Discs After Dynamometer Tests With Different Commercial Brake Pads, Paper EB2012-FM-01. EuroBrake2012, Dresden, Germany.

[4] Withers, G. (2012) Develoment of Low Cost High Performance Al MMC Materials for Brake And Clutch Applications, Paper EB2012-ABT-15. EuroBrake2012, Dresden.

[5] Storz, A. (2012) Development of Dispal-MMC Materials for Friction Applications, Paper EB2012-BC-15, EuroBrake2012, Dresden.

[6] Hassan, M.Z, Brooks, P.C. and Barton, D.C. (2009) A predictive tool to evaluate disc brake squeal using a fully coupled thermo-mechanical finite element model. *International Journal of Vehicle Design*, **51**, 124–142.

[7] Qi, H.S., Day, A.J., Kuan, K.H. and Rosala, G.F. (2004) A contribution towards understanding brake interface temperatures, in *Braking 2004: Vehicle Braking and Chassis Control* (eds D.C. Barton and A. Blackwood), PEP, London.

[8] Alsaif, M.A., Dahm, K.L., Shrestha, S. *et al.* (2010) Plasma Electrolytic Oxidation (PEO) Treated Aluminium Metal Matrix Composite Rotors for Lightweight Automotive Brakes. Proceedings of Sixth European Conference on Braking, Lille, France.

[9] Cox, R.L. (2012) *Engineering Tribological Composites: the Art of Friction Material Development*, SAE International, London, ISBN 978-0-7680-3485-1.

[10] Smales, H. (1994) Friction materials – black art or science? *Proceedings of IMechE Part D*, **209**, 151–157.

[11] Elzey, D.M., Vancheeswaren, R., Myers, S.W. and McLellan, R.G. (2000) Multi-criteria optimization in the design of composites for friction applications, in *Brakes 2000* (eds D.C. Barton and S. Earle), PEP, London.

[12] Eriksson, M., Bergman, F. and Jacobson, S. (2002) On the tribological contact in automotive brakes. *Wear*, **252**, 26–36.

[13] Bergman, F., Eriksson, M. and Jacobson, S. (1999) Influence of disc topography on generation of brake squeal. *Wear*, **225**, 621–628.

17

Low-Cost Carbon Fibre: Applications, Performance and Cost Models

Alan Wheatley[1], David Warren[2] and Sujit Das[2]

[1]*Department of Computing, Engineering and Technology, University of Sunderland, Sunderland, SR6 0DD, UK*
[2]*Oak Ridge National Laboratory, Oak Ridge, USA*

17.1 Current and Proposed Carbon Fibre Applications

Weight saving in automotive applications has a major bearing on fuel economy. It is generally accepted that, typically, a 10% weight reduction in an automobile will lead to a 6–8% improvement in fuel economy. In this respect, carbon fibre composites are extremely attractive in their ability to provide superlative mechanical performance per unit weight. That is why they are specified for high-end uses such as Formula 1 racing cars and the latest aircraft (e.g. Boeing 787, Airbus A350 and A380), where they comprise over 50% by weight of the structure However, carbon fibres are expensive and this renders their composites similarly expensive. Research has been carried out at Oak Ridge National Laboratories (ORNL; Tennessee, USA) for over a decade with the aim of reducing the cost of carbon fibre such that it becomes a cost-effective option for the automotive industry. Aspects of this research relating to the development of low-cost carbon fibre are reported in Chapter 3. In this chapter, the practical industrial applications of low-cost carbon fibre are presented, together with considerations of the performance and cost models which underpin the work.

The potential of carbon fibre reinforced polymer (CFRP) composites to reduce the weight of engineering components is tremendously high. CFRP components are typically 60% lighter than conventional steel and 30% lighter than aluminium components. CFRP composites are over three times stiffer than their glass fibre reinforced polymer (GFRP) counterparts. Therefore, compared to GFRP composites, CFRPs can increase the stiffness of the vehicle body, thereby further increasing the crash integrity of the passenger cell as well as improving ride comfort. CFRP use has, traditionally, been limited to luxury vehicles and racing cars in the automotive industry, along with finding adoption in some high-value applications within aerospace and sporting goods; however higher volume applications of carbon fibre composites in automotive structures are still only being planned. One example of a nearer term future application can be found at BMW, who will make extensive use of carbon fibre, primarily for the bodies of the battery-electric car i3 and plug-in hybrid i8 models on aluminium chassis in order to maximise performance and range while minimising the size of the battery packs [1].

Toyota Motor Corp. is currently using a lightweight CFRP body in its expensive Lexus LFA sports car and is planning to begin using it in more of its product line to cut weight and increase performance. According to Lexus, the main driver for the use of CFRP in the body is an improvement in the driving dynamics of the vehicle. Compared to aluminium the carbon fibre structure is a stiffer and higher resonant frequency structure yielding better handling ability for the driver [2]. Toyota is also planning to use a body frame made of carbon fibre composites in its 1/X concept car that will offer the same interior space as the Prius hybrid but only weigh one-third as much [3].

In the past two years, four major – and a growing host of minor – alliances between auto OEMs and/or tier suppliers and composite materials/processing companies have reflected automaker inclinations toward the use of CFRP in body and chassis structures in production passenger vehicles. Four major alliances announced recently include BMW and SGL Group, General Motors and Teijin, Toyota with Toray Industries and Fuji Heavy Industries and Ford and Dow Automotive Systems [4]. Even the Department of Energy's Oak Ridge National Laboratory has teamed with DOW (who is also in partnership with AKSA) to develop carbon fibre technologies [5]. The partnerships will provide increasing opportunities for the application of CFRP as an automotive material, but the fibre cost limits CFRP applications except in niche luxury and other low-volume platforms in the future. For those vehicle platforms OEMs are willing to pay a considerably higher premium for mass reduction, for example US\$6.25–17.50/kg saved for hybrid electric and extended range electric vehicles [4].

The latest 310 Vanquish by Ford Aston Martin due early next year will feature a full CFRP body, a 565-hp V-12 and a number of engineering revisions. Similarly, niche vehicles such as X-Power sports vehicles have demonstrated 70% weight reduction potential in body panels compared with steel. The Lamborghini Murcielago 12-cylinder sports car makes an extensive use of CFRP – 31% of total weight – including the body, floor, transmission tunnel, wheel housings and bumper section of the chassis.

CFRP applications in chassis components have so far been limited compared to body components, with the exception of the new Lexus LFA premium sports car where CFRP represents 65% of the chassis structure [6]. GMS Composites recently demonstrated a fabricated carbon/epoxy laminate cockpit chassis weighing only 80 kg for the Australian-designed and built two-seater concept car FR-1 [7]. Carbon Revolution, an Australian joint venture firm, recently developed a high-volume one-piece carbon fibre wheel, which is 40% lighter than existing products. The manufacturing process is based on an altered aerospace process for a high production volume and initially 6000 wheels are projected to be delivered for Porsche 911 by the end of 2012 [8].

Carbon/Amid/Metal Based Interior Structure with Multi-Material System Approach, a consortium supported by the German Ministry of Education and Research, is aiming to develop new lightweight structures for automotive seat applications using a multi-material approach [9]. This approach consists of a novel material distribution (i.e. 20% metals, 20% non-woven carbon fibre, 15% unidirectional carbon fibre, 45% injection moulded glass fibre-reinforced composites) for the final rib structure in a sports car front seat backrest structure. Both unidirectional tape and non-woven CFRP manufacturing technologies based on carbon fibres in PA12 matrix in the presence of activators and polymers are used. In addition, a thin seat concept coming from the mattress industry is bieng applied to a modern seat design, thereby facilitating the elimination of spacious foam parts and saving up to 20% in weight on a component level.

Carbon fibre production applications have often included roofs, hoods and other class A surfaces on low-volume sports cars. But an upcoming global production vehicle will have 75% of its body (including the hood, fenders and roof) comprised of carbon fibre. "It will be the first time that carbon fibre has been used this extensively on a base production car anywhere in the world," said Gary Lownsdale, Chief Technology Officer of *Plasan Carbon Composites* [10]. The Chevrolet Stingray will soon feature both a carbon fibre composite roof and a carbon fibre composite hood [11].

While the above applications are certainly good news for the potential applications of carbon fibre in automotive systems, they are mainly limited to high-cost, low-volume platforms due to the cost of the carbon fibre and the lack of resolution of some of the other technical obstacles previously outlined. For carbon fibre composites to make a significant dent in global petroleum consumption, they must become a standard material on high-volume vehicle platforms.

17.2   Carbon Fibre Polymer Composites: Cost Benefits and Obstacles for Automobiles

Oak Ridge National Laboratory recently completed an assessment of the cost effectiveness of a 25% vehicle weight reduction goal using a baseline 2002 midsize automobile cost model developed during 2011 [12]. The cost effectiveness of the proposed weight reduction goal was determined based on the vehicle retail price and vehicle life cycle costs of multiple lightweight material substitution scenarios in various body and chassis components to achieve the desired weight reduction goal. The baseline vehicle is compared to alternative vehicle scenarios involving lightweighting to evaluate the cost effectiveness of reducing the weight of a midsize passenger vehicle by 25%. The mass reduction analysis does not include the powertrain. The model's baseline vehicle was a 2002 average midsize vehicle manufactured by United States based OEMs having the specific vehicle characteristics as reported by EPA in its annual trend report of light-duty vehicles [13]. The 2002 midsize sedan considered as the baseline vehicle has a curb weight of 1477 kg, an interior volume of 3.25 m^3 (114.8 ft^3), a 2.9l, 185 HP, port fuel injected, V6 aluminium, four valves per cylinder, naturally aspirated engine, an automatic transmission, front wheel drive, 0–60 mph acceleration time of 9.4 s top speed of 216 kph (134 mph) and adjusted city, highway and combined fuel economy estimated to be 8.0 km/l (18.8 mpg), 11.3 km/l (26.5 mpg) and 10.0 km/l (23.3 mpg), respectively. For this analysis a carbon fibre cost of US$30.80/kg was assumed.

The three alternative lightweighting pathways considered for body and chassis components included: (i) lightweight metals (primarily advanced high strength steel, aluminium and magnesium), (ii) carbon fibre reinforced polymer (CFRP) composites and (iii) multi-materials.

Table 17.1 Potential lightweighting components using CFRP considered for 25% mass reduction scenarios.

Part	Baseline mass (kg)	Mass savings (%)	Mass savings (kg)	Comments and references (see also [12]
BIW	321	50	160.5	Refs. [6, 14]
Panels	60	50	30.0	BMW M3 CRT, Tesla Model S, Ref. [15]
Front/rear bumper impact module	5.5	55	3.0	Lamborghini Murcielago
Cradle	32.3	55	17.8	Ref. [16]
Suspension control arms	35	50	17.5	Ref. [16]
Steering knuckle	6.2	50	3.1	Ref. [16]
Instrument panel	29.4	57	16.8	Ref. [16]
Wheels	28.2	40	11.3	Ref. [17]
Seats	65.5	26	17.0	Ref. [9]
Total	583.1	47	277.0	

Lightweight metals and multi-materials are likely to be the most cost-effective options in the near future, but the focus lately has been on the latter approach which allows the best material for a given vehicle component application. Although CFRP is likely to be a long-term option, it is considered as one of the lightweighting options here to determine how its cost effectiveness could be improved in terms of vehicle life cycle cost by means of a trade off between lower fuel cost and higher carbon fibre and part manufacturing costs. The target 25% mass reduction is achieved through a combination of secondary and primary mass savings. Mass savings that result from lightweight material substitution at the component level vary for different technology scenarios because they have been estimated based on the component mass savings and the resulting secondary mass savings effect.

A total of nine components were considered for CFRP lightweighting, as shown in Table 17.1. The greatest mass reduction potential is anticipated in body in white (BIW) and demonstrated by the Focal Project 3 of the Automotive Composites Consortium (ACC), a collaborative, pre-competitive research and development partnerships of the three big United States OEMs and DOE. Focal Project 3 predicts a 66% weight reduction for the complete BIW, with total parts numbering less than 20, and based on the demonstrated weight savings potential of a CFRP fibre B-pillar [14].

The cost effectiveness of the target 25% mass reduction was considered initially in terms of lightweighting body and chassis components, followed by entire vehicle structure without the powertrain. The baseline line body and chassis structure weighed 656 kg and a 25% mass reduction would result in reducing the mass by 164 kg. Five material systems were chosen for consideration: advanced high-strength steel (AHSS), aluminium (Al) alloys, magnesium (Mg) alloys, carbon fibre reinforced polymer composites (CFC) and other material alternatives. In this analysis 53 vehicle components were considered for lightweighting. Based on a review of the literature, component level mass savings estimates from lightweight materials substitution were considered. References for each component substitution are included in the material system specific tables of the original report [12].

Based on the literature and for each of the material systems considered, a list of potential lighter weight components was constructed and the potential mass reduction associated with each component replacement was determined, along with the cost of replacing those components with the lighterweight alternative. Secondary mass savings were incorporated in the analysis. For each material system, the alternative material components were then prioritised, based on the cost penalty incurred (i.e. US$/kg saved). This information was then used to construct a table leading to a 25% mass reduction in the body and chassis structure for two vehicle options: (i) a light metals intensive solution and (ii) a carbon fibre composite intensive solution. Below are the results for the metals intensive (Table 17.2) and carbon fibre intensive (Table 17.3) body and chassis substitution scenarios.

Reducing the mass of the vehicle by 25% for the light metals solution carries a cost penalty of about US$0.48/kg due largely to several components being less expensive than the current incumbent materials once mass decompounding is considered. Even without mass decompounding the cost penalty for the optimised light metals solution is only US$1.94/kg. The carbon fibre intensive solution is significantly more expensive with the resultant cost being US$5.82/kg saved. Due to the low mass of carbon fibre composites, 160.5 kg can be saved with a carbon fibre intensive BIW which, with mass decompounding, can save a total of 184.6 kg. That substitution alone significantly exceeds the body and chassis mass reduction targets. Due to the reduced mass of the BIW, other components (not made from carbon fibre composites) could be down-gauged at significant cost savings which are included in the analysis. The down-gauging actually results in a cost saving for many of the components.

While the above scenarios considered achieving a 164 kg mass reduction by substitution of only body and chassis components, the same vehicle mass reduction can be achieved by the incorporation of alternative materials in non-body and non-chassis components of the vehicle. Additionally, since the validity of the extent of achievable secondary mass savings assumptions is debatable amongst industry experts, mass reduction scenarios that do not depend upon secondary mass savings were constructed. For both the light metals solution and the carbon fibre intensive solution, vehicle component combinations were considered in Tables 17.4 and 17.5 to achieve the same mass reduction without secondary mass savings. Using this approach, the cost penalty for an all metals solution is US$1.49/kg and for an all carbon fibre solution US$10.67/kg.

Finally, scenarios were considered for reducing the mass using all of the material options previously mentioned (Table 17.6). This was done for a 25% mass reduction in the body and chassis with and without mass decompounding and also for a 25% mass reduction in the entire vehicle (excluding the powertrain) with and without mass decompounding. Unfortunately, due to the cost of carbon fibre, no carbon fibre composite components were of sufficiently low cost to qualify for incorporation in these multi-material vehicle structures aimed at achieving the target mass reduction at minimum cost.

Vehicle lightweighting will incur a significant vehicle cost premium, in the range US$5.82–10.67/kg saved, particularly in the case of CFRP component substitutions, as noted above. The impacts of cost premium on vehicle retail price (due to OEM focus on sales and consumer heavy discounts on future fuel cost savings) can be estimated as 1.5 times higher [18]. Most vehicle operation cost savings will be from fuel cost, although higher financing, repair and end of life disposal cost due to vehicle lightweighting may reduce the overall vehicle life cycle cost savings only to a limited extent. Several recent studies have indicated that it is advisable to utilise mass difference rather than mass ratio when calculating the lightweight effect on fuel

Table 17.2 Lightweighting metals component substitutions (including estimated US$/kg saving) considered for 25% body and chassis mass reduction target scenarios (with mass decompounding).

Part	Baseline mass (kg)	LW mass (kg)		Mass saving (kg)	Baseline cost (US$)	LW cost (US$)		Cost saving (US$/kg)		Technology
		P	F			P	F	P	F	
BIW	321	224.7	206.8	114.2	1025	1213	1155	1.95	1.14	AHSS
Chassis										
Cradle	32.3	24.2	19.6	12.7	78	85	75	0.87	−0.22	AHSS
Corner suspension	134	125.2	105.9	28.1	627	649	589	2.50	−1.35	AHSS control arms
Braking system	70	68.6	67.8	2.2	294	303	301	6.16	3.08	Al master cylinder
Wheels and tyres	58	55.4	49.4	8.6	295	301	281	2.20	−1.60	AHSS Wheels
Steering system	41	38.7	38.3	2.7	350	351	349	0.48	−0.48	Al steering wheel column + Mg steering wheel
Total	656.3	536.8	487.8	168.5	2669	2901	2750	1.94	0.48	

Note: In all similar tables, P = final mass based on primary mass savings only, F = final mass savings including secondary mass savings.

Table 17.3 CFRP component substitutions (including estimated US$/kg saving) considered for 25% body and chassis mass target scenarios.

Part	Baseline mass (kg)	LW mass (kg)		Mass saving (kg)	Baseline cost (US$)	LW cost (US$)		Cost saving (US$/kg)		Technology
		P	F			P	F	P	F	
BIW	321	160.5	136.4	184.6	1025	2697	2455	10.42	7.75	CFRP
Chassis										
Cradle	32.3	32.3	26.1	6.2	78	78	69	NA	-1.45	Baseline
Corner suspension	134	134	108.1	25.9	627	627	554	NA	-2.81	Baseline
Braking system	70	70	69	1.0	294	294	291	NA	-2.52	Baseline
Wheels and tyres	58	58	50	8.0	295	295	271	NA	-3.05	Baseline
Steering system	41	41	40.4	0.6	350	350	347	NA	-5.12	Baseline
Total	656.3	495.8	430	226.3	2669	4341	3987	10.42	5.82	Baseline

Table 17.4 Lightweighting metal component substitutions for a 164 kg lighter vehicle incorporating components outside the body and chassis to achieve the same mass reduction goal (without mass decompounding).

Part	Baseline mass (kg)	LW mass (kg) P	F	Mass saving (kg)	Baseline cost (US$)	LW cost (US$) P	F	Cost saving (US$/kg) P	F	Technology
BIW	321	224.7		96.3	1025	1213		1.95		AHSS
Chassis										
Cradle	32.3	24.2		8.1	78	85		0.87		AHSS
Corner suspension	134	125.2		8.8	627	649		2.50		AHSS control arms
Braking system	70	68.6		1.4	294	303		6.16		Al master cylinder
Wheels and tyres	58	55.4		2.6	295	301		2.20		AHSS wheels
Steering system	41	38.7		2.3	350	351		0.48		Al steering wheel column + Mg steering wheel
Other										
Panels	60	46.8		13.2	197	197		0		AHSS
Instrument panel beam	9.3	4.6		4.7	68	43		−5.32		Mg
Front/rear bumpers	10	6.0		4.0	91	60		−7.75		Al
Seat frame	39.3	19.3		19.7	269	296		1.37		AHSS
Centre console	4.1	1.6		2.5	6	13		2.80		Mg
Steering knuckle	6.2	3.1		3.1	32	71		3.42		Al
Total	785.1	618.2		166.9	3332	3581		1.49		

Table 17.5 CFRP component substitutions (including estimated US$/kg saving) considered for 25% body and chassis mass target scenarios, including parts not considered part of the BIW or chassis subsystems (without mass decompounding).

Part	Baseline mass (kg)	LW mass (kg) P	F	Mass saving (kg)	Baseline cost (US$)	LW cost (US$) P	F	Saving (US$/kg) P	F	Technology
BIW	321	160.5		160.5	1025	2698		10.42		CFRP
Front/rear bumper impact module	5.5	2.5		3.0	33	100		22.33		CFRP
Total	326.5	163		163.5	1058	2798		10.67		

consumption during the vehicle use stage. Absolute or specific fuel savings [in l/(100 km × 100 kg)] represent the fuel savings rather than relative numbers (expressed in % fuel savings per unit % weight reduction) due to the independence of fuel savings from the absolute vehicle weight. A recent study estimated a specific fuel savings value of 0.38 l/(100 km × 100 kg) for a naturally aspirated engine based on an EPA drive cycle, including the consideration of powertrain adaptation resulting from the mass decompounding effect [19]. Table 17.7 indicates undiscounted life cycle cost savings for a lifetime vehicle driving distance of 250 000 km and a fuel price of US$3/gallon (1 gallon = 3.78 l).

Table 17.6 Multi-material component substitutions (including estimated US$/kg saving) considered for 25% body and chassis and 25% total vehicle (without powertrain) mass target scenarios with and without mass decompounding.

System	Baseline mass (kg)	LW mass (kg) P	F	Baseline cost (US$)	LW cost (US$) P	F	Saving (US$/kg) P	F
Body and chassis with mass decompounding	656	536.8	487.8	2669	2901	2750	1.94	0.48
Body and chassis without mass decompounding	656	492		2669	3201		3.24	
Entire vehicle (except powertrain) with mass decompounding	1029	846	771	7860	7342	7103	−2.83	−2.92
Entire vehicle (except powertrain) without mass decompounding*	1029	765		7860	9471		6.10*	

Note: An aluminium BIW was chosen to achieve the target mass reduction. However, if an AHSS BIW is chosen, vehicle mass reduction would be 22.5% but the cost penalty would be reduced to US$2.16/kg saved. An AHSS BIW is 32.1 kg heavier than aluminium but costs US$1077 less.

Table 17.7   Undiscounted life cycle cost savings for a lifetime vehicle driving distance of 250 000 km and a fuel price of US$3/gallon.

Scenario	Mass saving (kg)	Cost saving (US$/kg)	Vehicle retail price change (US$)	Vehicle life cycle cost savings (US$)
Lightweight metal body and chassis with mass decompounding	168.2	0.48	0.72	1266
CFRP body and chassis with mass decompounding	226.3	5.82	8.73	1695
Lightweight metals without mass decompounding	166.9	1.49	2.24	1254
CFRP body and chassis without mass decompounding	163.5	10.67	16.01	1215
Entire vehicle (except powertrain) with mass decompounding	258	−2.92	−4.38	1947
Entire vehicle (except powertrain) without mass decompounding	264	6.10	9.15	1979

17.3   Performance Modelling

A number of scenarios relating to materials migration in automotive applications have been considered above. In this section, some of the theory behind the prediction of weight savings is considered. It is clear that the weight saving achieved depends chiefly on the relative properties of the candidate materials. In turn, this depends upon the nature of the application where materials substitution is proposed. The analysis which follows considers what weight savings are achievable as a function of the relative properties of the materials concerned. Where carbon fibre composites are considered, the properties of these composites are considered as a function of their component fibres.

In general, it will be the *mechanical* properties of carbon fibre composites that we are looking to exploit. Hence we will restrict our considerations to this aspect of their performance. Within the spectrum of mechanical properties of the fibres themselves, the most critical tend to be strength and stiffness and the analysis presented is restricted to these specific mechanical properties. [It is, of course, acknowledged that composite *toughness* (in its various guises) is vital in determining the (say) impact performance of the composite, but this is a function of composite design and interfacial behaviour. It is therefore much more difficult to extrapolate from fibre properties to composite toughness than is true for strength and stiffness behaviour].

It is not the mechanical properties of carbon fibre composites alone that make them attractive in mechanical design. Rather, it is the mechanical properties *per unit weight* that is most important. For example, most carbon fibre composites are no stiffer (in terms of Young's modulus, E) than steel. Steel possesses a Young's modulus of around 205 GPa. Consider a steel rod and a pultruded carbon fibre rod of equal modulus and identical dimensions. Under identical tensile loads they will extend by equal amounts. However, because the carbon fibre rod possesses a density around one-fifth that of the steel rod, it weighs only one-fifth as much. Using the carbon fibre rod rather than the steel rod in a tensile application would result in an

Table 17.8 Performance criteria for mass saving in stiffness- or strength-critical applications.

Typical application/function	Stiffness-critical application	Strength-critical application
Rods in tension	E/ρ	σ_y/ρ
Beams in bending	$E^{1/2}/\rho$ or $E^{1/3}/\rho$	$\sigma_y^{2/3}/\rho$ or $\sigma_y^{1/2}/\rho$
Panels in bending	$E^{1/3}/\rho$	$\sigma_y^{1/2}/\rho$
Plates in compressive buckling	$E^{1/3}/\rho$	—
Columns in compressive buckling	$E^{1/2}/\rho$	—
Shafts in torsion	G/ρ, $G^{1/2}/\rho$ or $G^{1/3}/\rho$	σ_y/ρ, $\sigma_y^{2/3}/\rho$ or $\sigma_y^{1/2}/\rho$

80% weight saving. This argument can be extended to predict potential weight savings in a variety of automotive applications.

In the above example, the measure of performance which allowed us to determine potential weight saving was stiffness per unit density (E/ρ), rather than absolute stiffness (E). There are a range of similar measures which can be applied to a wide variety of engineering applications. The most commonly encountered measures used to assess performance per unit mass are as shown in Table 17.8 [20]. In this table, strength has been defined as yield strength (or elastic limit, σ_y) – most appropriate for metals, but any other relevant strength term would be equally valid (e.g. yield strength for metals and ductile polymers, ultimate tensile strength for brittle solids in tension). For torsional applications the relevant measure of stiffness is shear modulus, G.

On the basis of these performance measures, it is possible to calculate the potential mass savings arising from potential materials substitutions. The fundamental basis for these calculations is that of mass saving, given *equal mechanical performance*.

The calculations which follow are based on Table 17.9, which shows a comparison of the mechanical properties for a variety of potential automotive materials.

Table 17.9 Mechanical properties of various engineering materials.

Material	Density, ρ (g/cm^3)	Young's modulus, E (GPa)	Strength, σ_y (MPa)
Mild steel	7.9	205	350
High strength steel	7.9	205	600
Advanced high strength steel	7.9	205	1000
Aluminium (9-Si 3-Cu)	2.7	72	190
Magnesium (AM60)	1.8	45	140
Titanium (6-Al 4-V)	4.5	115	1000
MMC: magnesium/20% SiC (particulate)	2.1	75	420
MMC: aluminium/25% SiC (particulate)	2.9	115	400
Engineering/technical ceramics: silicon nitride	3.2	310	500
Thermoplastic PMC: 30% w/w GFPP	1.15	5	60
Thermoplastic PMC: PA66 30% w/w long glass fibre	1.38	8.5	120
Sheet moulding compound	1.9	15	150
Glass fibre composites (random continuous fibre)	2	25	300
Carbon fibre composites (random continuous fibre)	1.6	80	1300

Table 17.10 Mechanical performance per unit density measures for engineering materials relative to mild steel.

Material	E/ρ relative to mild steel	$E^{1/2}/\rho$ relative to mild steel	$E^{1/3}/\rho$ relative to mild steel	σ_y/ρ relative to mild steel	$\sigma_y^{2/3}/\rho$ relative to mild steel	$\sigma_y^{1/2}/\rho$ relative to mild steel
Mild teel	1.00	1.00	1.00	1.00	1.00	1.00
High strength steel (HSS)	1.00	1.00	1.00	1.71	1.43	1.31
Advanced high strength steel (AHSS)	1.00	1.00	1.00	2.86	2.01	1.69
Aluminium (9-Si 3-Cu)	1.03	1.73	2.06	1.59	1.95	2.16
Magnesium (AM60)	0.96	2.06	2.65	1.76	2.38	2.78
Titanium (6-Al 4-V)	0.98	1.31	1.45	5.02	3.53	2.97
MMC: Magnesium/20% SiC (particulate)	1.38	2.28	2.69	4.51	4.25	4.12
MMC: Aluminium/25% SiC (particulate)	1.53	2.04	2.25	3.11	2.98	2.91
Technical ceramics: silicon nitride	3.73	3.04	2.83	3.53	3.13	2.95
Thermoplastic PMC: 30% w/w GFPP	0.17	1.07	1.99	1.18	2.12	2.84
Thermoplastic PMC: PA66 30% w/w long glass fibre	0.24	1.17	1.98	1.96	2.80	3.35
Sheet moulding compound	0.30	1.12	1.74	1.78	2.36	2.72
Glass fibre composites (random continuous fibre)	0.48	1.38	1.96	3.39	3.56	3.66
Carbon fibre composites (random continuous fibre)	1.93	3.08	3.61	18.34	11.84	9.52

Table 17.10 shows some of the performance measures (as per Table 17.8) for the same materials *relative to mild steel*. The figures arising then allow potential weight savings to be calculated by migrating from any one material in this list to any other.

It is interesting to note that there is virtually nothing to choose between mild steel, HSS, AHSS, aluminium, magnesium and titanium in terms of the stiffness per unit density (E/ρ), the relevant measure for stiffness-critical components in tension. Hence tensile components of identical performance made from any of these materials would weigh much the same. The situation changes however if we alter the function of the component to one in which it supports bending or buckling loads. Consider the migration from mild steel to AHSS for a panel subject to bending loads. If stiffness is the prime requirement, there are no weight savings on offer ($E^{1/3}/\rho$ is identical for each). However, if strength is the prime requirement then AHSS outperforms mild steel ($\sigma_y^{1/2}/\rho$ for AHSS is 1.69× that of mild steel; i.e. better strength performance per unit density) and weight savings could be achieved. The question is how much weight?

17.3.1   Weight Saving Models

Recall the earlier example in which a steel rod in tension was replaced by a CFRP component of equal Young's modulus. The mass saving was calculated at 80%. In that case, the mass saving was calculated by comparing the stiffness per unit density (E/ρ) of each material. Because $(E/\rho)_{CFRP} = 5(E/\rho)_{Steel}$, the mass ratios of the two components were in the ratio of 1 : 5, with the steel component being five times heavier than the CFRP component. Hence use of the CFRP component would result in an 80% mass reduction.

Extending this argument, if we represent $(E/\rho)_{Steel}$ as M_1 and $(E/\rho)_{CFRP}$ as M_2, then the fractional mass reduction is given by

$$\text{Fractional mass reduction} = (1 - M_1/M_2) \qquad (17.1)$$

The % mass reduction is therefore

$$\% \text{ mass reduction} = 100\,(1 - M_1/M_2) \qquad (17.2)$$

Note that this expression remains valid if M_1 and M_2 are defined in their relative terms as per Table 17.10 rather than in their absolute terms. This simple but powerful expression is valid irrespective which of the performance measures listed in Table 17.8 is used. This allows us to calculate potential weight savings deriving from material substitution. Note that there are often subsidiary factors which may reduce actual weight savings. The derivation of the performance measures listed in Table 17.8 assumes unlimited design freedom in at least one geometric design variable (e.g. the ability to increase the thickness of a panel or beam to compensate for reduced stiffness or strength). Practical limitations on geometric design freedom may limit the weight savings achieved. Conversely, materials substitution accompanied by redesign may allow for even greater weight savings.

Table 17.11 lists the results of such weight saving calculations. Such calculations could be conducted with reference to any proposed material substitution, but Table 17.11 shows the situation with respect to migration from mild steel (assumed strength = 350 MPa). Where the substitution would actually cause a weight increase, the entry has been left blank. Perhaps the most noticeable thing here is the scale of the weight savings achievable. Values in excess of 50% weight saving are common, with some extreme cases exceeding 90%. Of course, the reference material considered here is mild steel of relatively low strength and high density, but similar calculations could be carried out to model any proposed materials switch. For example, the migration from an aluminium to a magnesium component in a bending application (where the relevant performance metric is $E^{1/3}/\rho$ or $\sigma_y^{1/2}/\rho$), then the weight saving would be around 22% irrespective of whether stiffness or strength were most critical. In actual fact, as a result of materials switching combined with a redesign, a weight saving of 35% (-5.5 kg) was reported when an aluminium engine cradle for the 2006 Z06 Corvette was replaced with a cast magnesium cradle [21].

Here, however, it is the potential weight savings which accrue due to the replacement of steel with low-cost CFRP which will be considered. In order to model this, a number of assumptions need to be made. For example, if the exercise is to replace a steel panel with a CFRP panel of equal mechanical performance (either strength- or stiffness-dominated), what

Table 17.11 Potential weight savings when migrating from mild steel to other materials in a variety of stiffness- or strength-critical engineering applications.

Material	\multicolumn{6}{c}{Mass saving (%) relative to mild steel for applications controlled by:}
	E/ρ	$E^{1/2}/\rho$	$E^{1/3}/\rho$	σ_y/ρ	$\sigma_y^{2/3}/\rho$	$\sigma_y^{1/2}/\rho$
Mild steel	0.0	0.0	0.0	0.0	0.0	0.0
High strength steel (HSS)	0.0	0.0	0.0	41.7	30.2	23.6
Advanced high strength steel (AHSS)	0.0	0.0	0.0	65.0	50.3	40.8
Aluminium (9-Si 3-Cu)	2.7	42.3	51.6	37.0	48.6	53.6
Magnesium (AM60)	—	51.4	62.2	43.0	58.0	64.0
Titanium (6-Al 4-V)	—	23.9	30.9	80.1	71.7	66.3
MMC: Magnesium/20% SiC (particulate)	27.3	56.1	62.8	77.8	76.5	75.7
MMC: Aluminium/25% SiC (particulate)	34.6	51.0	55.5	67.9	66.4	65.7
Engineering/technical ceramics: silicon nitride	73.2	67.1	64.7	71.6	68.1	66.1
Thermoplastic PMC: 30% w/w GFPP	—	6.8	49.8	15.1	52.8	64.8
Thermoplastic PMC: PA66 30% w/w long glass fibre	—	14.2	49.5	49.1	64.3	70.2
Sheet moulding compound	—	11.1	42.5	43.9	57.7	63.3
Glass fibre composites (random continuous fibre)	—	27.5	48.9	70.5	71.9	72.7
Carbon fibre composites (random continuous fibre)	48.1	67.6	72.3	94.5	91.6	89.5

mechanical properties are assumed for the steel being substituted (mild steel, HSS, AHSS)? Also, how are the equivalent mechanical properties of the CFRP composite calculated? The models and assumptions employed here relate CFRP properties directly to the properties of the carbon fibres it contains and are explained below.

17.3.2 Models for Density, Stiffness and Strength

As an aid to making meaningful comparisons between carbon fibre composites and traditional automotive materials, some simple models may be employed for calculating composite strength and stiffness as a function of the fibre and matrix employed. These models could be made more sophisticated but for our purposes a balance has been drawn and they are entirely adequate for the determination of trends and mass saving estimations. As an illustration, the matrix contribution to strength and stiffness of a continuous carbon fibre laminate is relatively insignificant. A minor increase in fibre modulus or strength can completely negate the matrix contribution. Nevertheless, for the sake of completeness, the matrix contribution to laminate strength and stiffness is included in the models below. Initially, we consider CFRP laminates, but later extend the arguments to deal with lower carbon content composites such as SMC.

In the sections which follow, subscripts c, f and m refer to composite, fibre and matrix, respectively. Subscript u refers to an ultimate property. Hence the symbol ρ_f refers to the fibre density, while σ_{uc} refers to the ultimate tensile strength of the composite.

Density: the composite density (ρ_c) is given by a simple rule of mixtures

$$\rho_c = V_f\,\rho_f + V_m\,\rho_m \tag{17.3}$$

In this expression, V_f is the volume fraction of fibres, V_m is the volume fraction of matrix, ρ_f is the fibre density and ρ_m is the matrix density.

Stiffness (Young's modulus): the stiffness of a carbon fibre laminate depends on many factors but mainly on: (i) fibre modulus, E_f, (ii) matrix modulus, E_m, (iii) fibre volume fraction, V_f, (iv) matrix volume fraction, V_m, and (v) fibre orientation distribution. The starting point for all of the subsequent analysis is that $V_f = 0.60$, $V_m = 0.40$ and $E_m = 3$ GPa. These are typical values for most CFRP laminated structures. For a continuous $0°$ aligned fibre composite, Young's modulus (E_c) may be reliably determined by a rule of mixtures expression

$$E_c = V_f\,E_f + V_m\,E_m \tag{17.4}$$

The composite modulus is the sum of the individual contributions from fibre ($V_f\,E_f$) and matrix ($V_m\,E_m$). However, the usefulness of unidirectional $0°$ aligned fibre composites is very limited – being largely restricted to unidirectional tensile applications (e.g. pultruded tie rods). For most real engineering applications, service stresses are not unidirectional and therefore fibres need to be aligned in more than one direction. This necessarily reduces the fibre contribution to composite stiffness. Equation 17.4 above may be rewritten to take account of the reduced fibre contribution to stiffness. An orientation efficiency factor (η, where $0 \leq \eta \leq 1$) may be introduced

$$E_c = \eta\,V_f\,E_f + V_m\,E_m \tag{17.5}$$

For the purposes of the following analysis, a realistic value of $\eta = 0.5$ has been assumed. Hence the Young's modulus of a quasi-isotropic CFRP laminate containing 60% by volume of fibres of $E_f = 250$ GPa would be $E_c = (0.5 \times 0.6 \times 250) + (0.4 \times 3) = 76.2$ GPa. Again, note that the matrix contribution to laminate stiffness is low ($<2\%$) and could be ignored for the purpose of this analysis without affecting the conclusions. In the analysis which follows, the mass savings resulting from replacing steel ($E = 205$ GPa) with CFRP laminates containing fibres with Young's moduli ranging from 150 to 600 GPa have been calculated.

Figure 17.1 shows the magnitude of mass reduction achieved when replacing steel ($E = 205$ GPa) with CFRP laminates prepared with fibres of Young's modulus ranging from 150 to 600 GPa.

Figure 17.1 shows a difference between what is achievable in tensile applications with what is achievable in other applications. For a component in tension, weight savings are relatively sensitive to fibre modulus. In contrast, weight savings for those components supporting bending loads are much less sensitive to fibre modulus with weight savings between 60 and 80% possible for all composites containing fibres of modulus >175 GPa.

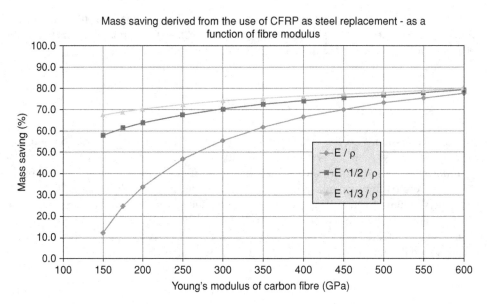

Figure 17.1 Mass reduction resulting from replacement of steel with CFRP laminates in stiffness-critical applications – as a function of fibre modulus.

Tensile strength: a similar approach is adopted here. The composite strength (σ_{uc}) is given by a modified rule of mixtures expression, as follows

$$\sigma_{uc} = \eta \, V_f \, \sigma_{uf} + V_m \, E_m \, \varepsilon_{uc} \tag{17.6}$$

In actual fact, the manner in which the matrix contribution to composite strength is evaluated makes little difference to the resulting composite strength and will not affect major trends or conclusions arising from the analysis.

In the above expressions, σ_{uf} is the tensile strength of the fibres and ε_{uc} is the failure strain of the composite. The mass savings resulting from replacing steel (σ_y = variously 300, 600, 900, 1200 and 1500 MPa) with CFRP laminates containing fibres with tensile strengths ranging from 1500 to 5000 MPa have been calculated. A matrix modulus of 3 GPa has been assumed along with a composite failure strain of 2% and a reinforcement efficiency of $\eta = 0.5$, as previously. If the matrix is assumed to be elastic within this range, the associated matrix contribution to strength turns out to be 24 MPa for a composite containing 40% matrix by volume. Note the relatively minor magnitude of this contribution (typically around 2–3% of the overall composite strength). Both strength and stiffness values are very much fibre-dominated in laminated CFRP composites.

Figures 17.2–17.4 show the achievable mass savings when replacing steel of varying strengths (300, 900 and 1500 MPa) with CFRP laminates containing carbon fibres of different strengths (from 1500 to 5000 MPa). Carbon fibre strengths have been chosen to be representative of those readily available. The United States DoE low-cost carbon fibre programme goals specify a minimum fibre modulus of 172 GPa and a minimum fibre strength

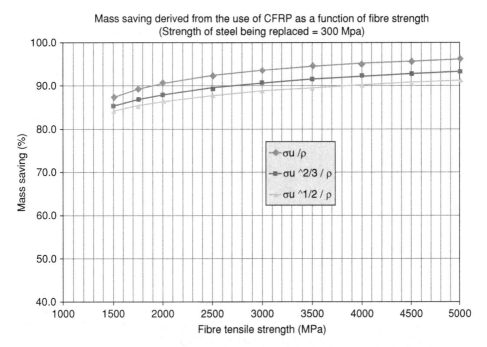

Figure 17.2   Mass reduction resulting from replacement of steel ($\sigma_y = 300$ MPa) with CFRP laminates in strength-critical applications – as a function of fibre strength.

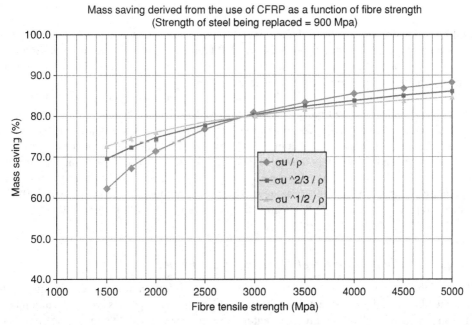

Figure 17.3   Mass reduction resulting from replacement of steel ($\sigma_y = 900$ MPa) with CFRP laminates in strength-critical applications – as a function of fibre strength.

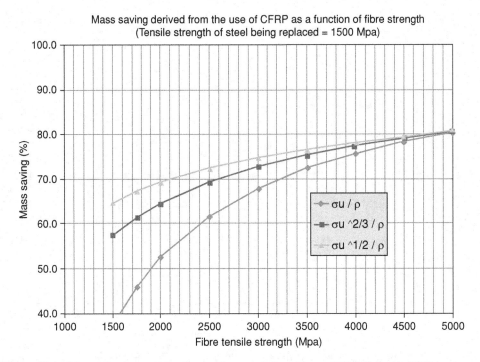

Figure 17.4   Mass reduction resulting from replacement of steel ($\sigma_y = 1500$ MPa) with CFRP laminates in strength-critical applications – as a function of fibre strength.

of 1720 MPa. These are toward the lower end of the range depicted here. Commodity grade carbon fibres possess moduli around 240 to 250 GPa and quoted strengths of around 4000 to 5000 MPa (though, being brittle solids, these will vary quite widely). Hence the property range of the target industrial market is covered in these figures.

There is strong current interest in the use of high strength steel (HSS), ultra-high strength steel (UHSS) and advanced high strength steel (AHSS) for weight reduction in automotive applications. While this approach is successful in strength-critical applications, the use of HSS materials in stiffness-critical applications is less successful since the high strength is not fully exploited and the stiffness of steel is much less amenable to "tailoring" than its strength. However, where strength is the critical performance factor, CFRP composites containing fibres of modest strength (1600–2500 MPa) are still capable of producing weight savings of 40–75% when compared with even very high strength steel (strengths of 1200–1500 MPa).

17.3.3   Carbon Fibre Sheet Moulding Compounds

The weight savings considered above all relate to carbon fibre laminates containing 60% by volume of carbon fibre. These are not the only carbon fibre composites which the automotive industry can exploit. There is great current interest in carbon fibre sheet moulding compound (CFSMC). CFSMC does have the disadvantages of not maximizing the material performance (and thus weight savings available) because practical considerations limit the fibre content to around 50–55% w/w (around 40% v/v). However, a SMC offers several advantages, namely:

- SMC is commonly used in non-crash critical components and therefore can be incorporated without adding the risk of changing energy management structures of the vehicle.
- SMC is a familiar material to designers and therefore working with it will afford them the opportunity to become more familiar with the material.
- Since most automotive plants are already tooled for manufacturing components from SMC, a minimum of capital investment will be necessary.
- Because of the improved material properties that are theoretically available by replacing glass in SMC with carbon fibre, a familiar material with familiar processing techniques may become available for more demanding applications.
- CFSMC can be a "drop-in" replacement for conventional SMC, yielding the opportunity to get field experience with real parts, without the need to redesign entire systems in order to exploit the benefits of composites (though note that redesign would help maximise weight savings).

Hence, one composite could simply be replaced with another, so the reluctance to use composites is not relevant. So, how would CFSMCs fare in relation to weight savings when used to replace steel and other material components – for example glass fibre-based SMC?

First, the same mechanical models employed to predict the properties of laminated composite can be employed. A typical SMC contains chopped discontinuous fibres of perhaps 25–50 mm (1–2 in) in length. As they are discontinuous, they suffer some performance shortfall with respect to continuous fibres (under load, discontinuous fibres exhibit some stress decay towards each fibre end). However, because the fibres are relatively long, this effect is relatively small. Essentially, the ineffective end portions of each fibre are very short in comparison to the overall fibre length. For these reasons the same strength and stiffness models have been used as a basis for weight saving calculations (Equations 17.5 and 17.6, in which reinforcement efficiency is assumed to be 50%, or $\eta = 0.5$, as previously). The carbon fibre modulus and strength ranges considered are as earlier (150 GPa $< E_f <$ 600 GPa and 1500 MPa $< \sigma_{uf}$ $<$ 5000 MPa). Typical fibre contents for SMCs are of the order of 40–55% by weight. For modelling, *volume fraction* rather than *weight fraction* is essential, but the two are easily converted:

$$V_f = W_f \, \rho_m / (W_f \, \rho_m + W_m \, \rho_f) \qquad (17.7)$$

In this expression, V_f is the volume fraction of fibres, W_f is the weight fraction of fibres, W_m is the weight fraction of matrix, ρ_f is the fibre density and ρ_m is the matrix density. On this basis, weight savings due to substituting steel with CFSMC can be calculated. Figure 17.5 shows the weight savings for stiffness-critical applications when replacing steel ($E = 205$ GPa) with CFSMC containing fibres with moduli ranging between 150 and 600 GPa at a fibre loading of 40% w/w (30.8% v/v). Figure 17.6 shows the equivalent graph, but at a fibre content of 50% w/w (40% v/v). These fibre contents are representative of what may be expected from CFSMCs.

As previously, purely tensile applications prove more difficult in achieving significant weight savings. For example at 40 and 50% w/w fibre content, weight savings do not occur until the

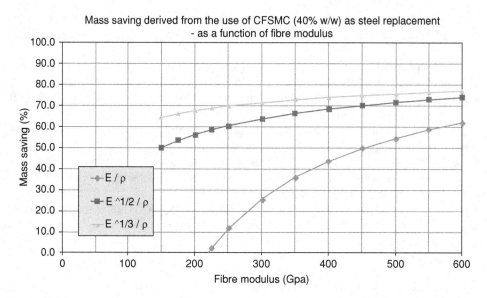

Figure 17.5 Mass savings resulting from replacing steel with 40% w/w carbon fibre SMC – as a function of fibre modulus.

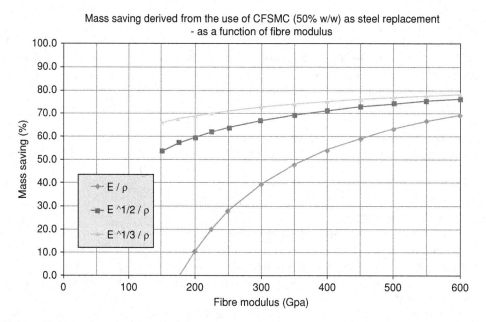

Figure 17.6 Mass savings resulting from replacing steel with 50% w/w carbon fibre SMC – as a function of fibre modulus.

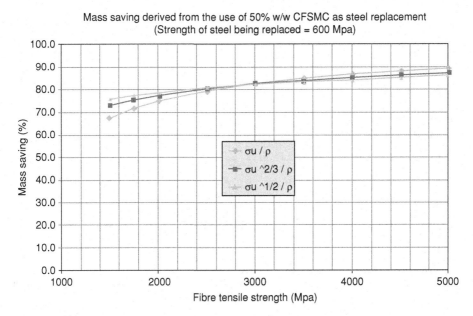

Figure 17.7   Mass savings resulting from replacing steel of strength 600 MPa with 50% w/w carbon fibre SMC.

fibre modulus exceeds 220 and 180 GPa respectively. For other applications (typically those where CFSMC will be used), potential weight savings are more significant – in excess of 50% across the range of realistic fibre moduli.

In strength-critical components, the achievable weight savings depend upon the strength of the steel being replaced and this can vary widely (e.g. 300–1500 MPa in our previous analyses). In order to give a representative portrayal of potential weight savings, Figures 17.7 and 17.8 show data with reference to steel of strength 600 and 1200 MPa when replaced by a 50% w/w CFSMC. For the lower strength steel case, weight savings are typically in the 75–85% range and reassuringly insensitive to fibre strength. For the higher strength steel case, weight savings remain high – in the 60–80% range in general. Even when a 40% w/w CFSMC is used, weight savings in the equivalent cases are in the 70–85% range (600 MPa steel) and 50–80% (1200 MPa steel). Further calculation shows that significant weight savings prove possible, right down to fibre contents of 25% w/w and below.

Finally in respect of CFSMC, what weight savings are on offer when substituting from traditional glass fibre SMC? These are useful comparison figures because they represent a simple like for like replacement in terms of design and process technology.

In terms of *strength critical* applications, the strength range exhibited by glass fibres is pretty much the same as that exhibited by carbon fibres. Therefore, for meaningful comparison, weight savings have been calculated on an equal strength for strength basis. Also because glass is denser than carbon fibre, it is fairer to compare on a volume fraction (rather than weight fraction) basis. The results of such calculations show that a 30% V_f (39.1% W_f) CFSMC would produce a weight saving of 13% over its glass fibre counterpart ($V_f = 30\%$, $W_f = 47\%$). This figure is independent of the application. For a 40% V_f (50% W_f) CFSMC

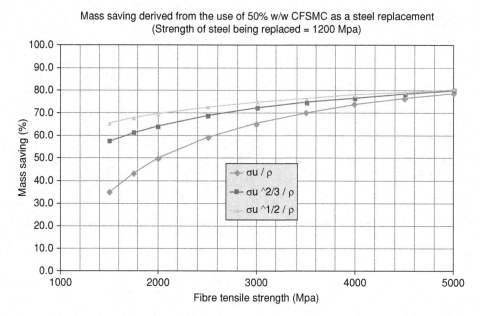

Figure 17.8 Mass savings resulting from replacing steel of strength 1200 MPa with 50% w/w carbon fibre SMC.

material, the equivalent figure would be a 16% weight saving over a 40% V_f (58% W_f) GFSMC material. These weight savings are relatively modest because the strengths of fibres within the composites have been assumed to be identical. The only weight savings which arise are due to the lower density of carbon fibre compared to glass fibre (1.8 g/cm^{-3} compared to 2.5 g/cm^3).

Turning to *stiffness critical* applications, the weight savings on offer are more substantial. Note, however, that these weight savings will only be realised if the opportunity is taken to reduce component dimensions (typically thickness) to exploit the higher stiffness of the CFSMC. A simple material substitution while keeping part dimensions constant will only produce minor weight savings. Most glass fibre used for reinforcement purposes is E-glass with a Young's modulus of around 70–80 GPa. For these calculations, a stiffness value of 75 GPa has been adopted. Achievable weight savings range from 30 to 70% in typical applications. At the industrial end of the carbon fibre stiffness scale, CFSMCs containing fibres of modulus 250 GPa offer 40–50% weight savings over GFSMCs. Even lower down the performance scale, CFSMCs containing fibres of modulus 175 GPa offer 35–43% weight savings. These figures represent significant weight savings and render CFSMCs very attractive.

17.3.4 Performance Modelling Summary

Whether based on strength-critical or stiffness-critical applications, carbon fibre composites offer very significant possibilities for weight reduction. Reassuringly, significant weight savings are possible with relatively low-performance fibres (e.g. $E_f = 175$ GPa, $\sigma_{uf} = 1750$ MPa) since (surprisingly) for many applications the potential weight savings are not particularly sensitive to fibre performance within the ranges associated with low- to medium-performance fibres. In addition, this potential for significant weight reduction is not limited to high fibre content laminated structures. It is also true for SMCs of moderate fibre content and even for composites with fibre contents as low as 20% volume fraction. This raises interest right down to thermoplastic matrix composites produced by (say) injection moulding. Hence carbon fibres meeting the DoE low-cost carbon fibre program goals will offer tremendous potential for weight savings in automotive components.

Finally, let us consider the strength and stiffness models employed and the conclusions from their use. Well, the critical factor in these equations employed (specifically Equations 17.5 and 17.6) is the reinforcement efficiency term, η. Thus far, we have assumed a value of $\eta = 0.5$. This may be an underestimate in certain cases (e.g. where judicious fibre orientated architecture is possible) or it may be an overestimate (e.g. due to fibre discontinuity or inefficient stress transfer due to interfacial frailties). In any event, even if the fibres are providing strength and stiffness contributions at only one-quarter of the efficiency of their continuous, fully aligned counterparts ($\eta = 0.25$), weight savings of 35–70% (stiffness) and 30–80% (strength) are still predicted for a 50% W_f CFSMC acting as a steel replacement.

17.4   Cost Modelling

17.4.1   Cost of Making Carbon Fibre

The baseline carbon fibre in the model is from an industrial grade precursor consisting of 50 000 filaments. An existing plant in the United States was used as the reference point for the model. The facility characteristics are:

- 2–1500 tonne/year carbon fibre lines per building
- 50 persons (including indirect) total staff (80 persons for two lines)
- Capital investment of just over US$119 million for two lines
- Precursor and carbon fibre manufactured by the same company
- 862 m heated length for the oxidation equipment
- 15 year lifetime of equipment
- 30 year lifetime of building
- 7% cost of capital
- 15–20% equipment installation cost
- Included cost of maintenance, insurance, taxes, start-up and contingency
- US$150/sq ft building cost
- US$0.0688/kWh electricity cost
- US$5.00/MMBTU natural gas cost (1 MMBTU = 1 million BTU)
- US$1.25/sq ft annual HVAC and lighting cost
- Baseline polymer is polyacrylonitrile with methyl acrylate co-monomer
- Precursor spinning method is solution spinning incorporating tank farm, solvent recovery and polymer drying.

Using this model, the precursor was found to cost US$2.37/lb to produce and the precursor cost was found to be most sensitive to the acrylonitrile price, which was US$2200/t for the baseline case. This information is depicted in Figure 17.9 below.

While the precursor cost was US$2.37/lb, it requires approximately 2.1 lb of precursor to make 1 lb of carbon fibre, resulting in a precursor investment of US$4.98 for 1 lb of carbon fibre. The model was extended to examine the cost of making 1 lb of carbon fibre and found

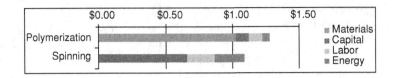

	Acrylonitrile	Natural gas	Direct labour	Methyl acrylate	Electricity
Minimum	$1000/tonne	$2/MMBTU	$20/hr	$1000/tonne	$0040/kWh
Baseline	$2200/tonne	$5/MMBTU	$28/hr	$2350/tonne	$0.0688/kWh
Maximum	$3400/tonne	$13/MMBTU	$36/hr	$3700/tonne	$0.120/kWh

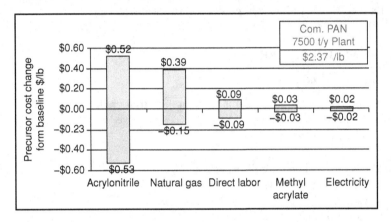

Figure 17.9   Baseline cost of producing carbon fibre precursor.

to be most sensitive to the precursor and electricity costs. For the baseline the resulting carbon fibre production would cost US$10.20/lb (Figure 17.10).

If the plant were scaled to produce 18 000 t/year, the resulting fibre manufacturing cost could be reduced to US$9.35/lb without the addition of any new technologies. The resulting costs are depicted in Figure 17.11.

17.4.2   Cost Model Results for Advanced Technologies

17.4.2.1   Carbon Fibre Cost Reduction Strategies

In response to the cost challenges, ORNL has been leading an effort, coordinated with automotive companies and carbon fibre producers, aimed at reducing the cost of carbon fibres. The current research portfolio includes the development of:

- Non-traditional, lower cost precursors
- Novel, lower cost fibre production methods
- A means for fully developing these advances in a demonstration user-facility which will soon be available to potential suppliers.

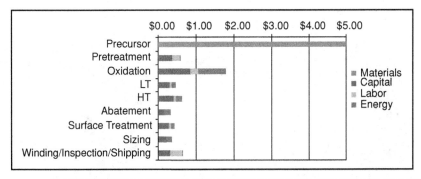

	Precursor	Electricity	Direct labour	LT / HT residence	Natural gas	Oxidation temperature	HT T_{max}
Minimum	$1.84/lb	$0.040/kWh	$20/hr	60s	$2/MMBTU	220°C	1200°C
Baseline	$2.29/lb	$0.0688/kWh	$28/hr	90s	$5/MMBTU	250°C	1400°C
Maximum	$2.89/lb	$0.120/kWh	$36/hr	120s	$13/MMBTU	300°C	1600°C

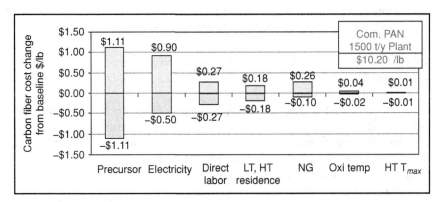

Figure 17.10 Baseline cost of producing carbon fibre (LT – low temperature carbonisation; HT – high temperature carbonisation).

The synergistic effects of reducing both precursor and manufacturing costs may bring carbon fibre costs within acceptable ranges for the automotive industry. Cost goals as coordinated with the automotive industry are to reduce carbon fibre costs to a range of US$11.00–15.40/kg (US$5–US$7/lb). Costs within or approaching that range would also have a significant impact on wind power, oil and gas, construction, power transmission and a variety of other industries.

17.4.2.2 Non-Traditional, Lower Cost Precursor

Below is a brief listing of lower cost precursor routes that are being evaluated at ORNL and by other private companies to develop structural carbon fibres. Other efforts are also being conducted to develop lower cost carbon fibre for non-structural applications. The current level

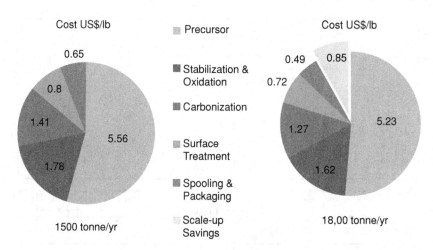

Figure 17.11 Baseline cost model using conventional technologies for two plant production volume levels: (a) 1500 t/year plant, (b) 18 000 t/year plant.

of development for each technology is either proprietary to those developing the technology or can be found in publically available presentations by the researchers:

- Textile-based precursors – This project developed a method for using carpet and yarn grade polyacrylonitrile as a carbon fibre precursor to produce carbon fibres with tensile strengths of up to 500 ksi (3450 MPa).
- High performance textile precursors – This project is using much of the same technology developed in the project above but refining the precursor chemistry and purity, while maintaining high throughput production economics to make a lower cost fibre with strengths above 650 ksi (4480 MPa) for use in compressed gas storage.
- Melt-spun PAN precursors – This effort is developing a low-cost, melt spinnable precursor for higher performance applications.
- High lignin content precursors – This research is still in the earlier stages but has a goal of developing a very low-cost, melt spinnable, lignin precursor with >85% lignin content.
- Lignin blended precursors – This project is aimed at developing a solution spinnable precursor that is blend of lignin and polyacrylonitrile [22].
- High polyolefin content precursors – This research is aimed at developing a melt spinnable precursor with >85% polyolefin content.
- Polyolefin blended precursors – Similar to the previous effort, this project is aimed at developing a precursor with a high polyolefin content but allows for the incorporation of other polymers into the precursor in significant quantities.

17.4.2.3   Non-Traditional, Lower Cost Conversion Technologies

In addition to the development of alternative precursors, ORNL has three programmes to develop alternative processing technologies. To the author's knowledge there is no other research in alternative processing technologies anywhere else with the goal of significantly reducing carbon fibre manufacturing costs:

- Advanced oxidative stabilisation – The technology in this effort could replace all four stages of oxidative stabilisation in the conventional process. Using a precursor that conventionally requires 90–120 min of residence time in thermal oxidative stabilisation ovens, the advanced modules have accomplished the same job in less than 30 min. Later carbonisation of those oxidised samples demonstrated strengths approaching 600 ksi.
- Microwave assisted plasma carbonisation – Microwave assisted plasma (MAP) technology could replace the low- and high-temperature carbonisation ovens in conventional processing of current lower cost carbon fibres. This technology is currently being scaled to a pilot line scale in a proprietary agreement with an industrial partner.
- Advanced surface treatment and sizing – This effort includes a plasma-based surface treatment method and the development of sizing systems focused on high volume resin systems. Many carbon fibres are not readily compatible with resin systems of interest to the automotive industry. The cost savings benefit of this technology is to increase adhesion between fibres and thus improve the load transfer efficiency resulting in the use of less fibre in the composites, thereby reducing part cost. (Example: 50% fibre / 50% resin part made of US$12/lb CF and US$1/lb resin would have US$6.50 in material costs. A 20% reduction in fibre use would yield a 40% fibre / 60% resin part which would have US$5.40 in material cost).

Some authors have questioned the prospect of US$5–7/lb, most often citing the US$5/lb target rather than the range which is truly the automotive target. Even those pundits who question whether or not that goal can ever be reached point out that carbon fibre composites are finding their way into more and more automotive applications, not because of price, but rather due to the advantages that carbon fibre composites offer [23]. The carbon fibre cost model, previously described, was then applied to many of these advanced precursors and alternative production methods to estimate the production cost savings that could be achievable if those technologies were fully developed and commercialised. The indications from that model are that US$7/lb carbon fibre is certainly quite feasible and US$5/lb may even be possible (Figure 17.12).

The ultimate goal is to combine technologies where possible: An example of this would be to combine savings from a new precursor produced by the advanced oxidation and MAP processes coupled with the fibre volume reduction achieved through interfacial optimisation.

17.4.2.4   Commercialisation of Advanced Technologies

As previously stated, the building of a two-line carbon fibre plant requires an investment of around US$100 million. With such a sizable investment, the risk of putting new technology into a commercial plant is significant. To reduce that risk, the technologies must be demonstrated to not only be technically feasible but also sufficiently robust and reliable prior to the decision to proceed with the investment. It must be demonstrated at or near a commercial scale that the equipment can run and operate continuously for weeks with minimal maintenance, minimal downtime and still produce carbon fibre with the desired properties. To address these

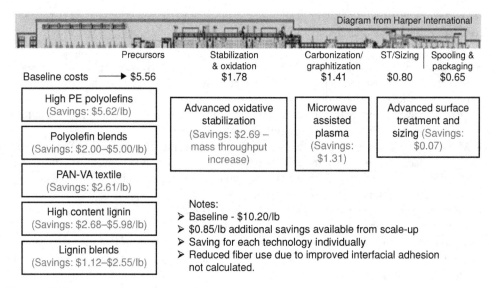

Figure 17.12   Potential cost savings advantage of individually cost lowering technologies if they are fully developed and commercialised.

manufacturing obstacles, the United States Department of Energy is funding the development of a manufacturing demonstration carbon fibre line, located in East Tennessee, named the Carbon Fibre Technology Facility (CFTF; Figure 17.13). The purpose of the CFTF is to scale technologies developed at ORNL and by other industrial producers to a preproduction scale in order to demonstrate a 7 days/week, 24 h/day implementation of new technologies.

The CFTF is designed to be readily accessible to domestic industry and universities for the development and commercialisation of carbon fibre technologies. Additionally, ORNL has developed and worked with an extensive network of industry, academia and government agencies to develop vertically integrated teams for commercialisation of carbon fibre technologies. The CFTF is intended to spur the development and growth of carbon fibre manufacturers and reduce the risk of implementing innovative, lower cost technologies by being a test bed for full commercialisation of those technologies.

Figure 17.13   Carbon fibre technology facility (CFTF) located in East Tennessee, USA.

The CFTF includes a thermal conversion line, a melt spun fibre production line, a fibre semi-production building and a composites research building. Specific goals for the CFTF are:

- Demonstrate the scalability of the science and technology for lowering the cost of carbon fibre.
- Produce and make available up to 25 t/year of conventionally converted fibres (made from low-cost precursors) to potential end users and upper-tier suppliers in multiple industries.
- Scale up advanced conversion technologies.
- Produce carbon fibres made from a variety of precursor materials, for example lignin and polyolefin, on a melt-spinning line in sufficient quantity to feed the conversion lines.
- Enable industrial and university collaborations in order to effectively leverage expertise of laboratory personnel in developing low-cost manufacturing techniques for carbon fibre components.
- Educate and train a highly skilled future workforce for lower cost carbon fibre implementation.

17.5   Conclusion

There is a tremendous and immediate need for affordable lightweight materials to reduce petroleum consumption in the automotive sector. That need is becoming more critical as petroleum prices rise and as government regulation tightens emission and fuel consumption standards. Carbon fibre composites offer a promising solution to meet that need. Carbon fibre composites are finding niche applications in low-volume, high-cost automotive applications and are being considered as a serious lightweighting option for future, higher volume vehicle platforms. The primary inhibitor to the incorporation of carbon fibre composites in high-volume automotive applications is the cost of the carbon fibre. Several research and development efforts are currently being conducted to reduce the cost of carbon fibre and pathways exist to be able to produce carbon fibre in the US\$5–7/lb range.

Acknowledgements

This work was sponsored, in part, by the United States Department of Energy, Office of Vehicle Technologies, Automotive Lightweighting Materials Program. Oak Ridge National Laboratory is operated by UT-Battelle, LLC, under contract DE-AC05-00OR22725.

References

[1] Brooke, L. (2010) BMW's 2013 EV is Pioneering Carbon-Fibre Technologies, Automotive Engineering International Online, 7th September 2010.

[2] Williamson, P. (2012) Q&A Lexus's Last Minute Switch to Composites (Part 1). E-mail Newsletter, American Composites Manufacturers Association, 2nd August 2012.

[3] Das, S. and Schexnayder, S.M. (2010) Carbon Fibre Supply Chain Analysis. ORNL Report prepared for the Strategic Planning Assessment Office of the US Department of Energy, May, p. 16.

[4] McKinsey and Company (2012) *Lightweight, Heavy Impact*, February. See: http://autoassembly.mckinsey.com/html/resources/publications.asp (Accessed 28th May 2012).

[5] Jost, K. (2012) Drastic Lightweighting on the Agenda. Automotive Engineering International Online, 22 June 2012.

[6] Stewart, R. (2010) *Carbon Fibre Producers Optimistic in Downturn*, ReinforcedPlastics.com, January 15 (Accessed 28th May 2012).

[7] Black, S. (2012) Out-of-autoclave prepreg enables concept sports car. High Performance Composites, 1 July 2012.

[8] KUSI News Channel Press Release (2013) *Australian-Based Carbon Revolution Successfully Opens North American Channels – Three Southern California Master Distributors Join U.S. Team As Suppliers Of World's First "One-Piece" Carbon Fibre Wheel*. See: http://www.kusi.com/story/20550616/australian-based-carbon-revolution-successfully-opens-north-american-channels-three-southern-california-master-distributors-join-us-team-as-suppliers (Accessed 10 January 2013).

[9] Koever, A. and Riedel, U. (2012) CAMISMA – Carbon/Amid/Metal based Interior Structure with Multi-Material System Approach. CFK-Valley Convention, June 12.

[10] Bucholz, K. (2012) Materials Lead the Way to Vehicle Mass Reduction. Automotive Engineering International Online, 30 August 2012.

[11] Brooke, L. (2013) Carbon fibre, new aluminium structure lighten 2014 Corvette Stingray. Automotive Engineering International Online, 14 January (2013).

[12] Das, S. and Warren, C.D. (2013) Cost Effectiveness of a 25% Body and Chassis Mass Reduction Goal in Conventional Powertrain Light-Duty Vehicles. Draft cost study prepared for the Vehicle Technologies Program Office of the US Department of Energy, January (2013).

[13] US Environmental Protection Agency (EPA) (2010) *Light Duty Technology Cost Analysis – Report on Additional Case Studies"* Report No. EPA-420-R-10-010 prepared by FEV Inc., April 2010. See: http://www.epa.gov/otaq/climate/420r10010.pdf (Accessed 28th May 2012).

[14] Iobst, S., Berger, L., Fernholz, K. *et al.* (2007) Fabrication of the Automotive Composites Consortium Carbon Fibre B-Pillar. International SAMPE Symposium and Exhibition (Proceedings), 52, SAMPE'07: M and P – From Coast to Coast and Around the World, Conference Proceedings, 2007.

[15] Marsh, G. (2006) Composites Conquer with Carbon Supercars. Reinforced Plastics, 20–24.

[16] Rehkopf, J. (2012) Report with Sujit Das at Oak Ridge National Laboratory, Oak Ridge, TN. Plasan Carbon Composites, Oak Ridge, TN, 16th April.

[17] Composites Manufacturing (2012) *Carbon Revolution Invents First One-Piece Carbon Fibre Wheel*, 29th June 2012. See: http://www.compositesmanufacturingblog.com/2012/06/one-piece-wheel/ (Accessed 28th July 2012).

[18] Rogozhin, A., Gallaher, M. and McManus, W. (2009) *Automobile Industry Retail Price Equivalent and Indirect Cost Multipliers* report by RTI International to Office of Transportation and Air Quality. U.S. Environmental Protection Agency, RTI Project Number 0211577.002.004, Research Triangle Park, NC, Feb 2009.

[19] Canadian Standards Association (CSA) (2012) Product Category Rules for Autoparts. CSA-PCR-2012:1, Nov. 2012.

[20] Ashby, M.F. (2010) *Materials Selection in Mechanical Design*, 4th edn, Butterworth-Heinemann.

[21] Osborne, R.J. (2006) Structural Cast Magnesium Development. Automotive Lightweighting Materials, FY2005 Progress Report, 2006 U. S. Department of Energy, Office of Energy Efficiency and Renewable Energy, pp. 57–67. See: http://www1.eere.energy.gov/vehiclesandfuels/pdfs/alm_05/2g_osborne.pdf (Accessed 28th May 2012).

[22] Husman, G. (2012) Development and Commercialization of a Novel Low-Cost Carbon Fibre", 2012 DOE Hydrogen and Fuel Cells Program and Vehicle Technologies Program Annual Merit Review and Peer Evaluation Meeting, Washington, DC, 14th May 2012. See: http://www1.eere.energy.gov/vehiclesandfuels/pdfs/merit_review_2012/lightweight_materials/lm048_husman_2012_o.pdf (Accessed 28th May 2012).

[23] Sloan, J. (2013) *Waiting for $5/lb Carbon Fibre?* See: http://www.compositesworld.com/columns/waiting-for-5lb-carbon-fibre (Accessed 2nd January 2013).

Index

Advanced Composite Materials for Automotive Applications: Structural Integrity and Crashworthiness, First Edition. Edited by Ahmed Elmarakbi.
© 2014 John Wiley & Sons, Ltd. Published 2014 by John Wiley & Sons, Ltd.

Printed in the United States
By Bookmasters